C0-ASZ-318

Engineering Materials and Processes

Springer
London
Berlin
Heidelberg
New York
Hong Kong
Milan
Paris
Tokyo

Series Editor
Professor Brian Derby, Professor of Materials Science
Manchester Science Centre, Grosvenor Street, Manchester, M1 7HS, UK

Other titles published in this series:

Fusion Bonding of Polymer Composites
C. Ageorges and L. Ye

Composite Materials
D.D.L. Chung

Titanium
G. Lütjering and J.C. Williams

Corrosion of Metals
H. Kaesche

Orientations and Rotations
A. Morawiec
Publication due October 2003

Computational Mechanics of Composite Materials
M.M. Kaminski
Publication due December 2003

Intelligent Macromolecules for Smart Devices
L. Dai
Publication due December 2003

Einar Bardal

Corrosion and Protection

Einar Bardal, Professor, dr.ing
Department of Machine Design and Materials Technology,
The Norwegian University of Science and Technology, Trondheim, Norway

British Library Cataloguing in Publication Data
Bardal, Einar
Corrosion and protection. – (Engineering materials and processes)
1.Corrosion and anti-corrosives
I.Title
620.1'1223
ISBN 1852337583

Library of Congress Cataloging-in-Publication Data
Bardal, Einar, 1933-
Corrosion and protection / Einar Bardal.
p.cm.
Includes bibliographical references and index.
ISBN 1-85233-758-3 (alk. paper)
1.Corrosion and anti-corrosives. I. Title.
TA418.74.B37 2003
620.1'1223--dc21 2003054415

Engineering Materials and Processes ISSN 1619-0181
ISBN 1-85233-758-3 Springer-Verlag London Berlin Heidelberg
a member of BertelsmannSpringer Science+Business Media GmbH
http://www.springer.co.uk

Typesetting: Electronic text files prepared by authors
Printed and bound in the United States of America
69/3830-543210 Printed on acid-free paper SPIN 10930090

Preface

Originally, this book was written in Norwegian, primarily for the teaching of corrosion theory and technology for students at the faculties of mechanical engineering and marine technology and for other interested people at various faculties of the Norwegian Institute of Technology (NTH) in Trondheim, now being a part of the Norwegian University of Science and Technology (NTNU). The book has also been used at some other universities and engineering schools in Scandinavian countries, and as a reference book for engineers in industry.

The book was written with the aim of combining a description of practical corrosion processes and problems with a theoretical explanation of the various types and forms of corrosion. Relatively much attention was paid to the effects upon corrosion of factors such as flow, heat, materials selection, design, surface conditions, and mechanical loads and impacts, as well as their roles in the development of different corrosion forms. The scope of the book is wet corrosion in general. However, because of the vital position of the offshore industry in Norway, several cases and aspects dealt with are related to marine technology and oil and gas production.

In general, this edition is based on my own work on corrosion and related subjects at NTH/NTNU and its associated research foundation, SINTEF, during 35–40 years. Results and experience from our research and engineering activities at SINTEF Corrosion Centre have deliberately been included because this work was done with the same objective in mind as was the teaching: to solve practical corrosion problems by more extensive use of theoretical tools and understanding, combined with empirical knowledge. My approach in this direction was particularly inspired by professor Almar-Næss. He started the first modern teaching of corrosion for students at the faculty of mechanical engineering and other typical engineering students at NTH in the early 1960s, and stimulated to my own engagement in the discipline. Considering the further work, I will like to acknowledge my nearest co-workers during many years, including the permanent staff at SINTEF Corrosion Centre as well as many former students. These people have carried out much of the research and engineering work that I have referred to. I hope that the many references that are made to their contributions show how important they have been.. Results from ones own research milieu are valuable directly as well as by the personal engagement they contribute to the teaching. These contributions to the book have, however, been balanced with a major proportion of general knowledge and research results from the world around. Several figures and tables are reproduced from external publications.

The preparation of the manuscript has been supported by a number of persons, with contributions such as comments to the content or language details of specific sections, assistance in providing literature and pictures, help in formatting the manuscript etc. Acknowledgement for the use of photographs is made in the captions. I wish to thank all for their kindness and help.

Particularly, I will take the opportunity to express my thanks to Detlef Blankenburg, head of Department of Machine Design and Materials Technology, NTNU, for permission to use the facilities of the department, and Siw Brevik for all her efforts and patience in preparing the manuscript. Last but not least, I will thank my wife, Liv, who originally typed the manuscript and who has supported me in many ways, my younger family members Eirunn and Sigmund Bardal, for, respectively, comments to the language, and indispensable computer expertise and efforts during the final preparation of the manuscript, and the rest of my family for their patience and understanding.

July, 2003
Einar Bardal

Contents

Preface v

1 Introduction 1

1.1 Definition and Main Groups of Corrosion – Terminology 1
1.2 Importance of Corrosion and Prevention Efforts 1
1.3 Corrosion Science and Corrosion Technology 3
References 4

2 Wet Corrosion: Characteristics, Prevention and Corrosion Rate 5

2.1 Description of a Wet Corrosion Process 5
2.2 Crucial Mechanisms Determining Corrosion Rates 6
2.3 Corrosion Prevention Measures 7
2.4 Expressions and Measures of Corrosion Rates 8
2.5 Basic Properties That Determine if Corrosion Is Possible and How Fast Material Can Corrode 10
References 10
Exercises 10

3 Thermodynamics – Equilibrium Potentials 13

3.1 Introduction 13
3.2 Free Enthalpy and Cell Voltage 13
3.3 The Influence of the State of Matter on Free Enthalpy and the Change of Free Enthalpy 16
3.4 Change of Free Enthalpy in Chemical Reactions. Reversible Cell Voltage 18
3.5 Electrode Reactions and Electrode Potentials 19
3.6 Series of Standard Potentials 23
3.7 Equilibrium Potentials of Reactions with Iron at 25°C 23
3.8 Pourbaix Diagram 26
3.9 A Simplified Presentation of Equilibrium Potential and Deviation from It 29
3.10 Possible Range for Real Potentials Under Corrosion Conditions 31
References 32
Exercises 32

4 Electrode Kinetics 35

4.1 Introduction: Anodic and Cathodic Reactions 35

4.2 Polarization and Overvoltage 36
4.3 Exchange Current Density 37
4.4 Activation Polarization 37
4.5 Concentration Polarization 38
4.6 Overvoltage Due to Concentration Polarization 42
4.7 Combined Activation and Concentration Polarization 43
4.8 Resistance (Ohmic) Polarization 44
4.9 Determination of Corrosion Potential and Corrosion Rate 44
4.10 Recording Polarization Curves with a Potentiostat 47
4.11 Conditions That Affect Polarization Curves, Overvoltage Curves and Corrosion Rates 50
References 50
Exercises 51

5 Passivity 53

5.1 Passivation and Passivity Described by Anodic Polarization and Overvoltage Curves 53
5.2 Passivity – Reasons and Characteristic Features 54
5.3 Breakdown of Passive Films 57
5.4 Stable and Unstable Passive State 58
5.5 Practical Utilization of Passivation and Passivity 61
References 62
Exercises 62

6 Corrosion Types with Different Cathodic Reactions 65

6.1 General 65
6.2 Corrosion under Oxygen Reduction 66
6.2.1 General Circumstances and Data 66
6.2.2 Effect of Temperature 68
6.2.3 Effect of Surface Deposits 70
6.2.4 Effect of Flow Velocity 70
6.2.5 Corrosion under Thin Films of Water 74
6.3 Corrosion under Hydrogen Evolution 75
6.4 Corrosion under Effects of Bacteria 77
6.5 CO_2 Corrosion 78
6.5.1 General 78
6.5.2 Mechanisms 78
6.5.3 Corrosion Rate 80
6.6 H_2S Corrosion 82
6.7 Other Cathodic Reactions 83
References 84
Exercises 86

7 **Different Forms of Corrosion Classified on the Basis of Appearence** 89

7.1 Introduction 89
7.2 Uniform (General) Corrosion 91
7.3 Galvanic Corrosion 94
7.3.1 Conditions That Determine Corrosion Rates 94
7.3.2 Prevention of Galvanic Corrosion 105
7.3.3 Application of Galvanic Elements in Corrosion Engineering 107
7.4 Thermogalvanic Corrosion 107
7.5 Crevice Corrosion 108
7.5.1 Occurrence, Conditions 108
7.5.2 Mechanism 109
7.5.3 Mathematical Models of Crevice Corrosion 113
7.5.4 Crevice Corrosion Testing 115
7.5.5 Practical Cases of Crevice and Deposit Corrosion 119
7.5.6 Galvanic Effects on Crevice Corrosion 120
7.5.7 Prevention of Crevice Corrosion 121
7.6 Pitting Corrosion 122
7.6.1 Conditions, Characteristic Features and Occurrence 122
7.6.2 Mechanisms 124
7.6.3 Influencing Factors 126
7.6.4 The Time Dependence of Pitting 127
7.6.5 Pitting Corrosion Testing 130
7.6.6 Prevention of Pitting Corrosion 131
7.7 Intergranular Corrosion 131
7.7.1 General Characteristics, Causes and Occurrence 131
7.7.2 Austenitic Stainless Steels 132
7.7.3 Ferritic Stainless Steels 134
7.7.4 Ni-based Alloys 135
7.7.5 Aluminium Alloys 135
7.8 Selective Corrosion (Selective Leaching) 135
7.9 Erosion and Abrasion Corrosion 138
7.9.1 Characteristic Features and Occurrence 138
7.9.2 Types and Mechanisms 139
7.9.3 Erosion and Erosion Corrosion in Liquid Flow with Solid Particles 140
7.9.4 Influencing Factors and Conditions in Liquids and Liquid–Gas Mixtures 144
7.9.5 Critical Velocities 146
7.9.6 Abrasion and Other Wear Processes Combined with Corrosion 149
7.9.7 Preventive Measures 150
7.10 Cavitation Corrosion 152

7.11 Fretting Corrosion (Fretting Oxidation) 154
7.12 Stress Corrosion Cracking (SCC) 156
7.12.1 Characteristic Features and Occurrence 156
7.12.2 Mechanisms 157
7.12.3 Fracture Mechanics Quantities 163
7.12.4 Cracking Course and Data for Some SCC Conditions 164
7.12.5 Prevention of SCC 170
7.13 Corrosion Fatigue 170
7.13.1 Definition, Characteristic Features and Occurrence 170
7.13.2 Influencing Factors and Mechanisms 171
7.13.3 Factors Most Important for Crack Initiation and Early Growth 175
7.13.4 Crack Growth Rate and Factors Affecting It 176
7.13.5 Calculation of Number of Cycles to Failure of Welded Steel Structures 179
7.13.6 Prevention of Corrosion Fatigue 180
References 181
Exercises 184

8 Corrosion in Different Environments 193

8.1 Atmospheric Corrosion 193
8.1.1 Environmental Factors and Their Effects 193
8.1.2 Atmospheric Corrosion on Different Materials 196
8.2 Corrosion in Fresh Water and Other Waters 198
8.3 Corrosion in Seawater 203
8.4 Corrosion in Soils 206
8.5 Corrosion in Concrete 210
8.6 Corrosion in the Petroleum Industry 212
References 215
Exercises 217

9 Corrosion Testing, Monitoring and Inspection 219

9.1 Corrosion Testing in General 219
9.1.1 Objectives 219
9.1.2 Test Methods 220
9.1.3 Testing Procedure 221
9.2 Electrochemical Testing 223
9.3 Corrosion Monitoring and Inspection 226
9.3.1 Monitoring of Cathodic Protection 227
9.3.2 Inspection and Monitoring of Process Plants 229
9.3.3 Monitoring and Testing in Other Environments 232
References 233
Exercise 234

10 Corrosion Prevention 237

10.1 Materials Selection 237
10.1.1 General Considerations 237
10.1.2 Some Special Aspects of Materials Selection for the Offshore Industry 240
10.1.3 Unalloyed and Low-alloy Steels and Cast Irons 241
10.1.4 High-alloy Cast Irons 242
10.1.5 Stainless Steels 243
10.1.6 Nickel Alloys 250
10.1.7 Copper and Its Alloys 250
10.1.8 Aluminium and Its Alloys 254
10.1.9 Titanium and Its Alloys 256
10.1.10 Other Metallic Materials 257
10.1.11 Non-metallic Materials 258
10.2 Change of Environment 259
10.3 Proper Design 262
10.4 Cathodic Protection 266
10.4.1 Principle 266
10.4.2 Protection Criteria and Specifications 269
10.4.3 Cathodic Protection with Sacrificial Anodes 273
10.4.4 Cathodic Protection with Impressed Current 276
10.4.5 Electrolyte Resistance, Potential Variation and Current Distribution in CP Systems and Galvanic Elements. 278
10.5 Anodic Protection 281
10.6 Corrosion Protection by Coatings 282
10.6.1 Metallic Coatings 283
10.6.2 Other Inorganic Coatings 292
10.6.3 Paint Coatings 293
10.6.4 Other Forms of Organic Coatings 299
10.6.5 Pre-treatment Before Coating 300
References 302
Exercises 304

Subject Index 309

1. Introduction

Thinking without learning is useless.
Learning without thinking is dangerous.
CONFUCIUS

1.1 Definition and Main Groups of Corrosion – Terminology

Corrosion is defined in different ways, but the usual interpretation of the term is "an attack on a metallic material by reaction with its environment". The concept of corrosion can also be used in a broader sense, where this includes attack on non-metallic materials, but such attacks are outside the scope of this book.

Corrosion of metallic materials can be divided into three main groups [1.1]:

1. Wet corrosion, where the corrosive environment is water with dissolved species. The liquid is an electrolyte and the process is typically electrochemical.
2. Corrosion in other fluids such as fused salts and molten metals.
3. Dry corrosion, where the corrosive environment is a dry gas. Dry corrosion is also frequently called chemical corrosion and the best-known example is high-temperature corrosion.

This book concentrates just on wet corrosion. The terminology used mainly corresponds to international standards and glossaries [1.2–1.4]. Equivalents to the English terms in other languages are given in References [1.4–1.6].

1.2 Importance of Corrosion and Prevention Efforts

Materials technology is a very vital part of modern technology. Technological development is often limited by the properties of materials and knowledge about them. Some properties, such as those determining corrosion behaviour, are most difficult to map and to control.

In general, the development of modern society and industry has led to a stronger demand for engineers with specialized knowledge in corrosion. There are a number of reasons for this:

a) The application of new materials requires new corrosion knowledge.
b) Industrial production has led to pollution, acidification and increased corrosivity of water and the atmosphere.
c) Stronger materials, thinner cross-sections and more accurate calculation of dimensions make it relatively more expensive to add a corrosion allowance to the thickness.
d) The widespread use of welding has increased the number of corrosion problems.
e) The development of industrial sectors like nuclear power production and offshore oil and gas extraction has required stricter rules and control.
f) Considering the future, it should be noticed that most methods for alternative energy production will involve corrosion problems.

The cost of corrosion in industrialized countries has been estimated to be about 3–4% of the gross national product [1.7, 1.8]. It has been further estimated that about 20% of this loss could have been saved by better use of existing knowledge in corrosion protection, design etc. In other words, there is a demand for applied research, education, information, transfer of knowledge and technology, and technical development. Teaching, where considerable emphasis is placed on the connections between practical problems and basic scientific principles, is considered to be of vital importance (see Section 1.3).

The cost of corrosion is partly connected with the efforts to give structures an attractive appearance, it is partly the direct cost of replacement and maintenance and the simultaneous economic loss due to production interruptions, and it includes extra cost of using expensive materials and other measures for the prevention of corrosion and the loss or destruction of products.

Besides the financial cost, a good deal of attention should be paid to safety risks and the pollution of the environment due to corrosion. Personnel injuries can occur due to the fracture of structures, the failure of pressure tanks and leakages in containers for poisonous, aggressive or inflammable liquids, for example.

The extent of corrosion damage in the process industry can be exemplified: of the material failures recorded in the DuPont Company during a two-year period, 57% were corrosion failures and 43% were mechanical failures. In offshore oil and gas production, Statoil's experience is that about half of the material failures are due to corrosion and corrosion fatigue.

The corrosion problems connected with oil production have been given much attention in many countries, not least those with extensive engagement in offshore oil production. This seems reasonable when one considers the huge offshore installations which consist of non-resistant materials in contact with highly corrosive environments, in deep water and under conditions that make inspection and control difficult. It is evident that these problems represent a great challenge and require a high level of knowledge in corrosion technology. Many research projects have been carried out to meet these challenges and requirements and considerable engineering activities have been established where careful attention is paid to modern, advanced methods for inspection, monitoring, control and calculation. Two examples of engineering improvements are dealt with below.

One of the first oilfields developed in the North Sea was Ekofisk. Up to the beginning of the 1980s, corrosion in the production tubing was the cause of 44% of all maintenance costs for "Greater Ekofisk" [1.9]. This maintenance cost due to corrosion amounted to the order of GBP 20 million during 1981–1982. In addition, the annual cost of inhibitors reached a maximum of about GBP 0.5 million in these years. From 1982–83, the maintenance cost has been reduced strongly by establishing a corrosion control programme calculated to cost approximately GBP 8 million [1.10].

Surface treatment in order to protect steel structures represents a large expense too. On the Norwegian continental shelf these expenses were about GBP 100 million annually until 1987–88. (This excludes the cost of preventing corrosion in process systems.) By means of increased know-how and conscious control by the oil companies, this amount was halved by 1994. A further reduction in cost can be obtained on the basis of life-cycle analyses and evaluation of the real needs for surface treatment on each part of the installations (selective maintenance) [1.11].

1.3 Corrosion Science and Corrosion Technology

It has been said that the aim of *science* is "knowing why", while *technology* deals with "knowing how". One of the famous corrosion scientists, T.P. Hoar, considered the technologist as a person who utilizes his scientific knowledge to solve practical problems [1.8]. This can stand as a wise and useful definition. However, it is the situation and the technological problem that one is facing that decides the extent to which the practical solution can be based on scientific knowledge. In corrosion technology many problems are solved more or less by pure experience because the conditions are too complex to be described and explained theoretically. On the other hand, it is evident that great progress in corrosion technology can be obtained by the application of corrosion theory to practical problems to a much higher extent than what has been done traditionally. This can be done i) in order to explain corrosion cases, to find the reasons and prevent new attacks, ii) in corrosion testing with the aim of materials selection, materials development etc., and in monitoring by electrochemical methods, iii) for the prediction of corrosion rates and localization, and iv) to improve methods for corrosion prevention in general, and to select and apply the methods more properly in specific situations. This point of view is considered useful as a basis for corrosion education and it has been a guideline for teaching corrosion to mechanical and marine engineering students at the university in Trondheim (NTH/NTNU) during the last 30 years. The present book has been written based on this opinion. It starts with a presentation of the most important corrosion theory. Then we apply this theory as much as possible to practical corrosion problems. Corrosion and corrosion prevention is more interdisciplinary than most subjects in engineering. Consequently, mastery of corrosion means that it is necessary to have insight into physical chemistry and electrochemistry, electronics/electrical techniques, physical metallurgy, the chemical, mechanical and processing properties of materials, fluid dynamics, the design of steel structures and machines, joining technology, and the materials market situation. These areas of knowledge constitute the foundation upon which corrosion technology is built.

References

1.1 Metals Handbook. 9th Ed. Vol.13 Corrosion. Ohio: ASM International, 1987. ASTM G15-86. Definition of terms relating to corrosion and corrosion testing, ASTM Annual Book of Standards, Vol. 03.02, 1987, pp. 134–138.
1.2 ISO/DIS 8044. Corrosion of Metals and Alloys – Basic Terms and Definitions.
1.4 Vukasovich MS. A Glossary of Corrosion-Related Terms Used in Science and Industry. Houston, Texas: NACE, 1997.
1.5 Costa JM, Cabot JLl (Eds). Glossari de Corrosio', Barcelona: Institut d'Estudios Catalans, 1991.
1.6 Korrosionsordlista (Glossary of Corrosion), TNC 67, Stockholm: Tekniska Nomenclaturcentralen, Stockholm, 1977.
1.7 Payer JH, Boyd WK, Dippold DG, Fisher WH. Materials Performance, NACE, May 1980.
1.8 Hoar TP (Chairm.). Report of the Committee on Corrosion and Protection, London: Dept. of Trade and Industry, H.M. Stationary Office, 1971.
1.9 Corwith JR, Westermark RV. The downhole corrosion control program at Ekofisk. Inspection, Maintenance and Repair, Febr. 1984.
1.10 Houghton CJ, Nice PJ, Rugtveit AG. The computer based data gathering system for internal corrosion monitoring at Greater Ekofisk, 2nd International Conference on Corrosion Monitoring and Inspection in the Oil, Petroleum and Process Industries, London, Febr. 1984.
1.11 Saltvold T. Norwegian Corrosion Consultants (NCC). Private communication. April 1994.

2. Wet Corrosion: Characteristics, Prevention and Corrosion Rate

Mighty ships upon the ocean
suffer from severe corrosion.
Even those that stay at dockside
are rapidly becoming oxide
Alas, that piling in the sea
is mostly Fe_2O_3.
And when the ocean meets the shore
You'll find there's Fe_3O_4,
'cause when the wind is salt and gusty
things are getting awful rusty.

T.R.B. WATSON

2.1 Description of a Wet Corrosion Process

The main features of corrosion of a divalent metal M in an aqueous solution containing oxygen are presented schematically in Figure 2.1. The corrosion process consists of an anodic and a cathodic reaction. In the anodic reaction (oxidation) the metal is dissolved and transferred to the solution as ions M^{2+}. The cathodic reaction in the example is reduction of oxygen. It is seen that the process makes an electrical circuit without any accumulation of charges. The electrons released by the anodic reaction are conducted through the metal to the cathodic area where they are consumed in the cathodic reaction. A necessary condition for such a corrosion process is that the environment is a conducting liquid (an electrolyte) that is in contact with the metal. The electrical circuit is closed by ion conduction through the electrolyte. In accordance with the conditions this dissolution process is called wet corrosion, and the mechanism is typically electrochemical.

In the example in Figure 2.1 the metal ions M^{2+} are conducted towards OH^- ions, and together they form a metal hydroxide that may be deposited on the metal surface. If, for instance, the metal is zinc and the liquid is water containing O_2 but not CO_2, the pattern in the figure is followed: Zn^{2+} ions join OH^- and form $Zn(OH)_2$. When

CO_2 is dissolved in the liquid a zinc carbonate is deposited. Corrosion of substances like iron and copper follow similar patterns with modifications: divalent iron oxide, $Fe(OH)_2$, is not stable, thus with access of oxygen and water it oxidizes to a trivalent hydrated iron oxide, $Fe_2O_3 \cdot nH_2O$, or an iron hydroxide, $Fe(OH)_3$, which also may be expressed as $FeOOH + H_2O$. FeOOH is the ordinary red (or brown) rust. If the access of oxygen is strongly limited, Fe_3O_4 is formed instead of the trivalent corrosion products. Fe_3O_4 is black (without water) or green (with water). Divalent copper hydroxide, $Cu(OH)_2$, is not stable either and tends to be dehydrated to CuO [2.1].

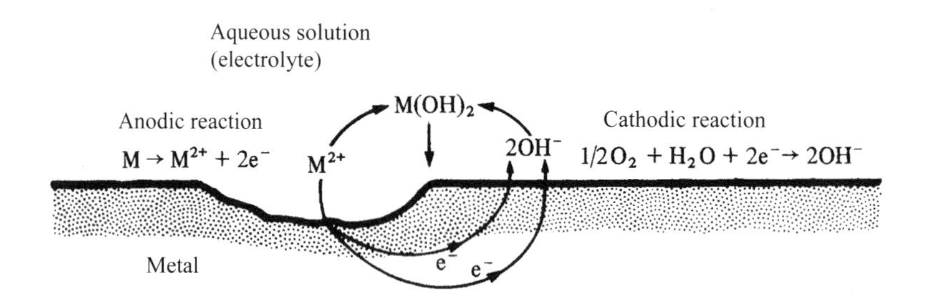

Figure 2.1 Wet corrosion of a divalent metal M in an electrolyte containing oxygen.

Reduction of oxygen is the dominating cathodic reaction in natural environments like seawater, fresh water, soil and the atmosphere. However, under certain conditions there are also other important cathodic reactions: the hydrogen reaction $2H^+ + 2e^- \rightarrow H_2$, reduction of carbonic acid (H_2CO_3) (in oil and gas production), reduction of metal ions etc. Various cathodic reactions are dealt with in Chapter 6.

Figure 2.1 illustrates an electrochemical cell, and the driving force for the electrochemical process (the corrosion) is the cell voltage, or in other words the potential difference between the anode and the cathode. This potential concept is developed further in Chapter 3.

2.2 Crucial Mechanisms Determining Corrosion Rates

It is seen that the corrosion process in Figure 2.1 depends on the availability of oxygen. When corrosion products, such as hydroxides, are deposited on a metal surface, they may cause a reduction in the oxygen supply because the oxygen has to diffuse through the deposits, which may form a more or less continuous layer on the metal surface. Since the rate of metal dissolution is equal to the rate of oxygen reduction (Figure 2.1), a limited supply and limited reduction rate of oxygen will also reduce the corrosion rate. In this case it is said that the corrosion is under cathodic control. This is a very widespread mechanism for corrosion limitation by nature. If the corrosion products are removed from the metal surface by mechanisms such as the corrosion medium flowing at high velocity and corresponding strong fluid dynamical forces, the corrosion rate may be greatly increased (erosion corrosion, see Section 7.9).

In certain cases, the corrosion products form a dense and continuous surface film of oxide closely related to the crystallographic structure of the metal. Films of this type prevent the conduction of metal ions from the metal–oxide interface to the oxide–liquid interface to a great extent so that the corrosion rates may be very low (anodic control). This phenomenon is called passivation and is typical for materials like stainless steel and aluminium in many natural environments. Ordinary structural steels are also passivated in alkaline waters. Passivation is promoted by ample access of oxygen on the material surface, which is obtained by high oxygen concentration in the liquid and by efficient transport of oxygen as a result of strong convection (high flow rates). Conversely, passivation may be hindered – or a passive film may be broken down – by the lack of oxygen. This often happens underneath deposits and in narrow crevices that obstruct the oxygen supply (crevice corrosion, Section 7.5). Aggressive species like chlorides are other major causes of the local breakdown of passive films that occurs in crevice corrosion, pitting (Section 7.6) and other forms of corrosion.

When more and less noble materials are placed in contact, the more noble material offers an extra area for the cathodic reaction. Therefore the total rate of the cathodic reaction is increased, and this is balanced with an increased anodic reaction, i.e. increased dissolution of the less noble material (galvanic corrosion, Section 7.3). If the more noble material (the cathodic material) has a large surface area and the less noble metal (the anodic metal) has a relatively small area, a large cathodic reaction must be balanced by a correspondingly large anodic reaction concentrated in a small area. The intensity of the anodic reaction, i.e. the corrosion rate (material loss per area unit and time unit) becomes high. Thus, the area ratio between the cathodic and the anodic materials is very important and should be kept as low as possible. It should be mentioned that in a galvanic corrosion process, the more noble material is more or less protected. This is an example of cathodic protection, by which the less noble material acts as a sacrificial anode (see next section).

2.3 Corrosion Prevention Measures

Corrosion prevention aims at removing or reducing the effect of one or more of the conditions leading to corrosion using the following measures.:

1 Selecting a material that does not corrode in the actual environment.
2 Changing the environment, e.g. removing the oxygen or adding anticorrosion chemicals (inhibitors).
3 Using a design that will avoid corrosion, e.g. preventing the collection of water so that the metal surface can be kept dry.

4 Changing the potential, most often by making the metal more negative and thus counteracting the natural tendency of the positive metal ions to be transferred from the metal to the environment.
5 Applying coatings on the metal surface, usually in order to make a barrier between the metal and the corrosive environment.

Corrosion prevention is partly dealt with in connection with the various forms of corrosion in Chapter 7, and more generally in Chapter 10.

2.4 Expressions and Measures of Corrosion Rates

There are three main methods that are used to express the corrosion rate:

a) Thickness reduction of the material per unit time.
b) Weight loss per unit area and unit time.
c) Corrosion current density.

Thickness reduction per unit time is the measure of most practical significance and interest. In the metric system this measure is usually expressed in mm/year. In some literature one can still find the unit mils per year (mpy) = 1/1000 inches per year, possibly also inches per year (ipy).

Weight loss per unit area and unit time was commonly used in earlier times, mainly because weight loss was usually the directly determined quantity in corrosion testing. Here the test specimens were weighed before and after the exposure to the corrosion medium. On this basis one could calculate the thickness reduction as weight loss per unit area/density.

From Figure 2.1 it can be understood that corrosion rate also can be expressed by corrosion current density. The dissolution rate (the corrosion rate) is the amount of metal ions removed from the metal per unit area and unit time. This transport of ions can be expressed as the electric current I_a per area unit, i.e. anodic current density i_a = corrosion current density i_{corr}.

If it is preferred to express the local corrosion current density in the anodic area in Figure 2.1, one has $i_{corr} = i_a = I_a/A_a$, where A_a is the anodic area. However, usually the average corrosion current density over the whole surface area A is given, i.e. $i_{corr} = i_a = I_a/A$. The most suitable measure to employ for calculating the corrosion rate depends on which form of corrosion one is dealing with (Chapter 7).

Corrosion current density is a particularly suitable measure of corrosion rate when treating corrosion theory and in connection with electrochemical corrosion testing. Current density is also directly applicable for cathodic and anodic protection (Sections 10.4 and 10.5). In corrosion testing the unit $\mu A/cm^2$ is most often used. When dealing with cathodic protection the units mA/m^2 and A/m^2 are used for the cathode (structure to be protected) and the anode, respectively.

The relationship between thickness reduction per time unit ds/dt (on each corroding side of the specimen/component) and the corrosion current density i_{corr} is determined from Faraday's equations:

$$\frac{ds}{dt} = \frac{i_{corr}\, M}{z\, F\, \rho} \text{ cm/s} \tag{2.1a}$$

or

$$\frac{\Delta s}{\Delta t} = 3268 \frac{i_{corr}\, M}{z\, \rho} \text{ mm/year ,} \tag{2.1b}$$

where i_{corr} is given in A/cm^2;

z	=	number of electrons in the reaction equation for the anodic reaction (dissolution reaction) (per atom of the dissolving metal);
M	=	the mol mass of the metal (g/mol atoms) (the numerical value of M is the atomic weight of the metal);
F	=	Faraday's constant = 96,485 coulombs/mole electrons = 96,485 C/mol $e^- \approx$ 96,500 As/mol e^- [2.2];
ρ	=	the density of the metal (g/cm^3).

Table 2.1 shows the conversion factors between the units of corrosion rates that are most frequently used in the literature. Note that, for most of the listed materials, a corrosion current density of 1 $\mu A/cm^2$ corresponds to a thickness reduction of roughly 0.01 mm/year. As an example of practical corrosion rates it can be mentioned that structural steels in seawater normally corrode by 0.1–0.15 mm/year $\approx$ 10–15 $\mu A/cm^2$ on average. The corrosion rate can be a few times higher locally.

Table 2.1 Corrosion rate conversion factors.

Material/reaction	Corrosion current density	Weight loss per unit area and unit time	Average attack depth increment per unit time	
	$\mu A/cm^2$	mdd*	mm/year	mpy*
$Fe \rightarrow Fe^{2+} + 2e^-$	1	2.51	1.16×10^{-2}	0.46
$Cu \rightarrow Cu^{2+} + 2e^-$	1	2.84	1.17×10^{-2}	0.46
$Zn \rightarrow Zn^{2+} + 2e^-$	1	2.93	1.5×10^{-2}	0.59
$Ni \rightarrow Ni^{2+} + 2e^-$	1	2.63	1.08×10^{-2}	0.43
$Al \rightarrow Al^{3+} + 3e^-$	1	0.81	1.09×10^{-2}	0.43
$Mg \rightarrow Mg^{2+} + 2e^-$	1	1.09	2.2×10^{-2}	0.89

* mdd = mg per dm^2 per day.
mpy = mils per year (1/1000 inches per year).

2.5 Basic Properties That Determine if Corrosion Is Possible and How Fast Material Can Corrode

The corrosion process in Figure 2.1 is an example of a so–called spontaneous electrochemical cell reaction. The driving force of the reaction is a reversible cell voltage. The cell voltage can be expressed as the difference between the potentials of the two electrodes (anode and cathode). Reversible cell voltage and the corresponding reversible potentials of the electrodes are determined by *thermodynamic properties*. These properties, which are dealt with in Chapter 3, mainly provide a possibility to determine the spontaneous *direction* which a given reaction tends to have. Applied to corrosion, the thermodynamics can tell us if corrosion is theoretically possible or not under given conditions.

The driving voltage of the cell reaction must cope with various types of resistance: a) resistance against charge transfer between a metal and the adjacent electrolyte at the anode and cathode, respectively; b) resistance due to limited access of reactants or limited removal of reaction products at the electrodes; c) ohmic resistance in the liquid and possibly in the metal between the anode and the cathode. The driving voltage and the sum of the resistances will together determine how fast the reactions will proceed, i.e. how fast a given material will corrode. Resistances against the reactions and the resulting reaction rates are described and explained under *electrode kinetics* (Chapter 4). In other words by using kinetic relationships we can determine *how fast* a material will corrode under certain conditions. Thermodynamics and electrode kinetics make up the two main parts of corrosion theory. These are the key to understanding and explaining the majority of practical corrosion problems. It is normal to apply this corrosion theory in modern corrosion testing, and it provides a more rational basis for corrosion prevention and monitoring.

References

2.1 Pourbaix M. Lectures on Electrochemical Corrosion. New York–London: Plenum Press, 1973.

2.2 Zumdahl SS. Chemical Principles. Lexington, MA: D.C. Heath and Company, 1992.

EXERCISES

1. Which reaction is the dominating cathodic reaction by corrosion in natural environments, such as seawater, fresh water, soil and condensation in the atmosphere? Write the reaction equation.

2. What types of corrosion products are usually deposited by corrosion in natural environments? What is rust? How can corrosion product deposits affect the corrosion rate?

3. Which property of the corrosion medium (the aqueous solution) is the most important prerequisite for electrochemical corrosion?

4. What can be said about the relationship between the anodic current I_a and the cathodic current I_c in a normal corrosion process?

5. A steel plate has corroded on both sides in seawater. After 10 years a thickness reduction of 3 mm is measured. Calculate the average corrosion current density. Take into consideration that the dissolution reaction is mainly $Fe \rightarrow Fe^{2+} + 2e^-$, and that the density and the atomic weight of iron are 7.8 g/cm^3 and 56, respectively.

6. What are the roles of thermodynamics and electrode kinetics in corrosion theory?

3. Thermodynamics – Equilibrium Potentials

There was opposed to Henry Le Chatilier a tradition that knew nothing of thermodynamics. He was reproached for encouraging the encroachment of physics and mathematics into a science which could perhaps ignore them. But in scientific matters, it is always a ticklish matter to make ignorance a weapon in your arsenal.

GEORGES URBAIN, 1925

3.1 Introduction

The above quotation from Georges Urbain has been reprinted from a book written by the corrosion scientist who more than any other person has contributed to the thermodynamic basis for modern corrosion science and technology, namely the Belgian professor Marcel Pourbaix (1904–1998) [3.1].

In corrosion theory and partly in practical corrosion engineering, the concept of electrode potential cannot be avoided. In order to understand corrosion and in many cases to take the appropriate steps to prevent it, a reasonably clear understanding of the term electrode potential is necessary. In this chapter it is the intention to establish such an understanding and present the formal quantitative expressions of electrode potentials, particularly under reversible or equilibrium conditions. Further, equilibrium diagrams of various metals are introduced. These diagrams show whether corrosion and other reactions are theoretically possible or not at given values of potential and pH of the environment.

3.2 Free Enthalpy and Cell Voltage

When a system is transferred from one state to another at constant pressure and temperature, it is subject to a change of free enthalpy (or Gibbs free energy) given by

$\Delta G =$ –(maximum work available from the system through the state transition), when this work is given a positive sign.

$$\Delta G = -(W_m - P\Delta V), \quad (3.1)$$

where W_m is the maximum work and $P\Delta V$ is used to press the atmosphere aside, so that the maximum available work is the difference between these terms.

We know already that the system tends to transform to a lower energy level by itself; i.e., a net state transition (a net reaction) that occurs without any external supply of energy, is accompanied by a negative ΔG. Such a reaction is called a spontaneous reaction. Thus we have:

ΔG negative:	*Spontaneous reaction, the system can carry out work, the system releases energy.*
$\Delta G = 0$:	*The system is in equilibrium.*
ΔG positive:	*Energy must be supplied to obtain this.*

In order to illustrate change of free enthalpy, a mechanical analogy, similar to that presented in Reference [3.2], can be used as shown in Figure 3.1.

We consider a stone that can move from position 1 on the top of a rock to position 2 in the valley following two possible paths: A, where the stone is immediately falling down the nearest cliff, and B, where the stone has to roll some distance with a little slope before falling. This is a spontaneous "reaction"; ΔG is negative. $|\Delta G|$ is the maximum available work Fh, where F is the force of gravity acting on the stone and h is the height of the step. The state transitions along the two different paths are both accompanied by the same ΔG. The example illustrates that ΔG is independent of the path between the states 1 and 2 (G is a state function). But reasonably, the transition along path A will occur faster than the one along path B. Correspondingly we also have for electrochemical reactions:

A change in free enthalpy does not tell us anything about the reaction rate.

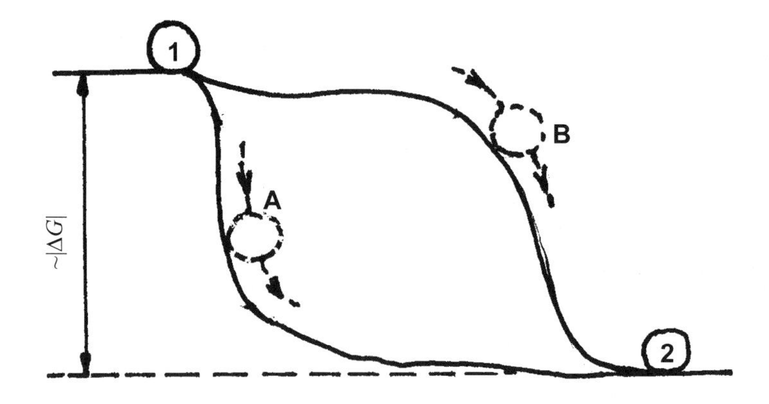

Figure 3.1 Mechanical analogy for change in free enthalpy.

Now we will consider an electrochemical cell with electrodes of metal A and metal B, respectively, as shown in Figure 3.2. A porous wall prevents mixing of the electrolytes in the two electrode compartments of the cell.

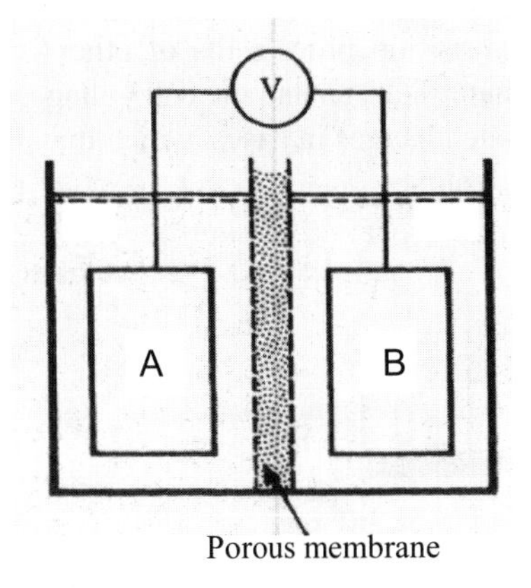

Figure 3.2 Reversible cell.

We presume that the internal resistance of the voltmeter is so high that the current in the external circuit is negligible, and further that the metals A and B are in equilibrium with their respective ions A^{z+} and B^{z+} in the liquid, i.e. that equilibrium has been established for the reactions:

$$A^{z+} + ze^{-} \leftrightarrow A,$$
$$B^{z+} + xe^{-} \leftrightarrow B.$$

This cell is called a reversible cell (there is no loss due to current flow, and equilibrium exists at the electrodes).

On the voltmeter we can read a voltage = E_o. The maximum electrical energy that can be achieved from the cell, which is equivalent to the maximum useful work, equals voltage · charge = E_oQ. At the same time, per definition, it is also = $-\Delta G$. Now, Faraday's laws [3.3, 3.4] tell us that per mole of a species (of valence z) reacting at an electrode, a charge of $Q = zF$ is circulating through the cell. F = Faraday's number ≈ 96,500 C/mol e^{-}. From this we have that the change of free enthalpy per mole of such a species reacting at an electrode, can be expressed by

$$\Delta G = -zFE_o. \tag{3.2}$$

By this, considerable progress has been made: the most important thermodynamic function has been related to the electric cell voltage.

By Equation (3.2) we have also defined the sign of cell voltage: a *spontaneous cell reaction corresponds to a positive cell voltage.* To force the reaction in the opposite direction, an external counter voltage of at least the same magnitude has to be applied. This counter voltage is defined as negative.

3.3 The Influence of the State of Matter on Free Enthalpy and the Change of Free Enthalpy

Since $|\Delta G|$ or E_o is the driving force of any corrosion process (as of other cell reactions), it is important to know which factors that affect ΔG and the cell voltage.

We forget corrosion for a while and consider state changes in gases, which may be known from the study of thermodynamics in other courses.

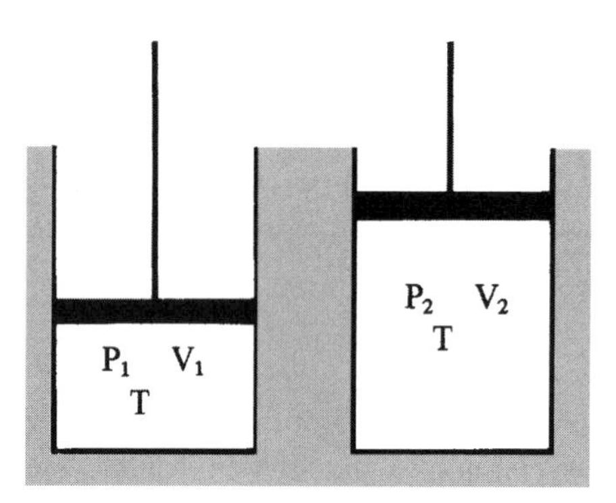

Figure 3.3 Change of state of an ideal gas.

For a quantity of n mol of an ideal gas that undergoes a change of pressure from P_1 to P_2 at constant temperature (see Figure 3.3) the following change in free enthalpy occurs:

$$\Delta G = nRT \ln \frac{P_2}{P_1} \qquad (3.3)$$

For the sake of simplicity, a *standard state* of the ideal gas has been defined, namely by $P = P^o = 1$ atm at the given temperature. The free enthalpy of this state is called the *standard free enthalpy* $= G^o$. For a pressure change from $P^o = 1$ atm to an arbitrary pressure P, at which the free enthalpy is G, we have according to Equation (3.3):

$$\Delta G = G - G^o = nRT \ln \frac{P}{P^o} = nRT \ln P.$$

If we define G and G^o per mole of the gas, the free enthalpy at pressure P is

$$G = G^o + RT \ln P. \qquad (3.4)$$

If instead of an ideal gas we have a liquid in equilibrium with the vapour above the liquid (Figure 3.4), we notice two conditions:

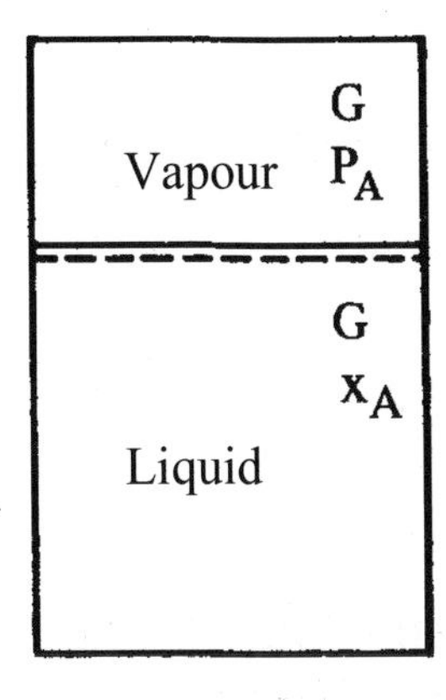

Figure 3.4 Liquid – vapour equilibrium.

a) At equilibrium, i.e. when there is no tendency of any net transfer of liquid to the vapour phase or vice versa, the free enthalpy in the liquid must equal the free enthalpy in the vapour. (If G_{liquid} deviated from G_{vapour}, we would have a driving force in one of the directions leading to a further levelling by either evaporation or condensation.)

b) For many liquids it has been shown that the partial pressure of a species in the vapour above the liquid is proportional to the mole fraction of the same species in the liquid. Such liquids are called ideal solutions.

With this background, and provided that the vapour above the liquid can be considered as an ideal gas, an expression of free enthalpy for ideal solutions is easily derived:

For the solvent: $$G = G^o + RT \ln x, \tag{3.5}$$

where x is the mole fraction of the solvent in the liquid.

For dissolved matter (in diluted solutions): $$G = G^o + RT \ln c, \tag{3.6}$$

where c = concentration of the dissolved species in mol/l (molarity).

Because not all gases and liquids can be considered ideal, Equations (3.4), (3.5) and (3.6) have been corrected by introducing the concept of activity, a, so that a general thermodynamic equation is obtained:

$$G = G^o + RT \ln a \tag{3.7}$$

that is valid both for gases and solutions. It is also valid for solid matter and pure liquids.

The standard state is defined by the pressure, the mole fraction or the molarity (depending on the state of matter) that gives a = 1 at the given temperature. When we are dealing with corrosion, we are usually so lucky that with satisfactory accuracy we can assume:

For gas:	$a \approx P$ (given in atm).
For solvent:	$a \approx x$ (usually ≈ 1).
For dissolved matter (in ionic form):	$a \approx c$ (given in mol/l).

For solid matter and pure liquids G varies very little over a large range of pressures, so that we can assume $G = G^o$, i.e. a = 1.

3.4 Change of Free Enthalpy in Chemical Reactions. Reversible Cell Voltage

Since we now have expressed G for any state of matter, these expressions can be used on chemical reactions. A general reaction can be written as:

$$aA + bB + \cdots\cdots \rightarrow lL + mM + \cdots\cdots$$

in which there are *a* mol of reactant A etc., and we obtain *l* mol of product L etc. (compare Figure 3.5).

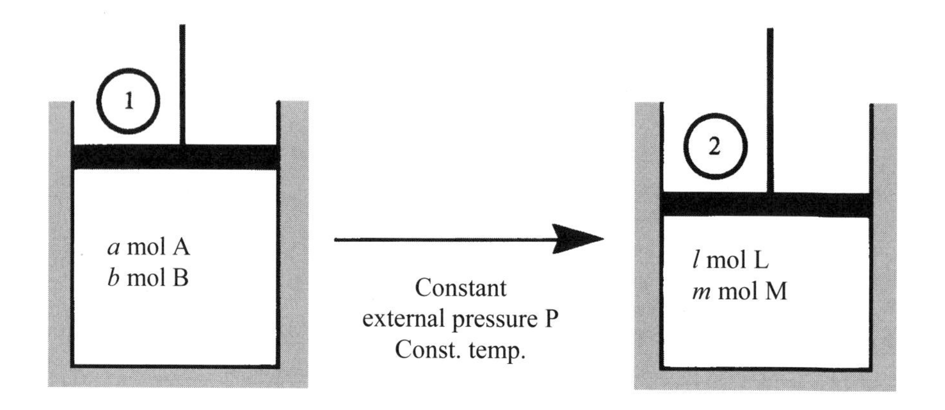

Figure 3.5 A general reaction at constant external pressure and constant temperature.

The total change in free enthalpy for the reaction is equal to the total free enthalpy of all the products minus the total free energy of the reactants:

$$\Delta G = lG_L + mG_M + \cdots - aG_A - bG_B - \cdots$$
$$= lG_L{}^o + mG_M{}^o + \cdots - aG_A{}^o - bG_B{}^o - \cdots$$
$$+ RT\,(l\ln a_L + m\ln a_M + \cdots - a\ln a_A - b\ln a_B - \cdots)$$

$$\Delta G = \Delta G^0 + RT\ \ln\frac{a_L^l\, a_M^m \cdots}{a_A^a\, a_B^b \cdots} \tag{3.8}$$

(In calculations, ΔG_O^O = standard free enthalpy of formation, and ΔG_f = free enthalpy of formation for the involved reactants and products.)

If this is a reversible electrochemical cell reaction, we have as previously mentioned:

$$\Delta G = -zFE_o, \tag{3.2}$$

and when the activity of all involved species = 1:

$$\Delta G^O = -zFE_O^O \tag{3.2.a}$$

where $E_o{}^\circ$ is *standard cell voltage.*

Inserting (3.2) and (3.2a) in (3.8) we obtain the important equation for the voltage of a reversible cell:

$$E_o = E_o^o - \frac{RT}{zF}\ln\frac{a_L^l\, a_M^m \cdots}{a_A^a\, a_B^b \cdots}. \tag{3.9}$$

3.5 Electrode Reactions and Electrode Potentials

We consider now a reversible cell consisting of a copper electrode and a zinc electrode, both in equilibrium with their respective ions of activity a = 1 at 25°C (Figure 3.6).

Such electrodes are called equilibrium or reversible electrodes or half-cells, and when the activities a = 1 as in this case, they are called standard electrodes or standard half cells.

We have the following electrode reactions:

$$Cu^{2+} + 2e^- = Cu \tag{3.10}$$
$$Zn = Zn^{2+} + 2e^- \tag{3.11}$$

Cell reaction: $Cu^{2+} + Zn = Cu + Zn^{2+}$.

If we connect the voltmeter (with large internal resistance) as shown in Figure 3.6, we will measure a voltage of

$$E_O^O = 1.1\text{ V}.$$

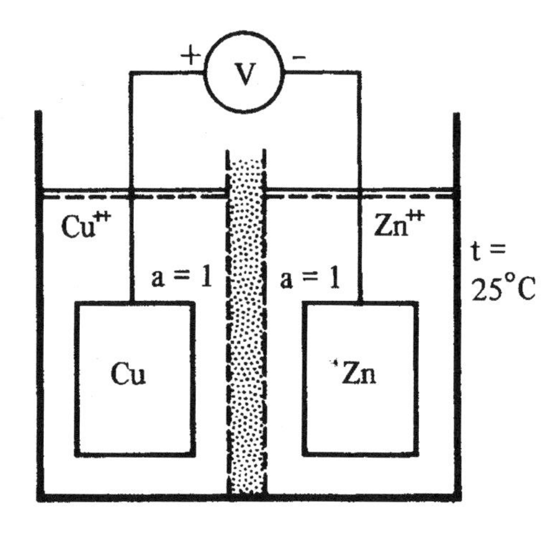

Figure 3.6 Reversible cell with standard electrodes.

Sometimes we are interested in knowing the reversible cell voltage of various electrode combinations. However, the number of such combinations is extremely large, and tables of cell voltages would make up a huge amount of data.

In order to simplify, *the quantified concept half-cell potential or electrode potential has been introduced.* This has been done by choosing a certain electrode reaction as reference, and defining the equilibrium potential of this electrode to be equal to zero. The numerical value of the equilibrium potential of any other electrode reaction X is given by the reversible cell voltage of the combination X–reference electrode, see Figure 3.7.

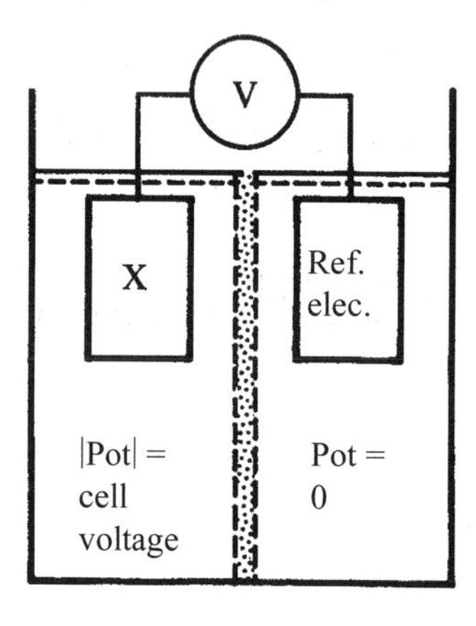

Figure 3.7 Reversible cell consisting of an arbitrary electrode X and a reference electrode.

An internationally accepted reference is the reversible reaction $H_2 = 2H^+ + 2e^-$ at activities a = 1, i.e. $a_{H^+} = 1$, $a_{H2} \approx P_{H2}(atm) = 1$.

This reaction equation is of the same form as the metal reaction equations we have considered (Cu and Zn reactions). For the hydrogen electrode we need, however, a metal as a base for the reaction. Platinum is the metal usually applied for this purpose, for one thing because of its noble and corrosion resistant character. Pt does not take part in the reaction itself, but serves as an electron conductor and base for electron transfer.

At equilibrium the hydrogen reaction occurs as illustrated in Figure 3.8. The electrode is called a standard hydrogen electrode (SHE).

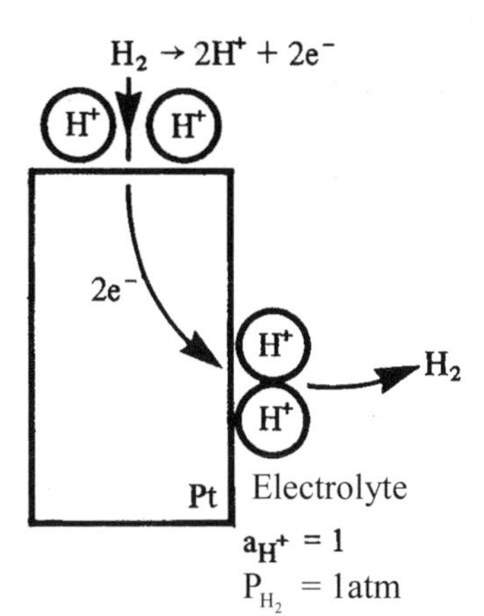

Figure 3.8 Standard hydrogen electrode (SHE).

If we replace the copper electrode in Figure 3.6 by a standard hydrogen electrode, we get a cell, as shown in Fig 3.9. The reactions in this cell are:

$$2H^+ + 2e^- = H_2, \quad (3.12)$$
$$Zn = Zn^{2+} + 2e^-, \quad (3.11)$$
$$2H^+ + Zn = H_2 + Zn^{2+}.$$

When the voltmeter is connected as shown in the figure a potential of $E_0^0 = 0.76$ V is measured. The Zn electrode is negative. *From this we say that the Zn electrode has a standard electrode potential of −0.76 V referred to the standard hydrogen electrode at 25°C. (E = −0.76 V Reference SHE, or $E_H = -0.76$ V.)*

As we have seen, the potential is defined (with sign) as an electrically measurable quantity. The numerical value is given by the measured voltage, and the sign is indicated by the polarity of the voltmeter terminals.

Since the numerical value of the Zn–electrode potential equals the voltage of the cell in Figure 3.9, we can also determine the potential by calculation, applying Equation (3.2). First we should, however, notice that, in the way we have written the reactions,

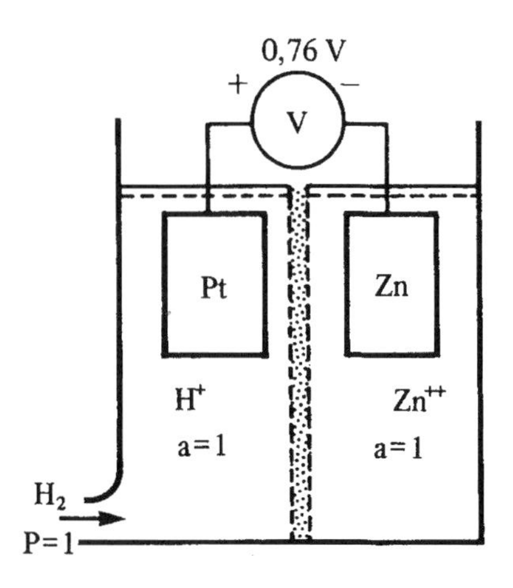

Figure 3.9 Cell consisting of standard hydrogen and standard Zn electrode.

the spontaneous reactions go the right, i.e. Zn is oxidized and hydrogen ions reduced. This is an example of a general rule:

In a spontaneous cell reaction, an oxidation reaction takes place at the negative electrode, and a reduction reaction takes place at the positive electrode.

Thus, the reaction to the right (oxidation of Zn) corresponds to a positive cell voltage = 0.76 V, while the Zn electrode potential is found to be −0.76 V. To get the correct sign when calculating the electrode potential we must use Equation (3.2) for the opposite reaction, i.e. for reduction of Zn^{2+}. Potentials with this sign rule are therefore called *reduction potentials.*

When $a_{Zn2+} \neq 1$, we can apply the same rule and use the general voltage equation (3.9) to determine the equilibrium potential for the Zn^{2+} electrode, writing the cell reaction equation such that Zn^{2+} is reduced, i.e. that reduced Zn is considered a reaction product, (right side of the reaction equation). (Compare the derivation of Equation 3.9.) This gives us the potential:

$$E_0 = E_0^o - \frac{RT}{zF} \ln \frac{a_{H^+}^2 \, a_{Zn}}{a_{H_2} \, a_{Zn^{2+}}}.$$

Since both a_{H+} and $a_{H2} = 1$, we obtain:

$$E_0 = E_0^o - \frac{RT}{zF} \ln \frac{a_{Zn}}{a_{Zn^{2+}}}.$$

Now it becomes clear that this equation could have been obtained directly by using Equation (3.9) on the electrode reaction $Zn^{2+} + 2e = Zn$ without going via the cell reaction. *Generally we can use Equation (3.9) to determine the potential of any electrode reaction when we consider this as a reduction reaction.* In other words we have the equilibrium electrode potential

$$E_0 = E_o^o - \frac{RT}{zF} \ln \frac{a_L^l \, a_M^m \cdots\cdots}{a_A^a \, a_B^b \cdots\cdots} \tag{3.9}$$

for the electrode reaction

$$aA + bB + \cdots\cdots + ze^- = lL + mM + \cdots\cdots$$

By this *calculation* the potential is determined as a thermodynamic function, while in connection with voltage *measurement* we defined and determined it as an electrical quantity. Both considerations are useful.

When we are dealing with corrosion we can usually replace the activities in Equation (3.9) with concentration, partial pressure etc., as mentioned on page no. 18. Equation (3.9) is well known as Nernst's equation, after Walther Nernst, a German physicist and chemist (1864–1941).

The conclusion of this section can be expressed as follows:

The electrode potential of a reaction X referred to standard hydrogen electrode = the voltage of the cell in which hydrogen reaction proceeds as oxidation and X as reduction.

3.6 Series of Standard Potentials

In the same way as we found the standard electrode potential of the reaction $Zn^{2+} + 2e^- = Zn$ to be −0.76 V, we can determine potentials for any other redox process. A series of such standard potentials is shown in Table 3.1. For example, the standard potential of the reaction $Cu^{2+} + 2e^- = Cu$ is 0.34 V. Thus, the reversible cell consisting of a standard Cu electrode and a standard Zn electrode has a cell voltage of E = 0.34 – (–0.76) = 1.1 V.

3.7 Equilibrium Potentials of Reactions with Iron at 25°C

By inserting

R = 8.3 J/K,
T = 298 K,
ln = 2.3 log

Table 3.1 Standard electrode potentials E_O^O at 25°C.

Electrode reaction	E_0^0 (V)
$Au^{3+} + 3e^- = Au$	1.50
$Cl_2 + 2e^- = 2Cl^-$	1.36
$O_2 + 4H^+ + 4e^- = 2H_2O$	1.23
$Pt^{2+} + 2e^- = Pt$	1.20
$Fe_3O_4 + 8H^+ + 2e^- = 3Fe^{2+} + 4H_2O$	0.98
$HNO_3 + 3H^+ + 3e^- = NO + 2H_2O$	0.96
$Ag^+ + e^- = Ag$	0.80
$Fe^{3+} + e^- = Fe^{2+}$	0.77
$Fe_2O_3 + 6H^+ + 2e^- = 2Fe^{2+} + 3H_2O$	0.73
$O_2 + 2H_2O + 4e^- = 4OH^-$	0.40
$Cu^{2+} + 2e^- = Cu$	0.34
$Hg^{2+} + 2e^- = 2Hg$	0.79
$AgCl + e^- = Ag + Cl^-$	0.22
$S + 2H^+ + 2e^- = H_2S$	0.14
$2H^+ + 2e^- = H_2$	0
$Fe_3O_4 + 8H^+ + 8e^- = 3Fe + 4H_2O$	–0.085
$Pb^{2+} + 2e^- = Pb$	–0.13
$Sn^{2+} + 2e^- = Sn$	–0.14
$Ni^{2+} + 2e^- = Ni$	–0.25
$Cd^{2+} + 2e^- = Cd$	–0.40
$Fe^{2+} + 2e^- = Fe$	–0.44
$Cr^{3+} + 3e^- = Cr$	–0.74
$Zn^{2+} + 2e^- = Zn$	–0.76
$2H_2O + 2e^- = 2OH^- + H_2$	–0.83
$Ti^{2+} + 2e^- = Ti$	–1.63
$Al^{3+} + 3e^- = Al$	–1.66
$Mg^{2+} + 2e^- = Mg$	–2.37
$Na^+ + e^- = Na$	–2.71

in Equation (3.9), a potential equation generally valid at 25°C is obtained:

$$E_0 = E_o^o - \frac{0.059}{z} \log \frac{a_L^l \, a_M^m \cdots\cdots}{a_A^a \, a_B^b \cdots\cdots} .$$

For the reaction

a) $Fe^{2+} + 2e^- = Fe$,

we get an equilibrium potential of:

$$E_o = -0.44 - \frac{0.059}{2} \log \frac{1}{a_{Fe^{2+}}} \quad (V)$$

$$E_o = -0.44 + 0.0295 \log a_{Fe^{2+}} \quad (V) ,$$

where $E_O^O = -0.44$ V is taken from Table 3.1. For another actual reaction on iron,

b) $Fe_3O_4 + 8H^+ + 8e^- = 3Fe + 4H_2O$,

we have

$$E_o = -0.085 - \frac{0.059}{8} \log \frac{a_{Fe}^3 \, a_{H_2O}^4}{a_{Fe_3O_4} \, a_{H^+}^8} \quad (V) .$$

Since $a_{Fe} \approx a_{Fe_3O_4} \approx a_{H_2O} \approx 1$, the expression is reduced to

$$E_o = -0.085 - \frac{0.059}{8} (-8 \log a_{H^+}) \quad (V)$$

and since $-\log a_{H^+} = \text{pH}$, we obtain

$$E_o = 0.085 - 0.059 \, \text{pH} \quad (V).$$

In addition to the reactions a) and b) we have also other reactions of importance for iron in water:

c) $Fe_3O_4 + 8H^+ + 2e^- = 3Fe^{2+} + 4H_2O$,

d) $Fe_2O_3 + 6H^+ + 2e^- = 2Fe^{2+} + 3H_2O$,

e) $O_2 + 4H^+ + 4e^- = 2H_2O$,

f) $2H^+ + 2e^- = H_2$.

The equilibrium potentials of these reactions are determined in the same way as for the reactions a) and b). Note that H^+ is involved in all reactions b)–f). Therefore, equilibrium potentials vary linearly with pH, as shown for reaction b).

3.8 Pourbaix Diagram [3.5]

pH is an important variable of aqueous solutions, and it affects the equilibrium potentials of a majority of the possible reactions that can occur. On this basis, Marcel Pourbaix derived and presented his pH potential diagrams, also called equilibrium diagrams and most frequently Pourbaix diagrams. These diagrams have become an import tool for the illustration of the possibilities of corrosion.
The Pourbaix diagram is a graphical representation of Nernst's equation for the various reactions. From the potential equations of the reactions a)–f) we obtain the Pourbaix diagram for iron in water at 25°C, as shown in Figure 3.10. Here, an activity of ferrous ions, $a_{Fe^{2+}}$, $= 10^{-6}$ is presumed. This order of magnitude is assumed to be a realistic lower level of $a_{Fe^{2+}}$ for situations where corrosion is of practical significance.

In addition, for reaction a) the equilibrium potentials for $a_{Fe^{2+}} = 10^{-4}$ and 10^{-2} have been given by dashed lines in Figure 3.10.

For the reactions e) and f) gas pressures (P_{O_2} and P_{H_2}) of 1 atm have been presumed.

The background and significance of the lines in the Pourbaix diagrams are now considered clear, but the regions between the lines are equally important. From the potential equation for the reaction $Fe^{2+} + 2e^- = Fe$ (and from the diagram in Figure 3.10) we can see that the equilibrium potential increases with increasing activity of ferrous ions. If we start with an activity of 10^{-6} and affect the potential to attain a level above the equilibrium potential at this activity, the system will (as any system that are moved away from equilibrium) try to re-establish equilibrium by itself. This can be done only by an increase of the Fe^{2+} activity, i.e. by the reaction going to the left. Fe^{2+} is therefore the dominating reaction product. In other words, we have a corrosion region above the equilibrium line of this reaction.

We see that the oxidized state of the material is the stable one when the potential is above the equilibrium potential. This is a general rule, and therefore we have Fe_3O_4 and Fe_2O_3 as stable products when we are above the lines b, c and d. Particularly, Fe_2O_3 is known for its ability to form a passivating film on the surface, i.e. an oxide film that prevents, to a great extent, Fe^{2+} ions from being transported from the metal to the liquid. Characteristic properties of such a film are low ionic conductivity, low porosity and good adherence to the substrate.

When the potential is below the lines a and b, the metallic state (Fe) is the stable one. Therefore, the metal is immune to corrosion in this region.

As it is seen, the pH–potential plane is clearly divided into a corrosion region, a passivity region and an immunity region (Figure 3.10). There is also a small corrosion region at high pH, where the dissolved corrosion product is $HFeO_2^-$.

Corresponding diagrams for other metals are shown in Figure 3.11.

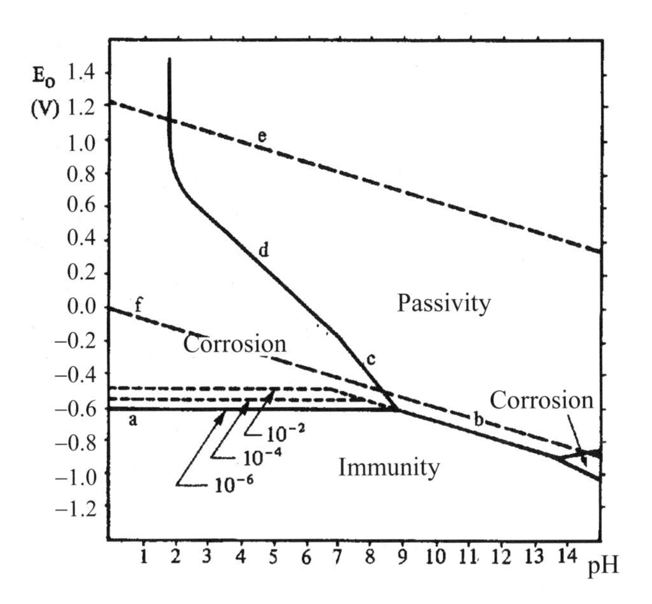

Figure 3.10 Pourbaix diagram of iron in water at 25 °C.

The main objectives of the Pourbaix diagrams are:

1. To show the directions of the various reactions at given pH and potential.
2. To make a basis for estimation of the corrosion product compositions at various pH and potential combinations.
3. To show which environmental pH and potential changes will reduce or prevent corrosion.

The validity of the diagrams is limited to reactions between pure metals, pure water and the species that can be formed from these. Small amounts of impurities and alloying elements in the metal and dissolved substances in the water do not necessarily influence strongly on the diagram, but in some cases they do. Typical examples of the latter are when iron base materials are alloyed with Cr and the water contains chloride. Chlorides destroy the passive films locally when the potential exceeds a certain level (pitting potential). It should also be noticed that the horizontal axis of the diagram refers to the pH close to the metal surface. This pH is affected by the local reaction and can therefore deviate, sometimes strongly, from the bulk pH.

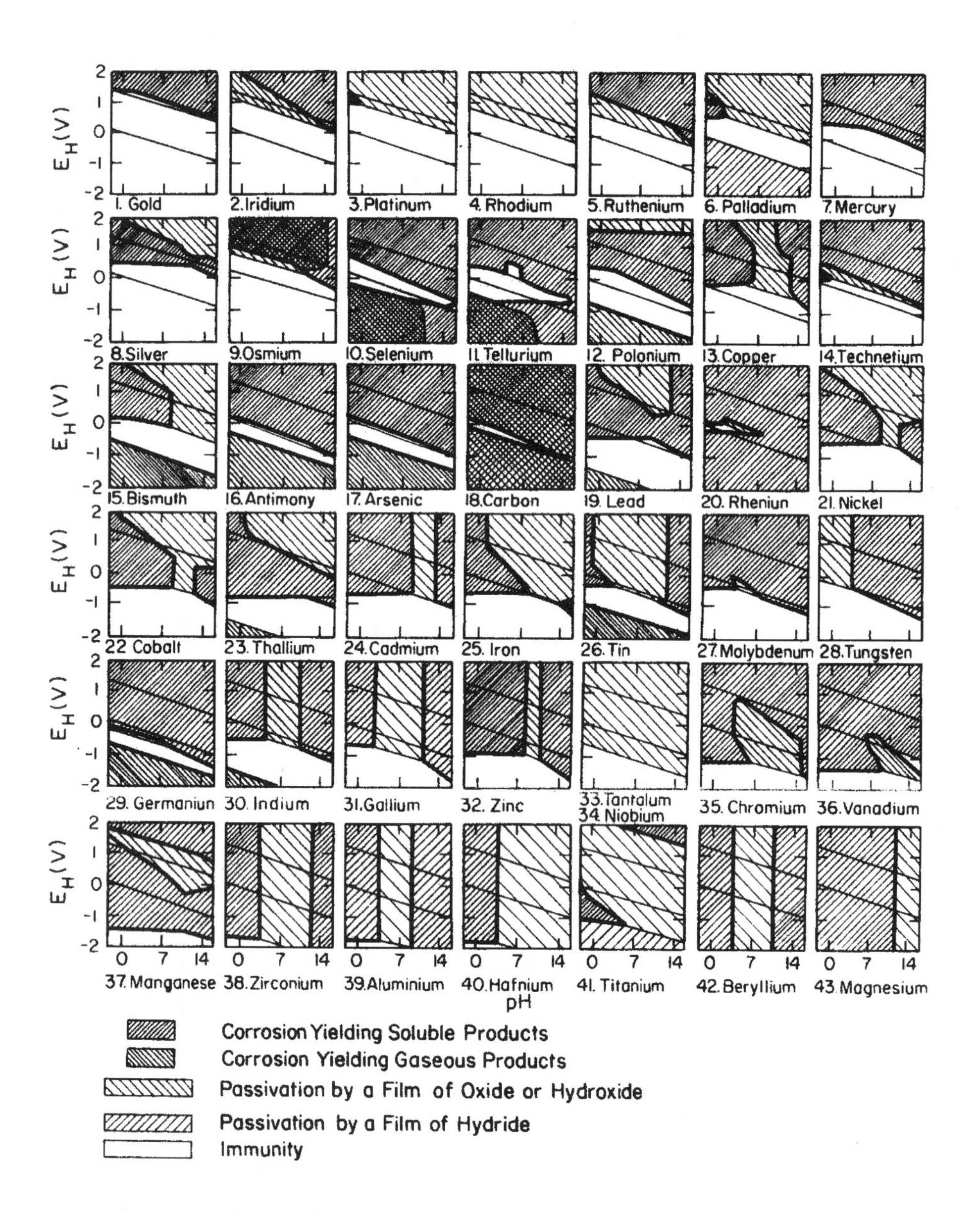

Figure 3.11 Pourbaix diagrams for various metals and metalloides arranged after their nobleness [3.1].

We have now completed the thermodynamic and formal presentation of equilibrium potential and equilibrium diagram. In addition to this knowledge it may often be useful to have a qualitative electrostatic image of the electrode potential, e.g. as shown in the next section.

3.9 A Simplified Presentation of Equilibrium Potential and Deviation from It

a) *In general*

A metal submerged in an electrolyte that does not contain the metal's ions has a tendency to release ions (Stage (1) in Figure 3.12). At this stage the metal is more positive compared with the nearest layer of the liquid than that corresponding to equilibrium.

If the release of ions (dissolution) is the only reaction, and the metal is not connected to an external circuit, a surplus of electrons is produced in the metal. It becomes more negative, and the dissolution rate is reduced. At the same time the concentration of positive metal ions in the liquid increases, causing an increasing

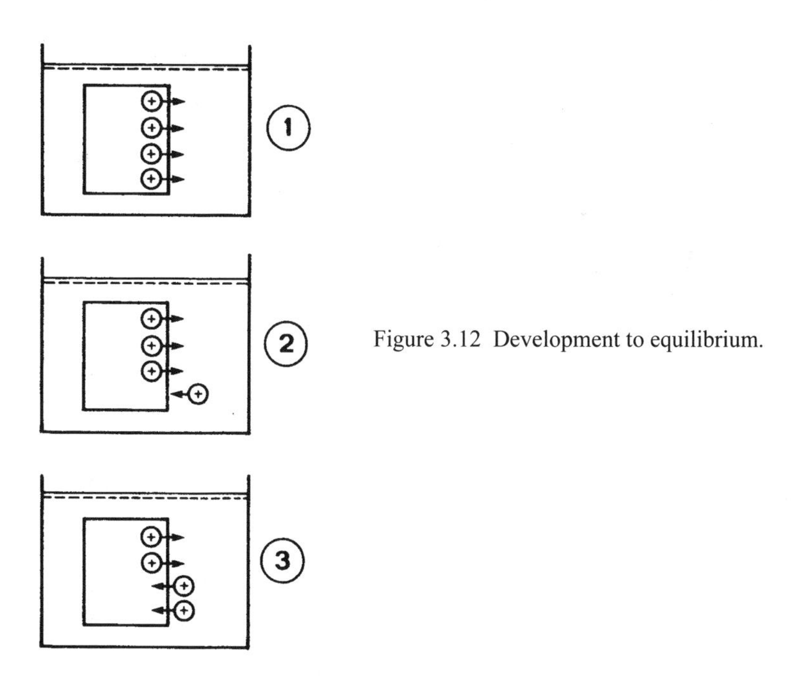

Figure 3.12 Development to equilibrium.

number of dissolved ions to be discharged on the metal surface again (2). This continues until the transport is equally large in both directions (3).

At his stage the metal is in equilibrium with its own ions in the solution under a potential difference between the metal and the liquid that is equal to the equilibrium potential at the ion concentration obtained in the liquid.

b) *The difference in standard potential* E_O^O *between different metals*

Zinc has a strong ability to dissolve. To balance the dissolution by ion discharge on the metal, the metal has to be charged relatively negative (strong attraction from negative charge, see Figure 3.13). In other words, the equilibrium potential in general, and hence the standard potential, is negative.

Gold has poor ability to dissolve, and gold ions in the liquid have strong liability to discharge on the metal. To obtain a balance, the atoms in the metals need a relative positive pressure, i.e. gold has a positive equilibrium potential, including a positive standard potential.

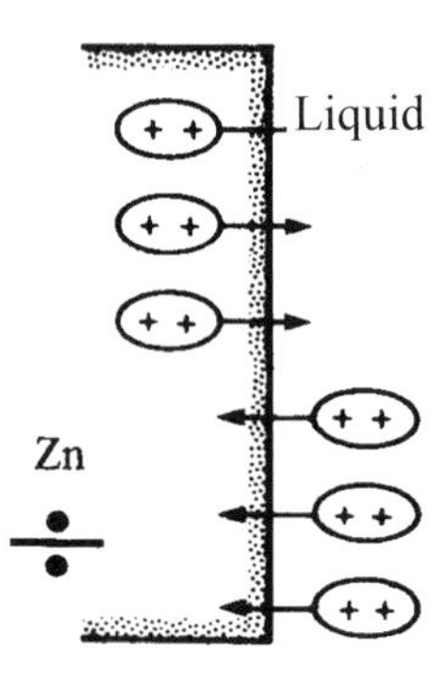

Figure 3.13 Difference between a reactive metal (Zn) and a noble one (Au).

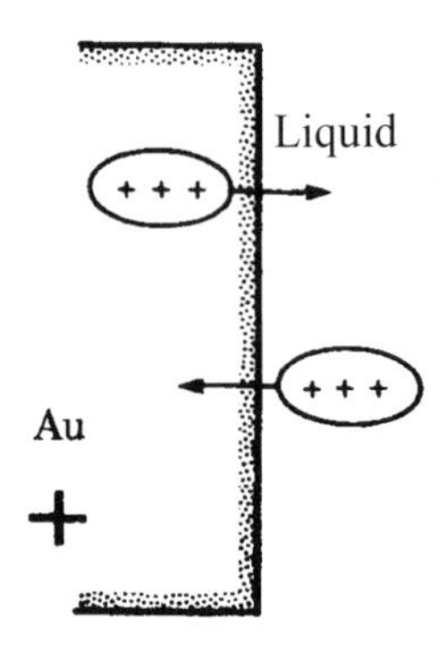

c) *The effect on the activity term in Nernst's equation*

The larger the activity of, e.g. Zn^{2+} in the solution, the more zinc ions are discharged. To balance this with a higher rate of ion release from the metal, a higher positive pressure is needed in the metal. That is, the equilibrium potential increases with increasing activity of metal ions in the liquid, in agreement with Nernst's equation:

$$E_o = E_o^o + \frac{RT}{zF} \ln a_{Zn^{2+}} .$$

d) *Deviation from the equilibrium potential*

If a metal is in equilibrium with its ions in the solution, and then is made more positive by external measures (Figure 3.14), i.e. the potential is made more positive than the equilibrium potential, we can expect to get a net dissolution of the metal.

As we can see, an oxidation reaction occurs if the potential is more positive and a reduction reaction occurs if the potential is more negative than the equilibrium potential, in agreement with the considerations of thermodynamics dealt with in Section 3.8.

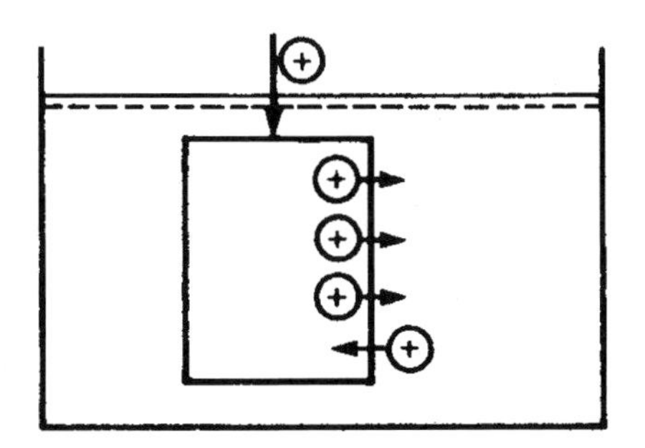

Figure 3.14 Deviation from the equilibrium potential.

3.10 Possible Range for Real Potentials Under Corrosion Conditions

As shown above, oxidation occurs when the electrode potential is higher, and reduction when it is lower, than the equilibrium potential. Hence, when we have, as in corrosion, two simultaneous reactions, of which one is oxidation (anodic reaction) and the other one is reduction (cathodic reaction), the real potential must lie between the equilibrium potentials of the two reactions. If we consider corrosion of iron in aerated water, with reduction of oxygen as the cathodic reaction, the potential has to be somewhere between the lines a and e in Figure 3.10. In acid (and usually in neutral) solutions the potential will lie in the corrosion region, in alkaline solutions in the passive region. With efficient oxygen supply, which for instance can be promoted by heavy convection of the solution, passivity may also be achieved in neutral water.

In acid solutions, iron can corrode without access to oxygen, because hydrogen can be reduced (reaction (f), hydrogen evolution) at a potential between the lines a and f in Figure 3.10. These relationships are explained more thoroughly in Chapters 4, 5 and 6.

It can be seen from preceding arguments that the electrode potential is in fact a potential difference imagined as an electrical potential difference between the metal surface and the adjacent electrolyte layer. This potential difference cannot be measured directly and absolutely, it has to be referred to (compared with) a corresponding potential difference for a reference electrode. What we define as electrode potential is therefore in reality a "difference between two potential differences". The electrode potential referred to a given reference electrode is measured as shown in principle in Figure 3.7. A practical set-up for measurements is dealt with in more detail in Chapter 4.

References

3.1 Pourbaix M. Lectures on Electrochemical Corrosion. New York–London: Plenum Press, 1973.
3.2 Fontana MG, Greene ND. Corrosion Engineering. New York: McGraw Hill, 1967, 1978.
3.3 Barrow GM. Physical Chemistry. New York: McGraw Hill, 1961.
3.4 Zumdahl SS. Chemical Principles. Lexington MA: D.C. Heath and Company, 1992.
3.5 Pourbaix M. Atlas of Electrochemical Equilibria, 2nd Ed. Houston, Texas: NACE, 1974.

EXERCISES

1. In the equation $\Delta G = -z\,FE_o$, E_o is the voltage of a reversible cell. Explain why E_o is positive for a spontaneous cell reaction. What has to be done to the cell to get a negative cell voltage, and what is the consequence on the reaction?

2. Free enthalpy of a species A in the vapour above a liquid is given by $G = G^o + RT \ln P_A$, where P_A is the partial pressure of A. Assume that the vapour is in equilibrium with the liquid and that the liquid is an ideal solution. Explain how we can conclude that the free enthalpy of species A in the liquid can be expressed by

   $$G = G^o + RT \ln x_A,$$

   where x_A is the mole fraction of A in the solution.

3. Explain the meaning of all the quantities in the formulae of the reversible cell voltage:

$$E_o = E_o^o - \frac{RT}{zF} \ln \frac{a_L^l \, a_M^m \cdots}{a_A^a \, a_B^b \cdots} .$$

Which quantity can usually be inserted instead of the activity a for a gas, a solvent, a dissolved substance and a pure liquid, respectively?

4. Why and how has the concept of electrode potential been established?

5. Which type of reaction occurs at the positive electrode and which type occurs at the negative electrode in a spontaneous cell reaction? Give reasons.

6. Explain why the equation for the reversible cell voltage also can be used for a reversible electrode potential. What must be noticed to get the sign right?

7. Calculate the standard potential E_o^o referred to the standard hydrogen electrode for the electrode reactions:

   I $Zn^{2+} + 2e^- = Zn$,
   II $ZnO + 2H^+ + 2e^- = Zn + H_2O$,
   III $HZnO_2^- + 3H^+ + 2e^- = Zn + 2H_2O$.

   The calculation is to be based upon change in standard free enthalpy (ΔG^o) of those cell reactions where I, II and III, respectively are combined with the standard hydrogen reaction, as shown in the following table:

Cell reaction	$\Delta G°$ per mol Zn reacting
$Zn + 2H^+ \rightarrow Zn^{2+} + H_2$	– 147.000 J
$Zn + H_2O \rightarrow ZnO + H_2$	– 87.700 J
$HZnO_2^- + H^+ + H_2 \rightarrow Zn + 2H_2O$	– 10.400 J

   The conclusion of Section 3.5 and the solution of Exercise 5 are useful with respect to the determination of the sign of the potential.
   Faraday's number F = 96,485 C/mol e^-.

8. Determine the equilibrium potentials E_o of the reactions I, II and III in Exercise 7 as functions of pH in the solution, when it is presumed that it contains 10^{-6} mole Zn^{2+} per litre for reaction I and the same concentration of $HZnO_2^-$ for reaction III. Draw the potential lines for these reactions in a potential pH diagram. The universal gas constant R = 8.3 J/K.mol. Assumed temperature T = 298 K.
9. Explain briefly the significance of the lines and of the regions between the lines in the Pourbaix diagrams and the usefulness of the diagrams.

10. State the most important conditions for the validity of the Pourbaix diagrams as well as some important deviations from these conditions in many practical cases.

11. Given a case where the two electrode reactions in a cell deviate more or less from equilibrium, in which range must the potential(s) be located? Give reasons.

4. Electrode Kinetics

One gets tired of everything
except to understand.
VIRGIL

4.1 Introduction: Anodic and Cathodic Reactions

While the Pourbaix diagram shows a qualitative picture of what can happen at a given pH and potential, we also need knowledge about the electrode kinetics to predict:

a) which values the potential and the local pH can attain, and
b) the reaction rates for the actual pH–potential combinations.

To begin with we look at the Zn–H cell again. We have previously considered this as a reversible cell (equilibrium at the electrodes, large resistance in the voltmeter, so that the current through the meter is close to zero). If instead of the voltmeter we only have a conductor that short-circuits the cell, electrons will flow through the conductor from the negative Zn electrode to the Pt plate. The Zn electrode and the Zn reaction are moved away from equilibrium by this removal of negative charges from the Zn plate. In an attempt to re-establish the equilibrium, positive ions are released from the Zn plate to the liquid. Thus, Zn is dissolved at the same rate as electrons are transported to the Pt plate, where they are consumed in the hydrogen reaction. Thus the following reactions occur at the electrodes:

$$Zn \rightarrow Zn^{2+} + 2e^-, \tag{4.1}$$
$$2H^+ + 2e^- \rightarrow H_2. \tag{4.2}$$

The same cell process can be totally obtained on a Zn plate submerged in a solution containing hydrogen ions and Zn ions. The reactions are accompanied by the same changes in free enthalpy and have the same equilibrium potentials as before. There is, however, a higher resistance against the hydrogen reaction on the Zn plate than on Pt, and thus the reaction rate will be lower on the Zn surface.

Here we have a typical corrosion process, with an anodic reaction = oxidation ($Zn \rightarrow Zn^{2+} + 2e^-$) on those parts of the surface that constitute the anode, and a cathodic reaction = reduction ($2H^+ + 2e^- \rightarrow H_2$) on the areas that constitute the cathode. The anode and cathode areas may change positions, such that a given surface area element is alternately the site for the Zn reaction and for the hydrogen reaction. If there is no such alternation or if it occurs very slowly, different forms of localized corrosion may be the result (see Chapter 7). This depends on whether there are homogeneous or inhomogeneous conditions along the surface. Inhomogeneous conditions cause permanent anodes and cathodes, while under homogeneous conditions arbitrary variations in the liquid may determine if a given area element is anode or cathode at a certain moment.

4.2 Polarization and Overvoltage

Usually, the electrode reactions (4.1) and (4.2) are moved away from equilibrium. For each of the reactions the potential is moved away from the equilibrium potential as a result of the net electrode reaction occurring, i.e. a net electric current flowing through the interface between metal and liquid. The deviation from equilibrium is called *polarization*, and we say that the electrode is polarized. A measure of polarization is the *overvoltage**, i.e. the difference between the real potential and the equilibrium potential. When a corrosion process takes place on a surface, the real potential adopts a value somewhere between the equilibrium potential of the cathodic and anodic reactions, respectively, as illustrated in Figure 4.1.

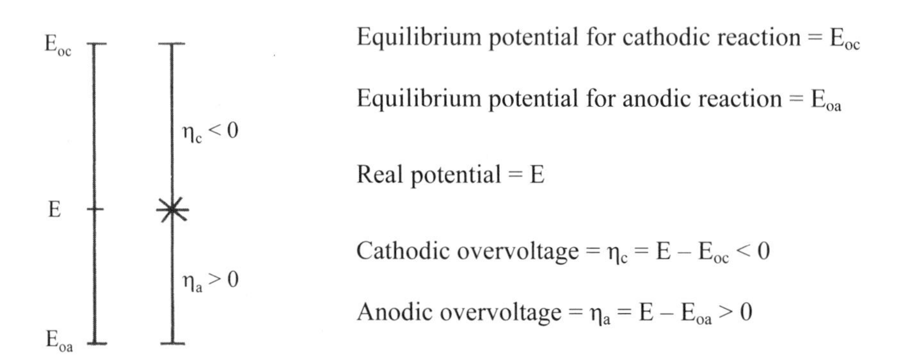

Figure 4.1 Equilibrium potentials, real potential and overvoltages.

* A more correct term would be overpotential. Overvoltage is used because this term is so well estabblished

4.3 Exchange Current Density

Before we look at the relationship between the overvoltage and the reaction rate, we return to the equilibrium state again. At the equilibrium potential of a reaction, a reduction and an oxidation reaction occur, both at the same rate. For example, on the Zn electrode, Zn ions are released from the metal and discharged on the metal at the same rate (Figure 4.2).

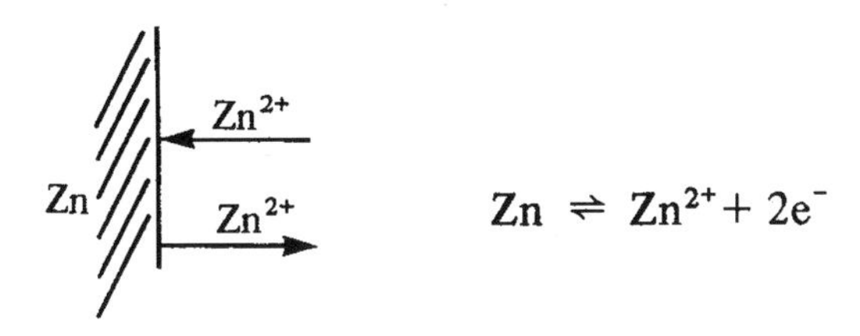

Figure 4.2 The Zn electrode in equilibrium.

The reaction rate in each direction can also be expressed by the transport rate of electric charges, i.e. by current or current density, called, respectively, exchange current, I_o, and (more frequently used) exchange current density, i_o. The net reaction rate and net current density are zero.

The exchange current density i_o depends strongly on the electrode reaction, electrode material and electrolyte (see Table 4.1) and significantly also on the temperature and concentrations. Data for some electrolytes and concentrations other than those included in Table 4.1 are given in References [4.1, 4.2]. For example, $i_o = 10^{-6}$ A/cm^2 is stated for the hydrogen reaction on Fe in 2 N H_2SO_4. For a reaction like $Fe^{2+} + 2e^- = Fe$ and correspondingly for other transition metals, i_o is often of the order of 10^{-5} A/cm^2 or less [4.3]. Later we shall see that i_o has strong influence on real corrosion rates, and thus it is of great practical significance. However, i_o data for various conditions are scarce.

4.4 Activation Polarization

When a reaction is moved away from its equilibrium, a net current is obtained in one of the directions. A larger or smaller resistance will always act against the current flow across the electrode interface, and as we have seen a certain overvoltage is required to cope with this resistance. Depending on the *type* of resistance that limits the reaction rate, we are talking about three different kinds of polarization, namely *activation polarization, concentration polarization* and *resistance (ohmic) polarization.* Activation polarization is caused by the resistance against the reaction itself at the metal–electrolyte interface. There is an energy barrier that the actual atoms or ions have to overcome to be transferred to a new state [4.1, 4.4, 4.5]. The rate-determining step may be ion or electron transfer across the interface, but it may also be some kind of conversion of a species involved in the reaction. For activation polarization the relationship between current density, i, and overvoltage, η, is given by the *Tafel equation* (after the German chemist Julius Tafel (1862–1918)):

$$\eta = \pm b \log \frac{i}{i_o}. \qquad (4.3)$$

For the anode: $$\eta_a = b_a \log \frac{i_a}{i_o}, \qquad (4.3\ a)$$

For the cathode: $$\eta_c = -b_c \cdot \log \frac{i_c}{i_o}, \qquad (4.3\ b)$$

where b_a and b_c are the anodic and cathodic *Tafel constants*. The Tafel constant is given by

$$b = \frac{2.3\ RT}{\alpha\ zF},$$

where α is determined by the shape of the energy barrier that must be overcome. The Tafel constant is most frequently in the range 0.05–0.15 V/current density decade. In certain cases, values of 0.3–0.6 V/decade have been observed. Table 4.1 presents values of b_c and overvoltages at $i_c = 1\ mA/cm^2$ in addition to i_o for various reactions.

The Tafel equation gives a straight line in potential–log current density diagrams (Figure 4.3), which are normally used in the description of reaction rates related to corrosion. The diagram is sometimes called a Tafel diagram.

4.5 Concentration Polarization

The electrode reactions involve both mass and charge transfer at the metal–electrolyte interface as well as transport of mass (ions and molecules) in the solution to and from the interface. Above we have introduced activation polarization, where the mass or charge transfer across the interface is rate determining. In other cases, the mass transport within the solution may be rate determining, and in this case we have concentration polarization. This implies either that there is a shortage of reactants at the electrode surface, or that an accumulation of reaction products occurs. We shall particularly consider the former case and use reduction of oxygen as an example primarily because the significance of this reaction is unique to corrosion in most natural environments:

$$O_2 + 4H^+ + 4e^- \rightarrow 2H_2O \tag{4.4}$$

(possibly superimposed dissociation of water, compare Section 6.2.1).

If the metal is submerged in an aerated solution, it can be assumed that the concentration of O_2 depends on the distance from the electrode surface, approximately as shown in Figure 4.4.

From a certain distance outwards the concentration is practically constant due to convection. At shorter distances the transport of oxygen occurs by diffusion. Under stationary conditions, we can describe diffusion by Fick's first law:

$$\frac{dn}{dt} = -D\frac{dc}{dx}, \tag{4.5}$$

where dn/dt is the mass transport in the x-direction in mol/cm^{2}s, D is the diffusion coefficient (cm^2/s), and c is the concentration in mol/cm^3. Thus, the mass transport towards the electrode surface in this case is dn/dt = D · dc/dx.

Under steady state, this mass transport rate equals the reaction rate at the electrode. By applying Faraday's laws we can therefore express the current density of oxygen reduction as follows:

$$i = D\,z\,F\frac{dc}{dx} = D\,z\,F\frac{c_B - c_0}{\delta}. \tag{4.6}$$

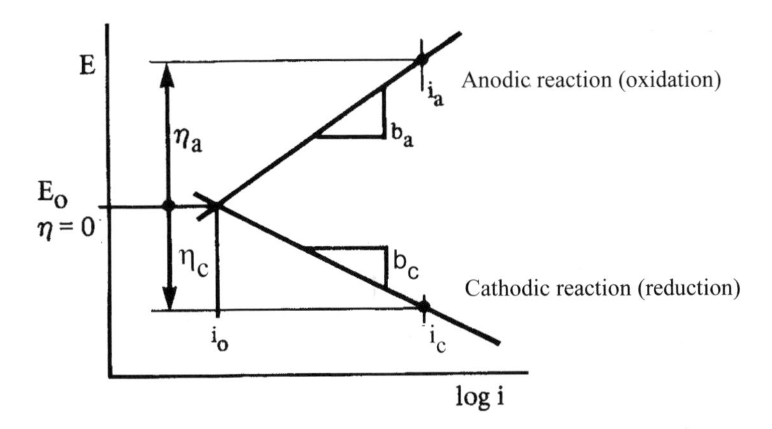

Figure 4.3 Potential–log current density diagram. Increase in the numerical value of the overvoltage by the value of b gives a tenfold increase in current density. $i_a = i_c = i_o$ corresponds to $\eta = 0$.

Table 4.1 Exchange current densities i_o, cathodic Tafel constants b_c and overvoltages at i = 1 mA/cm^2 [4.6]. (Reproduced from Uhlig HH. Corrosion and Corrosion Control. New York: John Wiley and Sons, 1971.)

Metal	Temp.,°C	Electrolyte	b_c,V	i_0, A/cm^2	$-\eta$ [1 mA/cm^2], V
Hydrogen reaction					
Pt	20	1 N HCl	0.03	10^{-3}	0.00
(smooth)	25	0.1 N NaOH	0.11	6.8×10^{-5}	0.13
Pd	20	0.6 N HCl	0.03	2×10^{-4}	0.02
Mo	20	1 N HCl	0.04	10^{-6}	0.12
Au	20	1 N HCl	0.05	10^{-6}	0.15
Ta	20	1 N HCl	0.08	10^{-5}	0.16
W	20	5 N HCl	0.11	10^{-5}	0.22
Ag	20	0.1 N HCl	0.09	5×10^{-7}	0.30
Ni	20	0.1 N HCl	0.10	8×10^{-7}	0.31
	20	0.12 N NaOH	0.10	4×10^{-7}	0.34
Bi	20	1 N HCl	0.10	10^{-7}	0.40
Nb	20	1 N HCl	0.10	10^{-7}	0.40
Fe	16	1 N HCl	0.15	10^{-6}	0.45
	25	4% NaCl pH 1–4	0.10	10^{-7}	0.40
Cu	20	0.1 N HCl	0.12	2×10^{-7}	0.44
	20	0.15 N NaOH	0.12	1×10^{-6}	0.36
Sb	20	2 N H_2SO_4	0.10	10^{-9}	0.60
Al	20	2 N H_2SO_4	0.10	10^{-10}	0.70
Be	20	1 N HCl	0.12	10^{-9}	0.72
Sn	20	1 N HCl	0.15	10^{-8}	0.75
Cd	16	1 N HCl	0.20	10^{-7}	0.80
Zn	20	1 N H_2SO_4	0.12	1.6×10^{-11}	0.94
Hg	20	0.1 N HCl	0.12	7×10^{-13}	1.10
	20	0.1 N H_2SO_4	0.12	2×10^{-13}	1.16
	20	0.1 N NaOH	0.10	3×10^{-15}	1.15
Pb	20	0.01–8 N HCl	0.12	2×10^{-13}	1.16
Oxygen reaction					
Pt	20	0.1 N H_2SO_4	0.10	9×10^{-12}	0.81
(smooth)	20	0.1 N NaOH	0.05	4×10^{-13}	0.47
Au	20	0.1 N NaOH	0.05	5×10^{-13}	0.47
Metal deposition					
Zn	25	1 M $ZnSO_4$	0.12	2×10^{-5}	0.20
Cu	25	1 M $CuSO_4$	0.12	2×10^{-5}	0.20
Fe	25	1 M $FeSO_4$	0.12	10^{-8}	0.60
Ni	25	1 M $NiSO_4$	0.12	2×10^{-9}	0.68

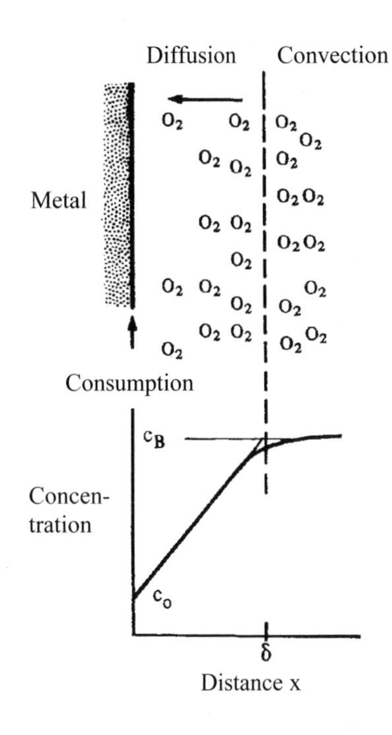

Figure 4.4 Oxygen concentration as a function of the distance from the electrode surface.

Maximum transport and reaction rate are attained when c_o approaches zero, and in this case the current density approaches the so-called *(diffusion) limiting current density:*

$$i_L = D z F \frac{c_B}{\delta}. \tag{4.7}$$

z (number of electrons per mole oxygen reacting) is 4 according to the reaction equation (4.4).* The concentration polarization means that the current density cannot increase beyond i_L no matter how large the overvoltage is. As we are going to see later, i_L depends both on the temperature and the flow rate of the solution.

For simple systems where exact formulae for δ exist (see Chapter 6), i_L can be calculated. Otherwise it can be determined experimentally.

The most typical concentration polarization occurs when there is a lack of reactants, and (in corroding systems) therefore most often for reduction reactions. This is the case because reduction usually implies that ions or molecules are transported from the bulk of the liquid to the electrode surface, while for the anodic (dissolution) reaction, mass is transported from the metal, where there is a large reservoir of the actual reactant.

---

* In reality not all the oxygen is reduced as shown in Eq. (4.4). Some O_2 is reduced to H_2O_2 with z = 2. Average z is therefore somewhere between 2 and 4, but measurements indicate that it often is close to 4.

Equations (4.5), (4.6) and (4.7) are valid for uncharged particles, as for instance oxygen molecules. If, on the contrary, charged particles are considered, migration (i.e. transport due to a potential gradient, in other words electrolytic conduction) will occur in addition to the diffusion. In this case Equation (4.7) must be replaced by

$$i_L = D\,z\,F\,\frac{c_B}{\delta\,(1-t)}, \tag{4.8}$$

where t expresses the actual ion's proportion of the total conductivity of the solution.

4.6 Overvoltage Due to Concentration Polarization

As described in Chapter 3, and expressed by Nernst's equation (3.9), the equilibrium potential depends on the activity, or practically on the concentration of the species that take part in the reaction. The reduction of the oxygen concentration that occurs close to the metal surface when the electrode is polarized in the cathodic direction, leading to the concentration profile shown in Figure 4.4, causes therefore a change in the equilibrium potential. This equilibrium potential shift is regarded as an overvoltage:

$$\eta_{conc} = E_{oc_o} - E_{oc_B},$$

where E_{oc_o} and E_{oc_B}, are the equilibrium potentials at the oxygen concentrations c_o and c_B, respectively. From Nernst's equation we can derive

$$\eta_{conc} = E_o^o - \frac{RT}{zF}\ln\frac{k}{c_o} - \left(E_o^o - \frac{RT}{zF}\ln\frac{k}{c_B}\right)$$

$$= \frac{RT}{zF}\left(\ln c_o - \ln c_B\right) = \frac{2.3RT}{zF}\log\frac{c_o}{c_B}, \tag{4.9}$$

where k is a factor representing the activities of the other substances taking part in the reactions, supposed to be negligibly influenced by the reaction (i.e. $k \approx$ constant). Further, we have from Equations (4.6) and (4.7), respectively

$$c_o = c_B - \frac{i\,\delta}{DzF} \quad \text{and} \quad c_B = \frac{i_L\,\delta}{DzF}, \quad \text{which give}$$

$$\frac{c_o}{c_B} = 1 - \frac{i\delta}{DzF}\,\frac{1}{c_B} = 1 - \frac{i}{i_L}.$$

By inserting this into Eq. (4.9), we obtain

$$\eta_{conc} = \frac{2.3\,RT}{zF}\log\left(1 - \frac{i}{i_L}\right). \tag{4.10}$$

The equation is shown graphically in Figure 4.5.

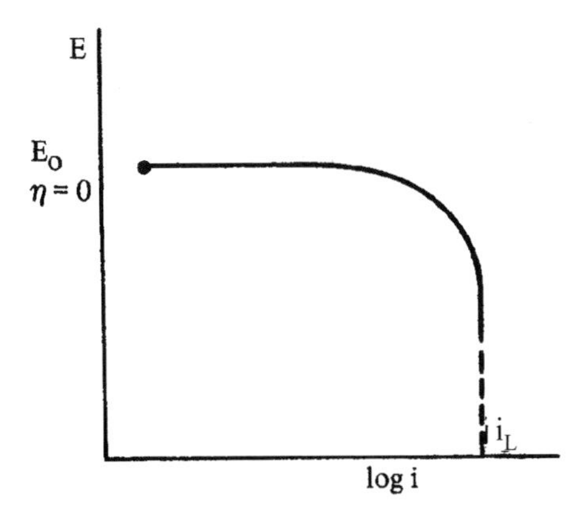

Figure 4.5 Overvoltage curve for pure concentration polarization.

4.7 Combined Activation and Concentration Polarization

In anodic reactions, activation polarization is often dominating (except when passivation effects occur). Then, the relationship between overvoltage and current density is, as previously shown:

$$\eta_a = b_a \cdot \log\frac{i_a}{i_0}. \tag{4.3a}$$

Conversely, in cathodic reactions (reduction processes), both types of polarization have to be taken into account. The relationship between overvoltage and current density will therefore be as follows:

$$\eta_c = -b_c \log \frac{i}{i_o} + 2.3 \frac{RT}{zF} \log \left(1 - \frac{i}{i_L}\right). \tag{4.11}$$

The corresponding overvoltage curve is shown in Figure 4.6.

The complete determination of position and shape of overvoltage curves requires knowledge about the quantities E_o, i_o, i_L, b_c, and b_a, of which the first one is given by Nernst's equation or the Pourbaix diagram. Through such knowledge, the best tool we have for description of the kinetics of electrode reactions is available. With a reservation for passivation effects (which we shall return to) we can with this background obtain a quantitative description of any corrosion process, in principle as shown in Section 4.9.

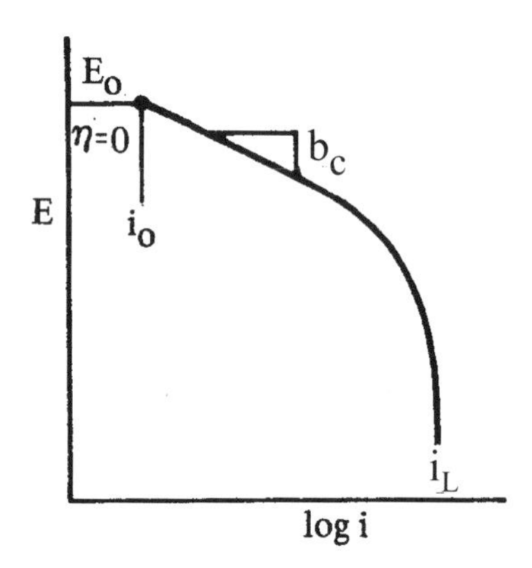

Figure 4.6 Overvoltage curve at combined polarization.

4.8 Resistance (Ohmic) Polarization

Surface layers on metals may have a considerable ohmic resistance. This is, for instance, the case for oxide films on stainless steels, aluminium, chromium etc., even if the film is very thin. When a current I (current density i) is flowing through the film, we will have an ohmic drop, i.e. a resistance polarization expressed by η = RI = ri, where R (Ω) and r (Ω cm^2) are the resistances in the film on the total electrode surface and in 1 cm^2 of it, respectively. Anodic resistance polarization affects the potential on surfaces passivated by oxides or other substances.

4.9 Determination of Corrosion Potential and Corrosion Rate

For a given metal exposed to an electrolyte with a certain pH value, the Pourbaix diagram tells us which potential ranges give immunity, corrosion and passivity, respectively. But what *determines* the potential in a real situation and what is the corrosion rate at this potential? From the description of the overvoltage curves we have now the basis for answering these questions.

We stick to two basic rules:

1. All electrochemical corrosion processes are composed of one or more anodic reactions and one or more cathodic reactions.

2. The electron release at the anodes must equal the electron consumption at the cathodes. Generally, this can be expressed by

$$\sum I_a = \sum I_c .$$

If the reactions occur on equally large areas (anode area = cathode area), we have

$$\sum i_a = \sum i_c .$$

In these equations, I_a and I_c are the anodic and cathodic currents, and i_a and i_c the corresponding current densities. We consider now a cell consisting of a steel or iron plate in an acid of known pH. The dominating reactions are dissolution of iron and development of hydrogen gas. We suppose that both reactions are activation polarized. With the knowledge about E_o, i_o and the Tafel constants b for both reactions we can draw the overvoltage curves corresponding to the Tafel equation (4.3) (the Tafel lines). The Tafel lines for both reactions are drawn in one E–log i diagram, as seen in Figure 4.7.

In the first instance, let us assume that the conductivity of the solution as well as of the metal is so large that there is no significant potential drop from one site at the electrode to another, neither through the electrolyte nor through the metal, i.e. the electrode potential must also be the same all over the electrode surface, which means $E_a = E_c$. We assume also that the anodic area A_a equals the cathodic area A_c, i.e. $i_a = i_c$.

It is seen that there is only one point in the diagram (Figure 4.7) that satisfies these two conditions, namely the intersection point between the anodic and cathodic overvoltage curves (the other two Tafel lines, i.e. the broken lines for the cathodic iron reaction and the anodic hydrogen reaction, contribute extremely little to the net cathodic respective anodic current density, so that in this case we can neglect both of them).

The potential and the current density at the intersection point are called corrosion potential and corrosion current density. The potential is also called the mixed potential.

As we have seen, if the equilibrium potentials, the exchange current densities, and the slope/shape of the overvoltage curves of the involved reactions are known, the corrosion current density and the (corrosion) potential can be determined by means of the E–log i diagram.

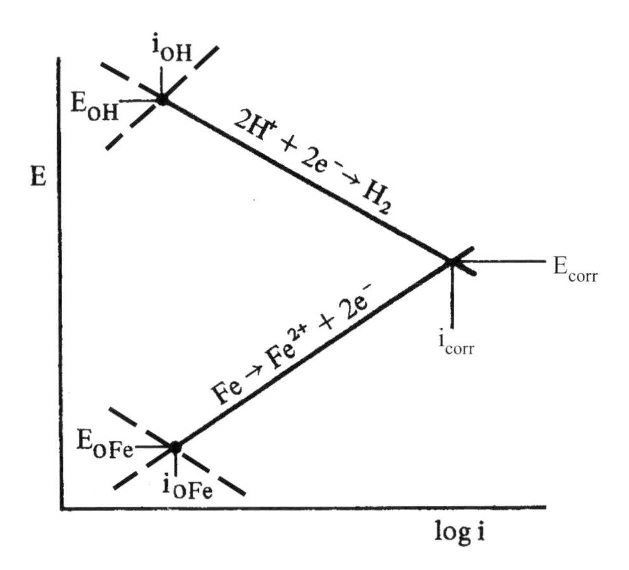

Figure 4.7 Potential–log current density diagram for determination of the corrosion potential E_{corr} and the corrosion current density i_{corr} in a case with one cathodic reaction and one anodic reaction.

When there are two cathodic reactions, e.g.

$$2H^+ + 2e^- \rightarrow H_2 \quad \text{and}$$

$$\frac{1}{2}O_2 + 2H^+ + 2e^- \rightarrow H_2O,$$

the cathodic current density will be

$$i_c = i_{H^+} + i_{O_2}$$

The corrosion current density and the potential are in this case most easily determined with a simple graphical analysis as shown in Figure 4.8. The thick curve shows the sum of cathodic current densities in the potential range where both reactions contribute.

If we deal with diluted strong acids saturated with air, the rates of the oxygen reduction and the hydrogen reaction are normally of the same order of magnitude at pH values around 3. The hydrogen reaction dominates at lower pH and the oxygen reaction at higher pH values.

Similar treatment is to be carried out if we have more than one anodic reaction.

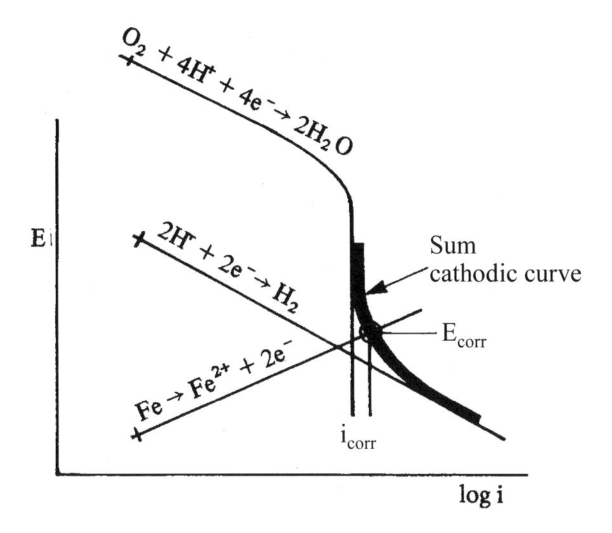

Figure 4.8 As Figure 4.7 but with two cathodic reactions.

Now we have seen how we can determine the corrosion rate when the anode area and the cathode area are equally large. But it should be remembered that this is certainly not a general condition. Thus, generally it is more correct to use the basic rule, i.e. $\sum I_a = \sum I_c$. Then the corrosion current and the corrosion potential are determined by means of E–log I diagram in the same way as we have shown by the E–log i diagram. As a basis for drawing the E–log I curve for each reaction, either the E–log i curve and the electrode area for each reaction, or experimentally recorded E–log I curves have to be known. We shall return to this, particularly in connection with galvanic elements (Section 7.3).

Combined potential–current diagrams for anodic and cathodic reactions, with intersecting points showing corrosion potential and corrosion current, were first introduced by the British corrosion scientist Ulick Richardson Evans (1889–1980), one of the greatest contributors to modern corrosion theory. The diagrams are often called Evans diagrams. Evans used originally a linear current scale. With a logarithmic current scale, the author of the present book has chosen to use only the designation E–log I diagram, and correspondingly E–log i diagram when dealing with current density.

4.10 Recording Polarization Curves with a Potentiostat

As we have seen, overvoltage curves can be used to determine corrosion rate. In order to draw the overvoltage curves, one must, however, know E_o, i_o, b_a, b_c and possibly i_L. E_o can be calculated by thermodynamics (or possibly measured on a reversible cell). The other quantities must be determined experimentally. For many reactions this has already been done, but both b and particularly i_o and i_L depend strongly on the conditions, and there is always a need for new data. The most important method of determining these is to record *polarization curves* with a *potentiostat*.

As the name indicates, the potentiostat is an instrument that keeps a set electrode potential (on the material we are investigating) constant, and it delivers the current which is instantaneously needed to hold the potential constant. A potentiostatic set-up is shown in Figure 4.9. There are three electrodes:

1. The working electrode W (the specimen to be investigated).
2. The counter electrode C.
3. The reference electrode R.*

The reference electrode usually has an electrolytic connection to the working electrode through a salt bridge with a narrow tube near the working electrode (a Luggin capillary), as shown in Figure 4.9.

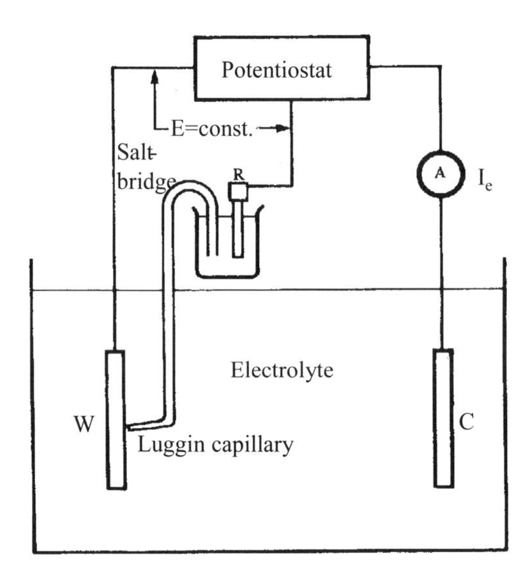

Figure 4.9 Wiring diagram for potentiostatic experiments.

* In the laboratory a saturated calomel electrode (SCE) consisting of Hg, Hg_2Cl_2/KCl (saturated) is commonly used. At 20°C this electrode has a potential 245 mV more positive than the standard hydrogen electrode potential.

As mentioned, the potential keeps the electrode potential constant at a set value, i.e. it keeps a constant voltage on the cell consisting of the working electrode and the reference electrode by delivering the necessary current between the counter electrode and the working electrode. For the working electrode this is experienced as an *external current*.

We assume that there are two possible electrode reactions on the working electrode, with overvoltage curves (potential–log current curves) and corrosion potential as shown by the solid curves in Figure 4.10. (We neglect still the dotted curves to the left in the diagram.)

When $E = E_{corr}$ no external current is supplied to the working electrode. If, by means of the potentiostat, we set another arbitrary potential E_1, an external current I_{1e} is supplied, which is the difference between the anodic and the cathodic reaction current at this potential. If we draw the logarithm of the external current as a function of potential, we obtain the dotted curves shown in the figure. These curves are called *polarization curves.*

It is seen that the overvoltage curves (the Tafel lines) are asymptotes to the polarization curves. When E deviates much from the corrosion potential, the polarization curves and the overvoltage curves merge, and this fact is utilized. In Figure 4.10 we started with the overvoltage curves and drew the polarization curves on this basis. But, as mentioned previously, the purpose is to determine the overvoltage curves from the polarization curves, i.e. to go the opposite way. We record I_e as a function of potential by means of the potentiostat, i.e. we record the polarization

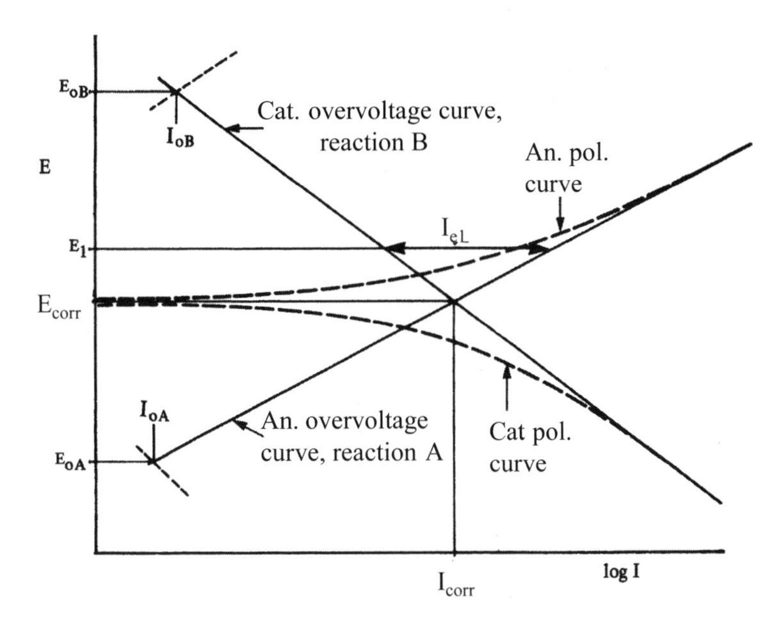

Figure 4.10 Overvoltage curves and corresponding polarization curves.

curves. The linear part of the polarization curves is extrapolated, and thus the overvoltage curves are determined. (This is easy when the polarization curves have definitely linear parts. This is not always the case; then a deep insight is necessary for interpreting the polarization curves properly.)

4.11 Conditions That Affect Polarization Curves, Overvoltage Curves and Corrosion Rates

The compositions of material and environment are crucial factors in corrosion. They determine the equilibrium potentials, with some influence of the temperature. Often they also determine qualitatively the shape and position of the overvoltage curves. Both qualitatively and quantitatively the contents of Cl^- and oxidizers in the electrolyte are of particularly great significance. This is dealt with in connection with passivity in Chapter 5, and also in other chapters. But quantitatively the overvoltage curves and thus the corrosion rates depend also strongly on other factors, such as surface state, temperature, electrolyte flow rate and geometry. The effects of these factors depend on several conditions and will be treated in various chapters. The state of the surface may have a most significant effect on the exchange current density. This is one of the reasons for the great variation of i_0 shown in Table 4.1. Deposits and liquid films on the surface, temperature and flow velocity are important factors for diffusion-controlled corrosion (Chapter 6), and the geometry (together with conductivity of the electrolyte) plays a decisive role in galvanic corrosion (Section 7.3) and cathodic protection (Section 10.4). Geometry and general design are also very important in many other cases (Chapters 7 and 10).

References

4.1 Bockris JO'M, Reddy AKN. Modern Electrochemistry, Vol. 2. New York–London: Plenum Press, 1970.
4.2 Fontana MG, Greene ND. Corrosion Engineering. New York: McGraw Hill. 1st Ed. 1967, 2nd Ed. 1978.
4.3 Kaesche H. Die Korrosion der Metalle. Berlin–Heidelberg–New York: Springer-Verlag, 1966.
4.4 Scully JC. The Fundamentals of Corrosion, 3rd Ed. Oxford: Pergamon Press, 1990.
4.5 Shreir LL, Jarman RA, Burstein GT. Corrosion, Vol. 1, 3rd Ed. Oxford: Butterworth–Heinemann, 1994.
4.6 Uhlig HH. Corrosion and Corrosion Control. New York: John Wiley and Sons, 1971.

EXERCISES

1. In what range of potential may a metal surface corrode?

2. What is wrong with the following statement:
"In a corrosion process, the sum of the numerical values of the anodic and cathodic overvoltages is considerably larger than the difference between the equilibrium potentials of the actual cathodic and anodic reactions"?

3. Which four quantities are necessary for determination of (the shape and position of) a cathodic overvoltage curve in a case with activation polarization at small overvoltages (low numerical values of η) and concentration polarization at higher values of $|\eta|$?

4. Which of the following relations are generally valid at free corrosion (corrosion without external current supply)?

$$I_a = I_c,\ E_c \geq E_a,\ E_c < E_a,\ \Sigma i_a = \Sigma i_c,$$

$$E_a = E_c,\ \Sigma I_a = \Sigma I_c,\ i_a = i_c.$$

Is any of the relations always wrong?

5. Iron corrodes in a diluted HCl/NaCl solution with pH = 3 and high conductivity. Determine the corrosion current density graphically when

a) The solution is saturated with air.
b) The solution is de-aerated (free of oxygen).

Assume that the reduction of H^+ and possibly of O_2 are the only reduction processes, that these reactions occur on surface areas equal to the areas with anodic dissolution, and use the following data:

Electrode reaction	Equilibrium potential, E_0	Exchange current density, i_0	Tafel constant, b	Limiting current density, i_L
$Fe^{2+} + 2e^- = Fe$	– 0.62 V	0.1 μA/cm^2	0.06 V	
$2H^+ + 2e^- = H_2$	– 0.18 V	0.1 μA/cm^2	0.12 V	
$O_2 + 4H^+ + 4e^- = 2H_2O$	– 1.05 V	0.05 μA/cm^2	0.12 V	40 μA/cm^2

6. Construct the cathodic and anodic polarization curves from the overvoltage curves in Exercise 5. How are polarization and overvoltage curves usually determined?

5. Passivity

If you drive nature out with a pitchfork,
she will soon find a way back.
HORACE

5.1 Passivation and Passivity Described by Anodic Polarization and Overvoltage Curves

Let us consider a case with ordinary structural steel or iron exposed to 1 N sulphuric acid which has been de-aerated, e.g. by bubbling nitrogen gas through it. If the metal is polarized to various potentials more positive than the corrosion potential E_{corr}, and the respective current density values are recorded, an anodic polarization curve like the solid one shown in Fig. 5.1 is determined. Other reactions than dissolution of iron are involved only at the bottom and top of the diagram. In these regions we also have reduction of hydrogen and development of oxygen, respectively, which cause the polarization curve to deviate from the overvoltage curve for dissolution of iron (----). In other regions these two curves are essentially identical.

When starting at the corrosion potential E_{corr} and increasing the potential, the corrosion current is increased, at first corresponding to activation polarization (with a straight overvoltage curve ----). As the potential is increased further the polarization curve will turn off to lower current than that corresponding to pure activation polarization. The reason for this behaviour is primarily that the large corrosion rate causes a high concentration of iron ions in the electrolyte close to the metal surface.* This in turn leads to:

i) Concentration polarization (increased equilibrium potential of the dissolution reaction, see Section 4.6).
ii) Deposition of ferrous sulphate on the surface and consequently reduction of the effective area for the dissolution process.

---

* The apparent deviation from a linear overvoltage curve is somewhat larger than the real one because there is a voltage drop between the reference electrode and the iron sample due to the high current density.

By further increase of the potential a *critical current density* i_{cr} (maximum corrosion current density) is reached at the potential E_{pass}, followed by a strong decrease in the current density, which is caused by *passivation.* The potential E_{pass} is designated by different terms in the literature, namely primary passivation potential, primary passive potential or simply *passivation potential.* By further increase of the potential through a value E_a, a low and approximately constant anodic current density, the *passive current density* i_p, is recorded. With sufficient potential increase a current increase is measured again, indicating *transpassivity* and/or oxygen development.

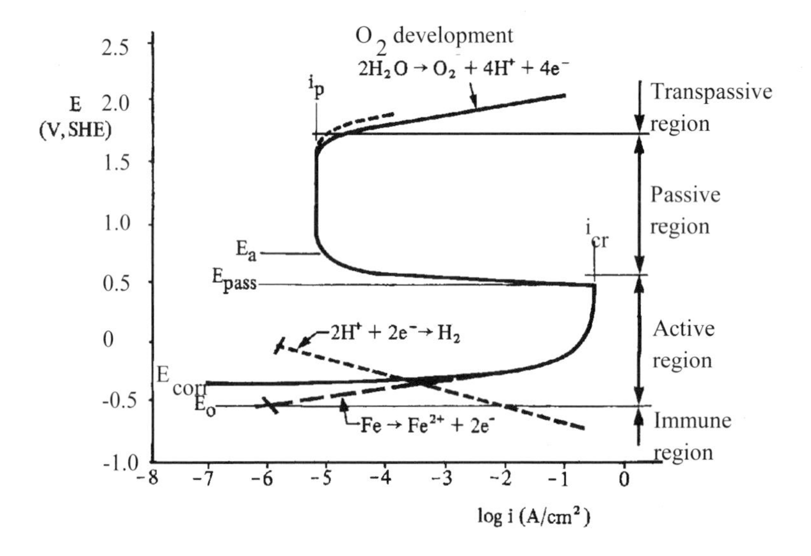

Figure 5.1 Anodic polarization and overvoltage curve for iron in 1 N H_2SO_4, mainly based upon data from Franck, Franck and Weil, and Hersleb and Engell, collected by Kaesche [5.1]. Overvoltage curve for hydrogen reduction estimated from other data [5.2, 5.3].

As we have seen, there are four characteristic potential regions, termed as shown to the right in the diagram. Metals which – in certain pH ranges – change from an active to a passive state when the potential is increased, as in the case in Fig. 5.1, are called *active–passive metals*. The Pourbaix diagrams indicate which metals belong to this group. Typical examples are Fe, Ni, Ti and their alloys.

5.2 Passivity – Reasons and Characteristic Features

As we have seen, we can show the regions of immunity, activity (corrosion) and passivity as potential regions by means of the anodic overvoltage curve and as pH–

potential regions in the Pourbaix diagram. A better view is obtained by parallel drawing of the Pourbaix diagram and overvoltage curves, as shown for iron with examples for two nearly neutral environments in Fig. 5.2. This way of presentation was used by Pourbaix and was extensively utilized for educational purposes by Almar-Næss [5.4]. As mentioned earlier, the Pourbaix diagram refers to pH values close to the metal surfaces. This must be remembered when using such presentations as the one in Fig. 5.2 for cases where the near-surface pH deviates considerably from the bulk pH as a result of electrode reactions. This is particularly relevant and important for various forms of localized corrosion (Chapter 7).

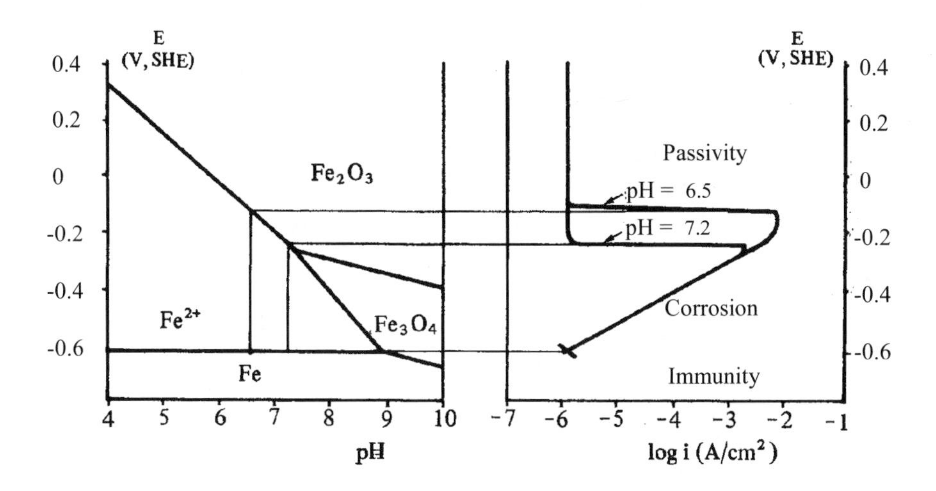

Figure 5.2 Pourbaix diagram and anodic overvoltage curves for iron in water.

Figure 5.2 illustrates that the passive state indicated by the overvoltage curve is characterized by thermodynamic stability of the oxide Fe_2O_3. In several other material–environment systems, other oxides have similar properties and effects.

It should be emphasized here that overvoltage curves based on experimentally recorded polarization curves do not always agree fully with theoretically determined Pourbaix diagrams. One reason for this is that the recorded polarization curves depend much on the experimental procedure, e.g. the rate at which the potential is changed. In addition there are practical deviations from the ideal conditions on which the Pourbaix diagram is based (see Chapter 3). For instance, there is some disagreement between the curve in Fig. 5.1 (for pH = 0) and the Pourbaix diagram for iron shown in Fig. 3.10. This can be explained by the formation of sulphate surface layers and larger concentrations of ferrous ions in the practical case with H_2SO_4 than what is assumed in the derivation of the Pourbaix diagram.

The reasons for passivity have been debated. In a publication as early as 1844, Faraday assumed that passivity was caused by an oxide film or a process that can be imagined as adsorption of oxygen on the metal surface [5.1]. These two possibilities have until recent decades been the basis for two competing theories of passivity. Strong arguments for the oxide film theory are found by the experimental demonstration of existing oxide films of a typical thickness up to 30 Å in connection with passivity, and a reasonably good correlation between passive regions shown with polarization curves and thermodynamic stability regions of oxides. The oxide films may, however, be very thin, and they can be formed via stages of oxygen adsorption, so that the controversy between the theories is not necessarily that large.

The properties characterizing a typical passivating oxide film are low ionic conductivity and low solubility. Due to these properties the oxide prevents to a large extent the transport of metal ions from sites in the crystal structure of the metal to the liquid, i.e. it prevents the anodic dissolution. However, there is a slow dissolution corresponding to the passive current density i_p. It is assumed that such oxide films are not formed by deposition of corrosion products from the liquid, as this usually gives more or less porous surface layers, but that the oxide is directly formed in close connection with the crystal structure of the metal. Because of this and because the films are very thin, many passivating films allow electrons to be transferred, so that electrochemical reactions can occur on the external face of the oxide. In order to understand practical corrosion cases it is often important to know the differences between various kinds of surface films.

The actual surface layers can be divided into three major groups:

a) Films that hinder the anodic dissolution effectively, but not the cathodic reaction. Examples of this type are typical passive films on Fe, Ni, Cr and their alloys. Cathodic reactions, e.g. oxygen reduction, can occur on the external side of the film because electron transfer across the film is allowed.

b) Films that hinder both the anodic and the cathodic reaction to a large extent. As a typical example we can mention passivating oxide films on aluminium, which, contrary to films on Fe, Ni and Cr, have high resistance against electron transfer. In natural oxides on aluminium there are, however, a large number of defects that allow some electron conductance and cathodic activity.

c) Deposits that reduce both anodic and cathodic reactions, but not sufficiently to give efficient passivity. This group comprises porous surface layers – e.g. rust and salts – deposited from the solution when the actual solubility product is exceeded. The oxygen reduction reaction is assumed to occur on the metal in the bottom of pores, and it is hindered more or less by the resistance of the deposit against oxygen diffusion. At the same time, the pores allow the anodic reaction, but this is also hindered more or less because of the deposit (compare the case with deposited ferrous sulphate in the active region, see Fig. 5.1).

In addition to low ionic conductivity and low solubility, a passive film must have sufficient *mechanical strength and adhesion* to the metal to avoid cracking, wear and disbonding.

5.3 Breakdown of Passive Films

Passive films may be destroyed mechanically, chemically, electrochemically or by undermining the film [5.5].

Mechanical breakdown occurs by formation of cracks under deformation of the material, or by wear due to high flow velocity of the liquid, solid particles or components touching the surface. Under such conditions it is important whether the film has sufficient chemical or electrochemical self-repairing ability. In case of heavy continuous wear, activation cannot be avoided, which is indicated by a larger or smaller potential drop (see Section 5.4 and Table 5.1). This is an example of corrosive–abrasive wear, which is dealt with in more detail in Section 7.9.

Chemical destruction occurs by direct dissolution of the film. The oxide has a certain solubility also in the passive state. By change of pH, or increase of temperature or concentration of aggressive species, the solubility may be so high that the passivity disappears completely or to some extent.

Electrochemical destruction occurs by potential changes from the passive region. This can happen by lowering the potential to the active region, or by increasing it to the transpassive region as illustrated in Fig. 5.1, or possibly only above a critical potential for localized corrosion (see below). In certain cases, e.g. for Fe in a moderately alkaline environment, a passive oxide can be reduced to other oxides or to the metallic state.

For stainless steels and Cr the transpassivity phenomenon is indicated by a branch on the anodic polarization curve as shown for an 18-8 CrNi steel in 1 N H_2SO_4 in Fig. 5.3. The reason for this indication of transpassivity is that the oxide is generally dissolved and $Cr_2O_7^{2-}$ and possibly Fe^{3+} are formed at potentials below the region where oxygen is developed.

If the same steel is exposed to a neutral NaCl solution, a lower critical potential – the pitting potential – is found. Pitting potentials for various mixtures of NaCl and Na_2SO_4 solution are indicated in Fig. 5.4 (the potentials at which there is a marked current increase). At potentials above the pitting potential, localized corrosion – pitting – occurs due to local attack and penetration of the oxide by Cl^- ions. The mechanism of this corrosion form is described in Section 7.6. However, it can be noticed already that salts like $NaNO_3$, $NaClO_4$ and Na_2SO_4 counteract NaCl, and thus lead to increased pitting potentials (see the example with Na_2SO_4 in Fig. 5.4).

The negative effect of increased Cl^- concentration on the pitting potential E_p is of great significance. For an 18-8 CrNi steel, E_p can be expressed by [5.3]:

$$E_p = 0.168 - 0.088 \log a_{Cl^-} \text{ (V, SHE)}, \qquad (5.1)$$

where a_{Cl^-} = activity ≈ concentration of Cl^-.

For stainless steels in neutral NaCl solutions there is no active potential range like that shown in Fig. 5.3. The alloys are passive all the way down from the region of localized corrosion to the immune region.

Localized destruction of oxides can also occur in crevices and underneath deposits (see Section 7.5).

If a discontinuity in the passive oxide film occurs for one reason or another, and the metal corrodes rapidly at this point, the oxide film may be *undermined by corrosion*. An extreme example is the behaviour of aluminium in the presence of mercury [5.5].

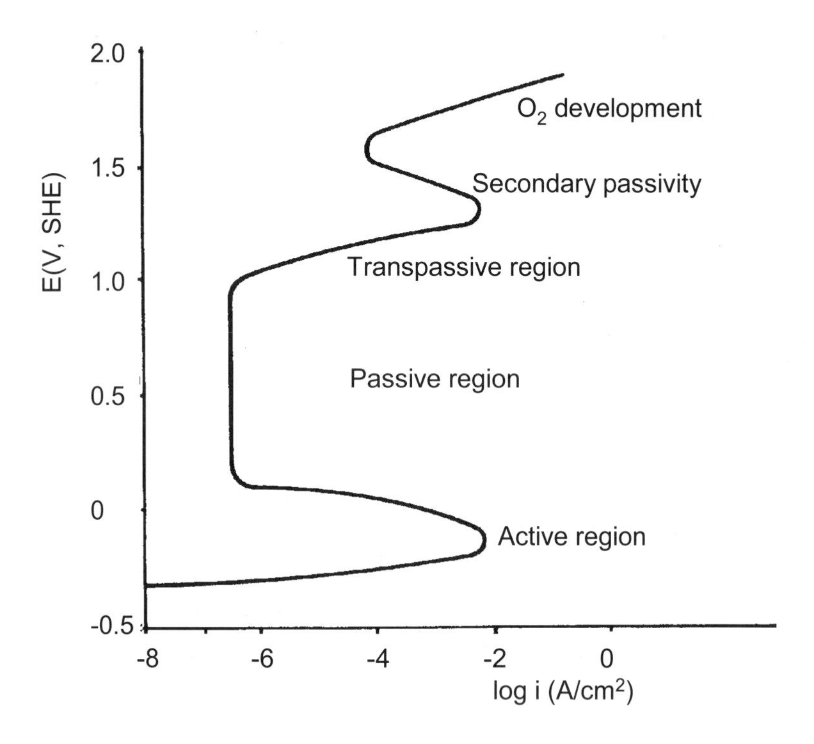

Figure 5.3 Anodic polarization curve (identical with the anodic overvoltage curve for metal dissolution in the potential range from –0.2 to +1.6 V) for an 18-8 CrNi steel in 1 N H_2SO_4 without oxygen at 50°C (from Kaesche [5.1] after Engell and Ramchandran).

5.4 Stable and Unstable Passive State

Depending on the availability of oxidizers, the passivity may be a) stable or b) conditional, practically unstable. A good demonstration of this is Faraday`s passivation experiment with iron in HNO_3 solutions [5.2]. In one case the piece of iron is immersed in 70% HNO_3, and the iron is passive. The passivity can be confirmed by weighing the iron sample before and after a certain period of exposure. If the oxide is scratched, heavy gas development occurs and stops again shortly after. It can be shown (again by weighing) that the sample does not corrode any more. This is an example of spontaneous passivation or self-passivation (1). If the acid is diluted with water, nothing happens as long as the sample is left undisturbed in the liquid (2). However, if the oxide film is injured mechanically, strong and permanent gas development occurs, indicating intensive and permanent corrosion (3). The same would have occurred if the iron piece originally was submerged directly in diluted acid.

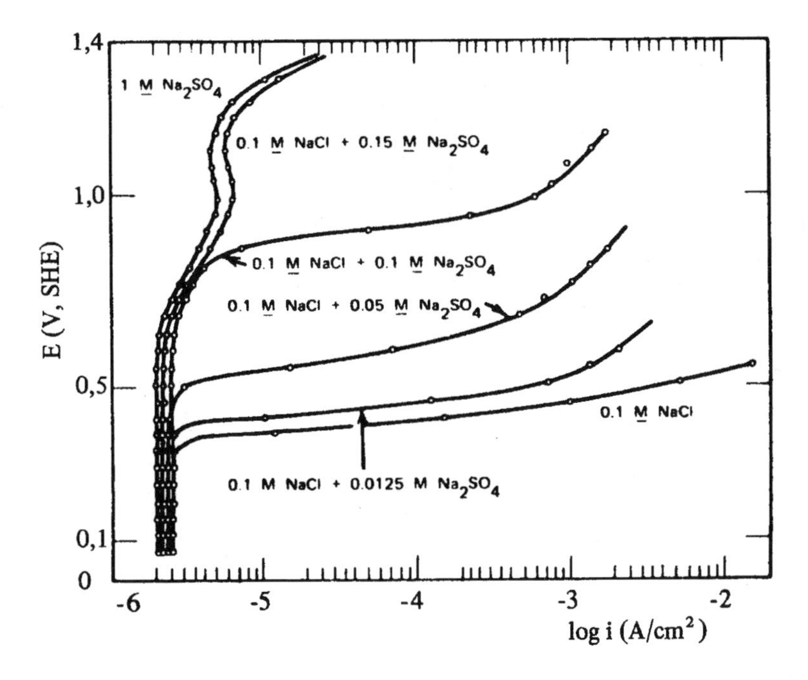

Figure 5.4 Anodic polarization curves for an 18-8 CrNi steel in a 0.1 M NaCl solution with different amounts of Na_2SO_4 added [5.3].

Faraday´s experiment can be explained by overvoltage curves as shown in Fig. 5.5. The intersection between the anodic and the cathodic curve shows that there is only one stable state (1) in concentrated acid (stable passive state). In diluted acid the reversible potential of the cathodic reaction (E_o) is more negative because of a lower concentration of oxidizer, i.e. nitric acid (in agreement with Nernst's equation (3.9), see also p. 65–66). The result is that there are three intersection points between the anodic and the cathodic curve in this case, giving the possibility of an unstable passive state (2)* or an active state (3). For comparison, a case (4) is shown where the active state is the only possible one.

* Electrochemically, 2 is a stable state, but the risk of mechanical damage of the oxide means that it is considered practically unstable.

Similar conditions can exist for a large number of material–environment combinations. For example, a strong increase in the limiting current density i_L of the oxygen reduction reaction on steel in neutral water may lead to stable passivity. The same state can be obtained if the critical current density is reduced by increasing pH, as shown in Fig. 5.2.

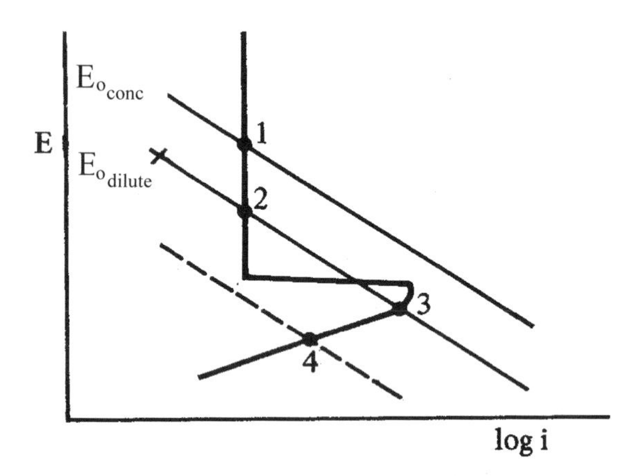

Figure 5.5 Illustration of stable passivity (1), and conditional, or practically unstable passivity (2), which can be transferred to active state (3). For comparison, a case with stable active state as the only possibility is shown (4).

As we have seen, a stable passivity depends on sufficient availability of an oxidizer. This means a high concentration and/or efficient transport of the oxidizer to the metal surface. Efficient transport is often a result of strong convection of the liquid. However, as long as the material is in the active state, increased supply of the oxidizer to the metal surface means increased corrosion rate. Hence, a situation that corresponds approximately to the line 2–3 in Fig. 5.5 is the worst case.

It follows from Fig. 5.5 that a transfer from the passive state (2) to the active state (3) implies a potential decrease. The greater the difference between the real corrosion resistance of the material based upon passivation (indicated by the rest potential in the passive state) and its thermodynamic nobleness (expressed by the EMF series in Table 3.1), the larger the potential decrease following the transfer from passive to active state. That is, this potential decrease indicates to what extent the real corrosion resistance of the material depends on passivity. Table 5.1 shows examples of such a potential decrease for several metal–environment combinations. The activation is in these cases caused by continuous grinding of the metal surfaces, i.e. the materials are kept active in spite of an otherwise spontaneous passivation tendency for many of the material–environment combinations represented in the table.

Table 5.1 Potential decrease, ΔE (mV), caused by continuous grinding of submerged metal surfaces (from ref. [5.6] after Akimov).

0.1 N NaOH		0.1 N HCl		0.5 N NaCl		0.1 N HNO_3	
Metal	ΔE	Metal	ΔE	Metal	ΔE	Metal	ΔE
Mn	827	Cr	556	Al	670	Al	424
Cr	680	Nb	463	Cr	577	Cr	350
Be	552	Al	438	Nb	449	Nb	345
Si	503	Mo	416	Mo	445	Mo	344
Fe	476	Ag	285	Ni	330	Si	183
Co	450	Cu	211	Ag	288	Ag	108
Ni	426	Si	95	Be	263	Be	68
Nb	380	Ni	22	Si	221	Mn	27
Mg	361	Fe	7	Co	216	Cd	1
Mo	330	Be	5	Mn	198	Sn	1
Ag	218	Zn	5	Sn	198	Åb	1
Cu	212	Pb	1	Cu	159	Fe	0
Cd	180	Mg	1	Fe	106	Co	0
Al	15	Sn	0	Zn	68	Ni	0
Zn	3	Mn	0	Mg	68	Cu	0
Sn	0	Co	0	Pb	36	Mg	0
Pb	0	Cd	0	Cd	6	Zn	0

5.5 Practical Utilization of Passivation and Passivity

Several of the commonly used materials depend on passivity to resist corrosive environments. To utilize the passivation properties of the materials, it is necessary to assure that the potential is kept in the passive region. This can be obtained by:

i) Such a combination of material and environment that spontaneous and stable passivation occurs automatically (self-passivation) (Section 5.4).

ii) Use of a passivating inhibitor causing stable passivity (Section 10.2).

iii) Use of anodic protection, which implies increase of the potential to the passive region (Section 10.5).

iv) Use of cathodic protection (cathodic polarization) to hold the potential below any critical potential for localized corrosion.

Material–environment combinations giving stable passivity are dealt with more extensively in later chapters (7, 8 and 10). Important examples are unalloyed steels in aerated alkaline aqueous solutions without significant contents of aggressive ions like Cl^-, stainless steels and titanium in oxidizing environments over a very wide range of pH (but stainless steels are also sensitive to chlorides) and aluminium in natural atmospheres and waters.

References

5.1 Kaesche H. Die Korrosion der Metalle. Berlin–Heidelberg–New York: Springer-Verlag, 1966

5.2 Fontana MG, Greene ND. Corrosion Engineering. New York: McGraw Hill Company, 1967, 1978.

5.3 Uhlig HH. Corrosion and Corrosion Control, 2nd Ed. New York: John Wiley and Sons, 1971.

5.4 Almar-Næss A. Korrosjon og korrosjonskontroll. Trondheim: Tapir, 1966 (in Norwegian).

5.5 Shreir LL, Jarman RA, Burstein GT. Corrosion, Vol. 1, 3rd Ed. Oxford: Butterworth-Heinemann, 1994.

5.6 Tomashov ND. Theory of Corrosion and Protection of Metals. New York: MacMillan, 1966.

EXERCISES

1. Draw an anodic polarization curve for an active–passive metal. Define in words and indicate on the figure all characteristic potentials, potential regions and current densities.

2. What properties are typical for a passivating oxide film?
   Mention examples of various types of surface films/surface layers.

3. A stainless 18-8 CrNi steel is exposed to a 0.1 M NaCl solution with various additions of Na_2SO_4. Draw a diagram showing the pitting potential as a function of the Na_2SO_4 concentration. As a criterion for pitting a fivefold increase of the corrosion current density is assumed for this exercise. (A quantitative criterion must always be defined when an exact pitting potential is to be determined.)

4. By immersion of a piece of iron in 70% (about 15 mol/l) nitric acid, spontaneous passivation is obtained (case I). With an HNO_3 concentration = 20% (about 4 mol/l), the condition for passivity is that the iron is passive from the beginning (due to previous exposure in a more concentrated nitric acid, for example) (case II). Table 5.1 gives some information about the conditions for iron in 0.1 N = 0.1 mol/l HNO_3 (case III).

An actual cathodic reaction is

$$HNO_3 + 3H^+ + 3e^- \rightarrow NO + 2H_2O$$

the standard reversible potential of which is E_0^0 =0.96 V (SHE).

a) In what way is the reversible potential for the given reaction affected when the conditions are changed from the standard state to the ones defined by cases I, II and III, respectively? (For simplicity, assume that the NO pressure = constant = the atmospheric pressure). Which terms in the reaction equation affect the reversible potential?

b) Assume the same exchange current density and Tafel gradient for the cathodic reaction in the three cases and illustrate schematically how the cathodic and anodic overvoltage curves are positioned in relation to each other in the three cases. (Presume that the information about Fe in 0.1 M HNO_3 given in Table 5.1 is valid independently of the kind of pre-exposure that might have been done.) What do we call the state in each of the three cases?

c) Which state do you think aluminium and chromium would adopt if they were exposed to the dilute acid dealt with in case III (see Table 5.1)? Which state would copper and nickel adopt?

6. Corrosion Types with Different Cathodic Reactions

All is flux, nothing is stationary
HERACLEITUS

6.1 General

Previously we have introduced two cathodic reactions of particularly great and wide importance:

Oxygen reduction:	$O_2 + 4H^+ + 4e^- \rightarrow 2H_2O.$	(4.4)
Hydrogen reduction:	$2H^+ + 2e^- \rightarrow H_2.$	(4.2)

Other cathodic reactions play important roles under specific conditions:

Reduction of metal ions, e.g.:	$Cu^{2+} + 2e^- \rightarrow Cu.$	(6.1)
Reduction of carbonic acid:	$H_2CO_3 + e^- \rightarrow H + CO_3^-.$	(6.2)
Reduction of hydrogen sulphate:	$H_2S + 2e^- \rightarrow 2H + 2HS^-.$	(6.3)
Reduction of chlorine:	$Cl_2 + 2e^- \rightarrow 2Cl^-.$	(6.4)
Reduction of nitric acid:	$HNO_3 + 3H^+ + 3e^- \rightarrow NO + 2\ H_2O.$	(6.5)

Whether significant corrosion is to occur or not is often determined by the possibilities for cathodic reactions. Therefore we shall have a closer look at the most common cathodic reactions. In addition to reactions in natural waters we shall also consider those typical for oil and gas production environments.

When there is an oxidizer in the liquid and the concentration of this is increased, the corrosion potential is generally increased, as shown in Figures 6.1 and 6.2. The shift is due to the activity term in the reversible potential equation (Nernst's equation, (3.9)): increased activity of oxidizers gives increased reversible potential for the cathodic reaction. This may lead to an increase in corrosion rate, as shown on the right in Figures 6.1 and 6.2 for active and active–passive metals, respectively.

A general relationship should be noticed: due to the reactions, concentration differences are established along the metal surface. These differences often affect the corrosion pattern and local corrosion rates. For instance, pH differences are frequent and important. The increase of pH on cathodes as a result of the oxygen reduction is an example of particular importance (see Section 6.2, Equation (4.4b)).

Exchange current densities i_0 and Tafel gradients b_c for some reduction reactions are shown in Table 4.1.

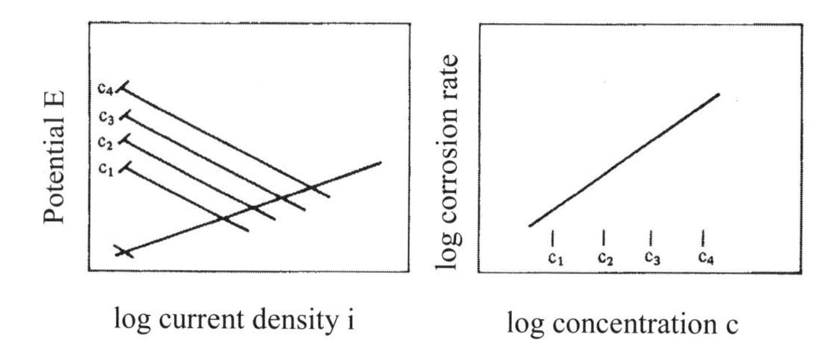

Figure 6.1 Oxidizer concentration c affects the reversible potential of the cathodic reaction and consequently also the corrosion rate of an active metal.

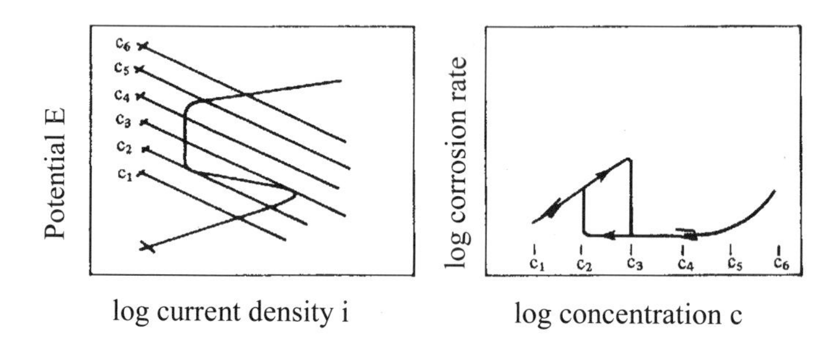

Figure .6.2 As for Figure 6.1, but now for an active–passive metal.

6.2 Corrosion under Oxygen Reduction

6.2.1 General Circumstances and Data

From the reaction equation (4.4) it is seen that access of oxygen and hydrogen ions is necessary. Sufficient direct supply of H^+ ions occurs only in acid media, where the H^+ concentration is high. However, oxygen reduction is possible also in neutral and alkaline media because reaction (4.4) is supported by continuous dissociation of water:

$$4H_2O \rightarrow 4H^+ + 4OH^-$$

$$O_2 + 4H^+ + 4e^- \rightarrow 2H_2O \tag{4.4a}$$

Total reaction: $$O_2 + 2H_2O + 4e^- \rightarrow 4OH^- . \tag{4.4b}$$

Thus, the oxygen reduction in acid liquids is described by Equation (4.4a) and in neutral and alkaline solutions by Equation (4.4b). The reversible potential of the two reactions as a function of pH is represented by the same straight line in the Pourbaix diagram. At a given pH value and 25°C the reversible potential is

$$\begin{aligned} E_0 &= E^0_{0a} - \frac{0.059}{4} \log \frac{1}{p_{O_2}\, a^4_{H^+}} \\ &= E^0_{0b} - \frac{0.059}{4} \log \frac{a^4_{OH^-}}{p_{O_2}} . \end{aligned} \tag{6.6}$$

It should be noticed that E^0_{0a} (= 1.23 V (SHE)) is the reversible potential at pH = 0 ($a_{H+} = 1$) and $p_{O_2} = 1$, while E_{0b} (= 1.23 – 0.059 × 14 = 0.40 V) is the reversible potential at pH = 14 ($a_{OH^-} = 1$) and $p_{O_2} = 1$.

Exchange current density and Tafel gradient for O_2 reduction on steel in air-saturated 0.5 N Cl^- electrolyte was determined by Haagenrud [6.1] on the basis of data from the literature:

$$i_o = 2 \times 10^{-8}\ \text{A/cm}^2.$$
$$b_c = 0.22\ \text{V/decade}.$$

Other published values of i_{0O_2} are between 10^{-10} and 10^{-13} A/cm^2 [6.2], which once again shows that i_0 depends strongly on the conditions.

Of conditions that are particularly important for the exchange current density, here we shall mention the activity of bacteria in the corrosion medium. On stainless steels and other relatively corrosion-resistant materials in seawater a microbiological slime layer is usually formed, and below 30–40°C this slime layer may have a dramatic effect on the cathodic properties. The oxygen reduction reaction is strongly catalyzed; during a few weeks of exposure it may accelerate 1–3 orders of magnitude depending on the conditions [6.3, 6.4]. Under certain conditions the local corrosion rate can increase correspondingly (see Section 7.3, Galvanic Corrosion and Section 7.5, Crevice Corrosion). The free corrosion potential increases strongly as a consequence of the catalyzed cathodic reaction, which also promotes initiation of localized corrosion. The exact mechanism of the bacterial effect is not fully clarified, but research results indicate that the exchange current density of the oxygen reduction is the electrochemical property mainly affected; it can increase a few decades.

As mentioned earlier, the oxygen reduction will normally dominate over the hydrogen reaction in air saturated aqueous solutions when pH is, roughly speaking, above 3. Therefore in most natural environments with access of air, such as the atmosphere, seawater and fresh waters, the oxygen reduction is the crucial cathodic reaction.

Distribution and intensity of corrosion is therefore usually determined by the access of oxygen to the various parts of the surface. Uneven oxygen access causes concentration cells (Section 7.5). Very often the corrosion is diffusion controlled, i.e. the diffusion-limiting current density is of special interest. Several examples of corrosion rate determined by the oxygen reduction rate are dealt with in Chapter 7.

6.2.2 Effect of Temperature

The temperature affects the exchange current density and the Tafel gradient, but because of relationships just mentioned above, the effect of temperature on the diffusion-limiting current density i_L is often the most significant one in natural environments. On surfaces without diffusion-limiting deposits, i_L is expressed by:

$$i_L = D_{O_2} \, zF \frac{c_B}{\delta} . \tag{4.7}$$

The diffusion coefficient D_{O_2} depends strongly on the temperature:

$$D_{O_2} = A \, e^{-Q/RT} , \tag{6.7}$$

where A and Q can be considered as constants in water, R the universal gas constant and T temperature in K.

The thickness δ of the diffusion boundary layer depends on various factors. For cases where the liquid flows at 90° towards the edge of a plane electrode and parallel to the electrode surface, Vetter [6.5] gives, after Eucken and Levich, for laminar flow:

$$\delta_{lam} = 3\ell^{1/2} \, v_\infty^{-1/2} \, \nu^{1/6} \, D^{1/3} . \tag{6.8}$$

and after Vielstich, for turbulent flow:

$$\delta_{turb} \propto \ell^{0.1} \, v_\infty^{-0.9} \, \nu^{17/30} \, D^{1/3} \tag{6.9}$$

In these equations, ℓ is the distance from the edge of the plate in the flow direction, v_∞ is relative velocity between the electrode and the liquid at infinite distance from the electrode surface, ν is kinematic viscosity and D the diffusion coefficient.

Increased temperature gives reduced viscosity ν and increased diffusion coefficient D, and the total effect of temperature on the boundary layer thickness δ is therefore relatively small.

For open systems where water is in equilibrium with the atmosphere, the oxygen concentration c_B is determined by the solubility, i.e. the saturation concentration, of air in the water. This concentration depends on the temperature, as shown in Figure 6.3.

The effects on D and c_B together lead to a corrosion rate as a function of temperature as shown in Figure 6.4, curve a) for an open steel tank, with a marked maximum at about 80°C. If the system is closed, a certain supersaturation of air arises when the temperature is increased, with the result indicated by curve b). If the oxygen concentration was constant, the corrosion rate increase would approximately follow the increase in the diffusion coefficient, like that shown by the dashed curve c). In all three cases the corrosion is diffusion controlled, i.e. $i_{corr} = i_L$.

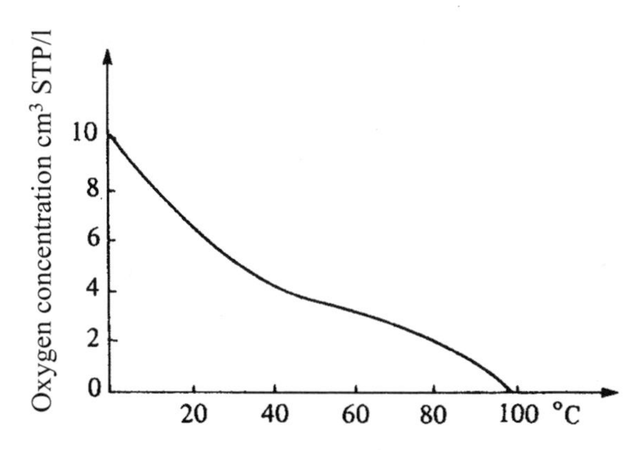

Figure 6.3 Concentrations of oxygen in air-saturated water as a function of temperature [6.6]. (Reproduced from Tomashov ND. Theory of Corrosion and Protection of Metals. New York: Macmillan, 1966.)

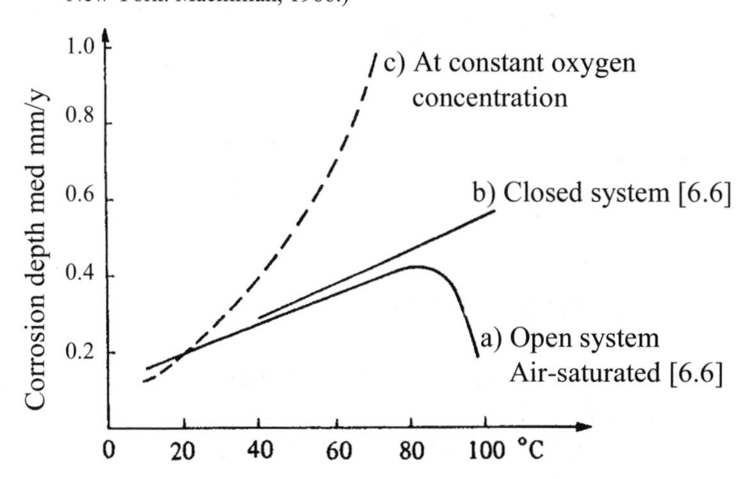

Figure 6.4 Corrosion rate of steel in water as a function of temperature for an open system (a), a closed system (b), and a system where the oxygen concentration is kept constant (c).

6.2.3 Effect of Surface Deposits

Equation. (4.7) is valid only when there are no diffusion-limiting deposits on the surface. Under real service conditions, however, surface layers of more or less porous corrosion products, salts or other substances that hinder the diffusion, often exist. These layers are not usually electron conducting, and thus the cathodic reaction has to take place at the metal surface underneath the deposits, possibly in holes or pores in the surface layers. This will give a diffusion stance in series with the usual diffusion resistance in the boundary layer.

Equation (4.7), valid for a clean surface, can be written in the following form:

$$i_L = zF \frac{c_B}{\delta/D},$$

in which δ is an expression for the diffusion resistance in the boundary layer. For a surface with a diffusion-hindering deposit the corresponding equation will be [6.7, 6.8]:

$$i_L = zF \frac{c_B}{\delta/D + t_B/P_B}, \qquad (6.10)$$

where t_B is the thickness and P_B the oxygen permeability of the deposit. t_B/P_B expresses the diffusion resistance of the layer of deposited substances, and the total resistance against diffusion of oxygen to the metal surface is given by the sum in the denominator.

Such surface layers can reduce the real corrosion rates considerably. One example will be mentioned here: a clean iron or unalloyed steel surface in stagnant air–saturated water would corrode at a rate of 0.6 mm/y at 20°C,* and in flowing water up to several times more, while the average corrosion rate in free seawater really is only 0.1–0.15 mm/y due to deposition of rust and salts, particularly $CaCO_3$.

6.2.4 Effect of Flow Velocity

As shown by the examples for a plane metal surface in Equations (6.6) and (6.7), the diffusion boundary layer thickness depends on the flow velocity. Generally, Nernsts diffusion boundary layer (Figure 6.5) can be expressed by

---

*This is based on a boundary layer thickness of 0.5 mm, which represents conditions with only free (not forced) convection, and further on a saturation oxygen concentration at 20°C of 2.5 mol/cm^3, a diffusion coefficient of 2.5 cm^2/s, z = 4, and F = 96,500 C/mol e^-.

$$\delta = C \ f(L) \, Re^{-m} \ Sc^{-n} \ \propto v^{-m}, \tag{6.11}$$

where

Re is Reynolds' number = vL/ν,
Sc is Schmid's number = ν/D,
v is the free flow velocity (= v_∞ in Equations (6.8) and (6.9)),
L is a characteristic length and f(L) is a simple geometrical function (cf. Equations (6.8) and (6.9)).

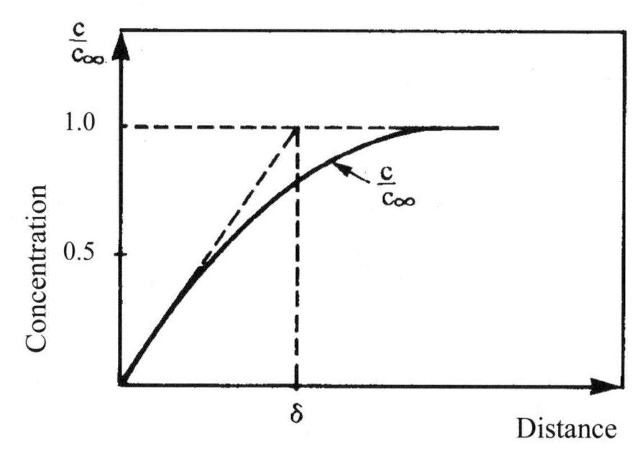

Figure 6.5 Nernst's diffusion boundary layer δ defined by the slope of the concentration profile tangent through origin and its intersection with the line $c/c_\infty = 1$.

Values of k, f(L), m and n are given in the literature [6.9–6.11] for various conditions. Usually, m = 0.5 for a laminar boundary layer, while it is 0.8–0.9 for a turbulent boundary layer.

Thus, for diffusion-controlled corrosion on clean metal surfaces we have according to Equation (4.7):

$$i_{corr} = i_L \propto v^{0.5} \quad \text{for laminar boundary layer,}$$
$$i_{corr} = i_L \propto v^{0.8-0.9} \quad \text{for turbulent boundary layer.}$$

These relationships are indicated by the left part of the solid curve in Figure 6.6. If there is a deposit layer on the metal surface whose diffusion resistance dominates in relation to the diffusion resistance of the boundary layer, the corrosion rate will be low and hardly affected by an increase in the flow rate, as shown by the left part of the dashed curve in Figure 6.6. When, however, the flow rate becomes high enough, erosion and successive removal of the deposit layer takes place, or it dissolves, and the corrosion rate increases strongly. When the velocity is so large that all deposits are removed, the two curves coincide as shown in the figure. For the steepest part of the curve, the exponent in the relation $i_{corr} \propto v^n$ will have a value between 0.8 and 2 depending on the combination of material and environment. At even higher flow velocities, there is a transition to pure activation control, which means that the corrosion rate becomes independent of velocity (Figure 6.6). The conditions at the various velocities v_1, v_2 and v_3 in Figure 6.6 correspond to the overvoltage curves in Figure 6.7.

An example of measured corrosion rates on steel in seawater is shown in Figure 6.8. If this diagram is redrawn with $v^{1/2}$ as the variable along the horizontal axis, we will get an S-shaped curved in this case also, as we have in Figure 6.6.

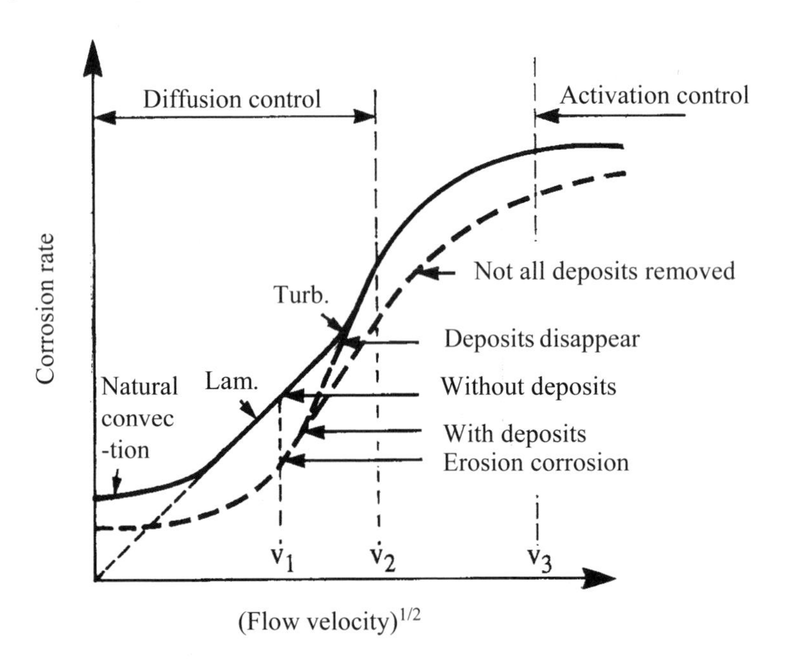

Figure 6.6. Corrosion rate as a function of (flow velocity)$^{1/2}$ (active metal).

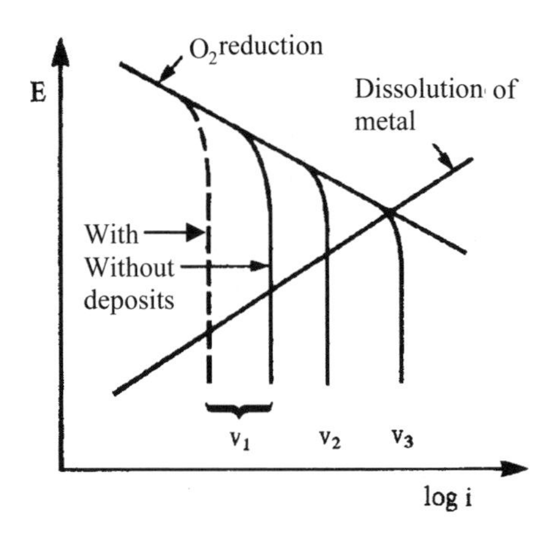

Figure 6.7 Schematic overvoltage curves corresponding to Figure 6.6 (active metal).

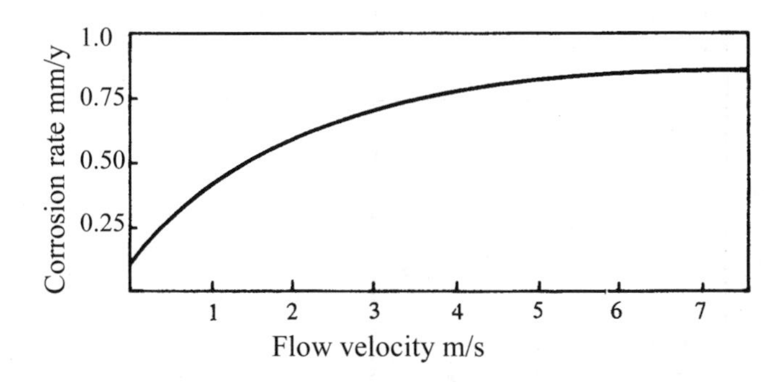

Figure 6.8 Corrosion rate as a function of flow velocity. Steel in seawater at 23°C

In Figures 6.6 and 6.7 it was assumed that the material is active in all the actual situations, which is the case, e.g. for steel in more or less acidic aqueous solutions. Conversely, if we consider steel in neutral or slightly alkaline environments, passivation may occur when we increase the flow rate sufficiently. The reason for such a behaviour is that i_l exceeds the critical current density for passivation, i_{cr}. This is illustrated schematically in Figures 6.9 and 6.10. Under real conditions the transition to the passive state is not as abrupt as in Figure 6.10.

The relationships shown between flow velocity and corrosion rate are not the only possible ones. Other forms of functions, depending on combination of material and environment, have been illustrated by Lotz and Heitz [6.13]. Lotz [6.14] has stated different velocity exponents (n) for different corrosion mechanisms. Thus, values of n determined experimentally for different cases may indicate which mechanism is acting in each case.

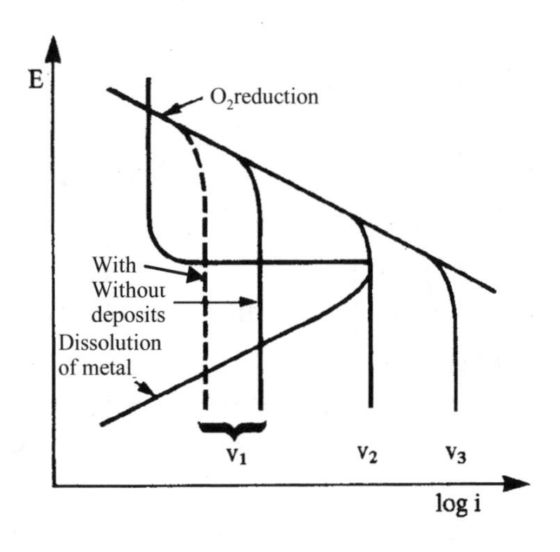

Figure 6.9 As Figure 6.7, but for an active–passive metal.

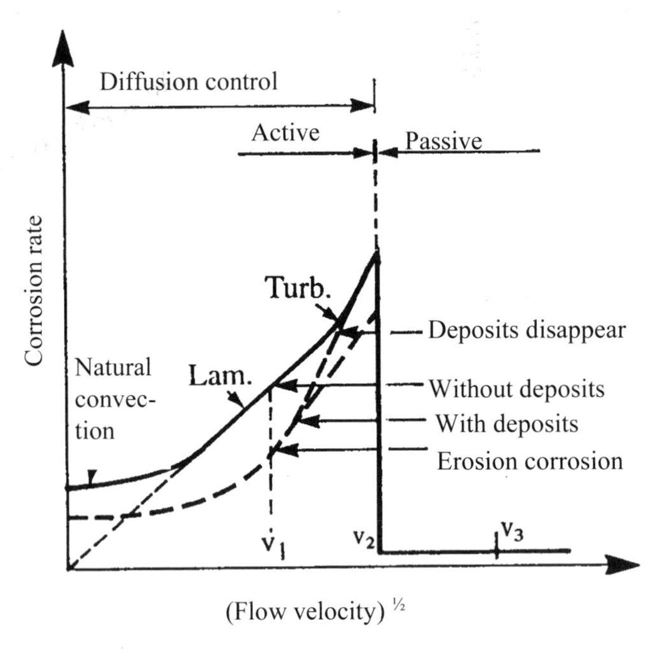

Figure 6.10 As Figure 6.6, but for an active–passive metal.

6.2.5 Corrosion under Thin Films of Water

In the atmosphere, thin films of water are often formed, which leads to electrochemical corrosion, in principle wet corrosion. Tomashov [6.6] considers such cases as wet atmospheric and humid atmospheric corrosion. The significance of water film thickness is shown schematically in Figure 6.11.

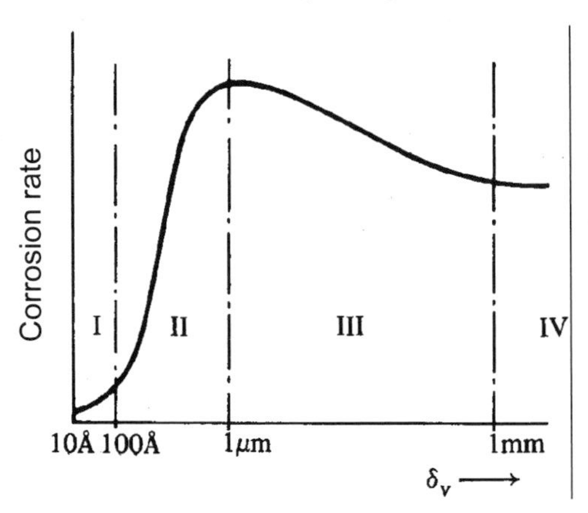

Figure 6.11 Rate of atmospheric corrosion as a function of water film thickness (after Tomashov [6.6]).

According to Tomashov, $\delta_v > 1$ mm (region IV) gives ordinary wet corrosion, with corrosion rates like those for submerged metal. In region III, $\delta_v = 1$ mm to 1 μm, the corrosion is characterized as wet atmospheric (visual water film), and when $\delta_v = 1$ μm to 100 Å (region II) as humid atmospheric corrosion, with a film which is at least partly invisible (chemically adsorbed). Maximum corrosion rate, indicated by Tomashov at a water film thickness of about 1 μm, is due to easy access to oxygen while the corrosion is still electrochemical. When the film thickness increases from this level the corrosion rate decreases because the diffusion layer increases. On the other hand, when δ_v is reduced from about 1 μm, the anodic polarization increases successively, and the corrosion is reduced as a consequence [6.6].

The values of δ_v given above should not be taken too seriously, since they depend much on the conditions. According to Shreir [6.11], the maximum corrosion rate has been found at a water film thickness of about 150 μm. It must also be taken into account that real atmospheric corrosion is strongly affected by deposits on the surface. Atmospheric corrosion is treated more from a practical point of view in Section 8.1. Figure 8.1 shows a photo of corrosion attack on a 14-year-old ventilation outlet from a bathroom in a private house. Here a water film has been formed by condensation on the inner surface.

6.3 Corrosion under Hydrogen Evolution

Like the oxygen reduction reaction described by Equation (4.4a), the hydrogen reaction corresponding to Equation (4.2) requires sufficient access of hydrogen ions. This is satisfied at low pH values, where reaction (4.2) can occur directly. Considerable corrosion of iron and steel, with the hydrogen reaction as the only cathodic reaction, is limited to this low-pH region. In any case, for completeness we shall look at a wider pH region in this connection. In near neutral and alkaline liquids, hydrogen ions can be produced by dissociation of water, so that the hydrogen gas evolution occurs by Equation (4.2b) in these cases:

$$\begin{array}{r} H_2O \rightarrow 2H^+ + 2OH^- \\ 2H^+ + 2e^- \rightarrow H_2 \\ \hline \end{array} \qquad (4.2a)$$

$$2H_2O\ 2+ 2e^- \rightarrow H_2 + 2OH^-. \qquad (4.2b)$$

The reversible potentials of reactions (4.2a) and (4.2b) are defined by the same line in the Pourbaix diagram.* However, the nature of hydrogen ion access affects the electrode kinetics, thus the overvoltage curve of the two reactions are different. This

* $E^0_{0a} = 0$ V (SHE) ($pH = 0$, $p_{H2} = 1$).

$E^0_{0b} = 0 - 14 \times 0.059 = -0.83$ V vs. SHE ($pH = 14$, $p_{H2} = 1$).

(Compare equilibrium potentials for the oxygen reduction reaction, p. 67.)

means that the cathodic overvoltage curve and polarization curve in an oxygen-free solution depend on the pH level, as shown in Figure 6.12.

The lower straight line is the overvoltage curve of reaction (4.2b), while the cathodic curves above this line represent reaction (4.2a). It appears that (4.2b) has a stronger activation polarization than (4.2a). On the other hand, reaction (4.2a) is limited by concentration polarization (diffusion control) at higher current densities. The consumption of H^+ is in these cases determined by the diffusion rate of H^+ ions to the metal surface.

From the background described, the reaction course at increasing cathodic polarization will be as follows: at low current densities, reaction (4.2a) occurs under activation polarization. By increasing the current, this reaction becomes successively concentration polarized, and finally the reaction (4.2b) takes over. Since this reaction does not depend on H^+ diffusion, a current increase is allowed again.

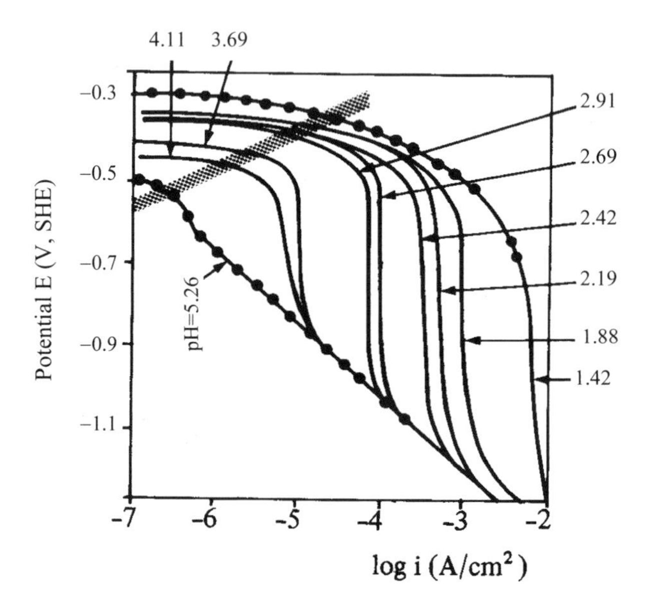

Figure 6.12 Polarization curves for iron in oxygen-free 4% NaCl solution, at various pH values (adapted from Kaesche [6.15]).

The exchange current density for the hydrogen reaction depends strongly on the conditions, while the Tafel gradient in most cases is 0.10–0.12 V/decade (see Table 4.1).

As previously mentioned, the hydrogen reaction is normally the dominating cathodic reaction on steel in air-saturated solutions with pH less than about 3, and in liquids with low oxygen concentration also at somewhat higher pH values. However, the hydrogen reaction may be significant at far higher bulk values of pH in cases of localized corrosion (local acidification), which we will come back to in

Section 7.5. The reaction also plays an important role under anaerobic conditions involving bacteria, and hydrogen is an important element under CO_2 corrosion as well, as described in the following sections.

6.4 Corrosion under Effects of Bacteria

If there is no access to air, there is usually no significant corrosion on ferrous materials when pH > 5. However, in some environments, heavy localized corrosion may occur due to bacteria that thrive in near neutral environment (pH ≈ 5–8) at temperatures of 10–40°C in the absence of air (anaerobic bacteria). These bacteria promote reduction of present sulphate to sulphide. Hence, they are usually called sulphate–reducing bacteria (SRB). The "classical" mechanism comprises the following reactions at/on the cathodes [6.16]:

Dissociation of water: $8H_2O \rightarrow 8H^+ + 8OH^-$.

Cathodic reaction: $8H^+ + 8e^- \rightarrow 8H$. (4.2c)

Cathodic depolarization: $SO_4^{2-} + 8H \xrightarrow{SRB} S^{2-} + 4H_2O$. (6.12)

The former two reactions are like the usual ones that constitute the total hydrogen reaction in near neutral or alkaline environments (compare Section 6.3), except that the hydrogen atoms are not combined with hydrogen molecules in the case where bacteria are active, according to this "classical" mechanism theory. The cathodic depolarization reaction that occurs due to the bacteria plays an important role. SRB catalyze this reaction, hence it consumes efficiently the uncharged H atoms produced in the cathodic reaction. Because of this the cathodic reaction can proceed at high rates at low overvoltages. To understand the significance of this relationship one can look at the Pourbaix diagram for iron in water in Figure 3.10. Under near neutral conditions (pH = 5–8) the region between the reversible potential of the reaction $Fe = Fe^{2+} + 2e$ (line a) and that of the hydrogen reaction (line f) is rather narrow. The polarization of the hydrogen reaction makes high local corrosion rates possible within this narrow potential range. In this connection it must also be taken into consideration that the area ratio between cathodes and anodes can be relatively large, so that the corrosion potential may be very close to the reversible potential of the hydrogen reaction. Without the SRB, the reaction at these low overvoltages would essentially stop as soon as the steel surface was covered with a monolayer of hydrogen atoms.

Ferrous ions from the anodic reaction $Fe \rightarrow Fe^{2+} + 2e^-$ react with S^{2-} from the cathodic depolarization reaction and with OH^- from the water dissociation reaction and form ferrous sulphide, FeS, and hydroxide, $Fe(OH)_2$. FeS can play an important role. Where the sulphide forms a continuous film on the surface it acts as protection and as an effective site for the cathodic reaction. If the film is injured or there is a lack of continuity in the film for other reasons, local galvanic corrosion will occur. Experiments and experience indicate that also the anodic reaction ($Fe \rightarrow Fe^{2+} + 2e^-$) is depolarized as a result of the SRB environment. This is of interest in connection with cathodic protection (Section 10.4). Depending on the conditions, other products can be formed as well, e.g. H_2S.

Some SRB may cause localized corrosion on stainless steels, nickel alloys, aluminium, zinc and copper alloys. Mechanisms of sulphur–assisted corrosion, with emphasis on Fe– and Ni–based materials, have recently been reviewed by Marcus [6.17]. The review includes the fundamentals of enhanced dissolution, retarding or blocking of passivation, and passivity breakdown.

Corrosion due to SRB may occur in oil and gas wells, and in cases of anaerobic conditions in soils, sediments on sea and river beds, and beneath organic layers on metal surfaces in seawater. The anaerobic conditions may in many cases be limited to small areas in crevices, beneath deposits or injured coatings, etc.

Examples of corrosion due to SRB and other bacteria are dealt with in Chapter 8. SRB must be taken into account in design of systems for cathodic protection of steel (Section 10.4).

6.5 CO_2 Corrosion

6.5.1 General

The most serious corrosion problem in oil and gas production is caused by CO_2. In production tubing and pipe systems there is some CO_2 in the gas phase. In addition, water is often precipitated. The water phase dissolves CO_2, which causes its own cathodic reactions (see below). The water phase contains also considerable amounts of dissolved salts, and thus its conductivity is high.

Pictures of CO_2 corrosion attacks in production tubing and a flow line after a few years of service at a field in the North Sea are shown in Figures 8.13 and 7.43. Cases and corrosion data are dealt with in Section 8.6.

6.5.2 Mechanisms [6.18]

In the oil/gas system the partial pressure of CO_2 equals the mol fraction of CO_2 in the gas multiplied by the total pressure. The total pressure is often of the order of 100 bar, while the CO_2 partial pressure may be, e.g. about 1 bar, although it varies considerably from one site to another.

A precipitated water phase in such a system dissolves CO_2 up to a concentration proportional to the partial pressure of CO_2 in the gas phase. The solubility depends also on the temperature.

CO_2 dissolved in water gives carbonic acid:

$$H_2O + CO_2 \rightarrow H_2CO_3, \tag{6.13}$$

which is a weak acid, i.e. it dissociates to a minor extent:

$$H_2CO_3 \rightarrow H^+ + HCO_3^-. \qquad (6.14)$$

At higher pH values (> 6) HCO_3^- dissociates further to $H^+ + CO_3^{2-}$. The pH value is usually lower, so the latter dissociation occurs normally to a very small extent. The cathodic reaction is assumed primarily to follow the equation

$$H_2CO_3 + e^- \rightarrow H + HCO_3^-. \qquad (6.15)$$

The hydrogen atoms that are adsorbed on the surface are there combined with hydrogen molecules and form hydrogen gas:

$$2H \rightarrow H_2 \qquad (6.16)$$

At lower pH values (2–4), reduction of H^+ from reaction (6.14) may give a certain contribution, while reduction of HCO_3^- is significant at pH > 6 [6.19, 6.20].

The cathodic reaction in solutions containing CO_2 may be very efficient and give high current densities, as shown in Figure 6.13 for different rates of convection in the water.

When carbon steel corrodes in water containing CO_2, the anodic reaction is

$$Fe \rightarrow Fe^{2+} + 2e^-. \qquad (6.17)$$

The total reaction is

$$Fe + H_2CO_3 \rightarrow FeCO_3 + H_2. \qquad (6.18)$$

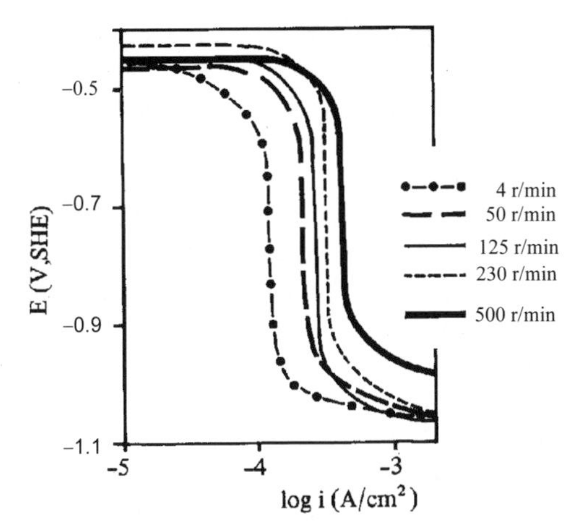

Figure 6.13 Cathodic polarization curves of steel in CO_2-saturated 0.5 M Na_2SO_4 solution with CO_2 concentration = 0.027 mol/l and pH = 4.2, for different revolution numbers (r/min) of the electrode (after Schmitt and Rothmann [6.20]).

The solubility of $FeCO_3$ is low and decreases with increasing temperature. $FeCO_3$ is therefore deposited when the temperature exceeds a limit that depends on the CO_2 partial pressure (often 60–80°C). In the solution, $Fe(HCO_3)_2$, the formation of which is described by combination of Equations (6.2) and (6.16), appears to a considerable extent. At higher temperatures it is decomposed to $FeCO_3$ and H_2CO_3.

When $FeCO_3$ deposits, a protecting film is formed. According to SEM investigations at the Institute for Energy Technology in Oslo, the film also contains Fe_3C in a skeletal form [6.21, 6.22]. Increasing flow velocity causes increasing mass transport and consequently a higher extent of dissolution of corrosion products with increased corrosion rate as a result [6.18].

6.5.3 Corrosion Rate

Partly on a theoretical and partly on an experimental basis, de Waard and Milliams developed a model that has been used frequently for determination of CO_2 corrosion rates on steel [6.23]. A newer version of de Waard–Milliams equation is

$$\log V_{corr} = 5.8 - \frac{1710}{273 + t} + 0.67 \log(pCO_2) \tag{6.19}$$

where V_{corr} is corrosion rate in mm/year, t is temperature (°C) and pCO_2 the partial pressure of CO_2 in bar [6.18, 6.24].

The rate of the cathodic reaction is proportional to the CO_2 partial pressure. Thus, the factor 0.67 in front of log (pCO_2) indicates that the corrosion is not completely under cathodic control. (If it was, the factor would be = 1.) In the equation, mass transport limitation due to deposits of corrosion products has not been taken into consideration. Therefore it represents a so-called "worst case".

The de Waard–Milliams equation is often presented graphically in a nomogram, as shown in Figure 6.14. Here a scale for a deposit factor (scale factor) is also included. This "scale factor scale" has been determined by comparison of measured corrosion rates at various temperatures above 60°C with those calculated from Equation (6.18). The read corrosion rate at a certain combination of CO_2 pressure and temperature has to be multiplied by the read deposit (scale) factor.

To be more accurate, the actual pHvalue of the water phase has to be taken into account. This has been done in different ways [6.18, 6.25]. In the latter publication the effects of other environmental conditions have been described as well. For instance in a gas-dominated system where water may be condensed, the water-to-gas and temperature-to-dewpoint ratios are of great significance. It may also be necessary to correct for high flow rates, particularly when the flow velocity is higher than a critical velocity, which according to API RP 14E [6.26] can be expressed by

$$V_c = \frac{C}{\sqrt{\rho}} \tag{6.20}$$

where V_c is critical velocity leading to erosion (removal of scale), given in ft/s, ρ is density of the fluid (lb/ft^3) at the actual pressure and temperature, and C is an empirical constant. Various values of C have been proposed, depending on the continuity/intermittence of service. The cited edition of API RP 14E (1991) recommends a C-factor of 100 for corrosive systems and 150–200 for inhibited systems. If there is sand in the fluid the value of C must be reduced. The relevance and justification of Equation (6.19) have been questioned, e.g. by Salama [6.27].

The velocity effects on CO_2 corrosion have been studied in several projects dealing with multiphase flow. Mechanistic models based on electrochemistry, reaction kinetics, and mass transfer effects have been developed, e.g. by Nesic et al. [6.28]. A semi-empirical model was presented by de Waard et al. [6.29]. This model and the corresponding experimental results are valid for cases without carbonate scales. Inhibition can be accounted for by inserting an inhibitor efficiency factor. A model mainly based on the same data as the one developed by de Waard et al. is included in the standard NORSOK M-506 [6.30]. The application of the NORSOK model is based on a computer program. Common to these models is that they are valid for cases with a bulk phase of water. For mist flow and dewing conditions the calculation basis is inferior, but such conditions give low corrosion rates.

The effect of velocity on CO_2 corrosion in a real gas–condensate well is shown in Figure 8.12, which demonstrates how the critical velocity depends on geometry and inhibition.

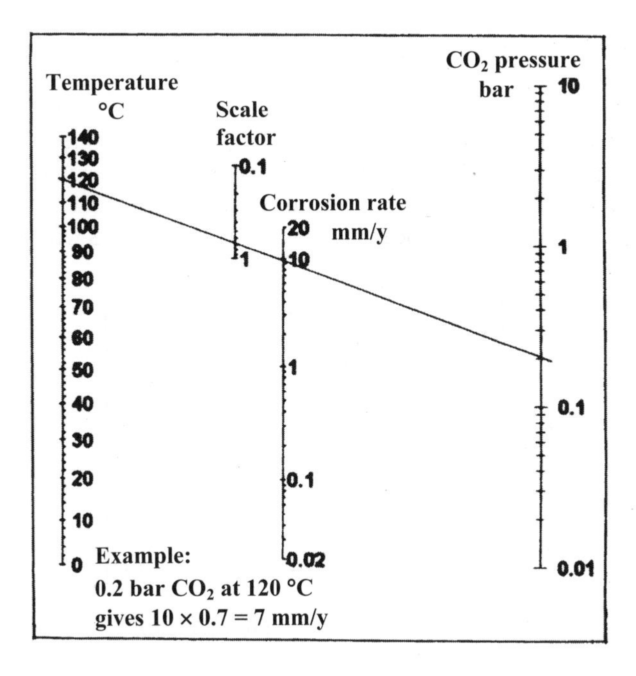

Figure 6.14 Nomogram for CO_2 corrosion (after de Waard and Lotz [6.18]). Reproduced with permission of NACE

6.6 H_2S Corrosion

In some oil/gas wells there is so much hydrogen sulphide (H_2S) that the sulphide reacts significantly with the structural materials. It is said that the oil/gas is sour in these cases. According to the NACE standard the gas is considered sour when the partial pressure of H_2S is higher than 0.003 bar. However, in the guidelines from the European Federation of Corrosion (EFC) [6.31], the pH is taken into account, and the limits between sweet and sour gas are therefore defined by curves. According to the EFC guidelines a higher H_2S pressure is in most cases acceptable compared with the limits given in the NACE standard. In most fields in the North Sea the hydrogen sulphide concentration has been low, but it is assumed that it will increase during the years of production because seawater injection in the wells may cause increased H_2S content. The reason for this is that bacteria can reduce sulphate (originating from the injected water) to sulphide. There is therefore a risk of more material failures due to H_2S in the future (see also Sections 7.12.4 and 8.6).

Like CO_2, H_2S dissolves in water, forming a weak acid:

$$H_2S(g) \rightarrow H_2S(aq) \rightarrow H^+ + HS^- . \tag{6.20}$$

At higher pH (> 6) HS^- dissociates further to $H^+ + S^{2-}$. The most important cathodic reaction in a sulphide–containing environment can be expressed by

$$2H_2S + 2e^- \rightarrow 2H + 2HS^- . \tag{6.3}$$

H atoms are combined with hydrogen gas, but some hydrogen in the atomic state diffuses into the material. This may cause hydrogen embrittlement, particularly in high-strength steels. The result may be hydrogen-induced cracking, which is dealt with under stress corrosion cracking (Section 7.12). Together with a much higher concentration of CO_2, low or moderate concentrations of H_2S will not contribute

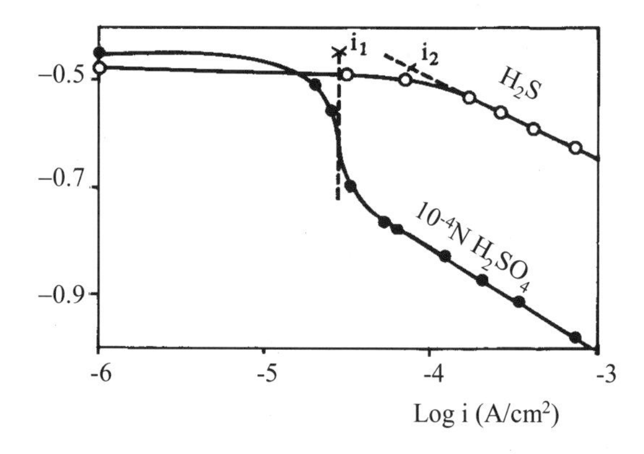

Figure 6.15 Cathodic polarization curves of iron in a diluted strong acid and in water containing H_2S, in both cases at pH = 4 (after Neumann and Carius [6.32]).

significantly to the surface corrosion of carbon and low-alloy steels in oil production plants, even though the reaction (6.3) can be relatively efficient, as indicated by the cathodic curves in Figure 6.15. Sulphur originating from H_2S can affect the corrosion properties of passive materials [6.17]; sometimes it causes localized corrosion, e.g. at welds in low-carbon martensitic stainless steel [6.33].

6.7 Other Cathodic Reactions

Reduction of metal ions must be taken into consideration in some cases. A typical example is reduction of copper, Equation (6.1). (This is further dealt with under galvanic corrosion, Section 7.3.)

An important reaction in some cases is reduction of chlorine:

$$Cl_2 + 2e^- \rightarrow 2Cl^- \tag{6.4}$$

The oxidizer, chlorine, may in larger concentrations lead to breakdown of the passive oxide on stainless steels and cause heavy corrosion. This has happened around swimming pools, where evaporation has led to enrichment of Cl_2 in water films. In smaller concentrations it is beneficial, preventing organic growth, and chlorination (addition of sodium hypochlorite) is used for this purpose in seawater pipe systems.

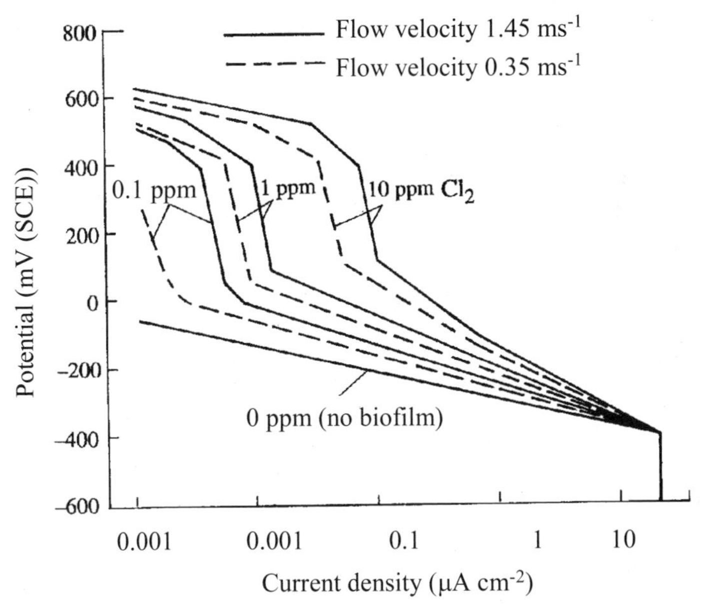

Figure 6.16 Smoothed cathodic overvoltage curves for a high-alloy stainless steel in aerated and chlorinated seawater at 25°C (From Gartland and Drugli [6.34]).

Figure 6.16 shows how different concentrations of chlorine can affect cathodic overvoltage curves for stainless steel in seawater. The corrosion risk for stainless steel at higher Cl_2 concentrations arises because the increased cathodic reaction rate lifts the potential so that critical potentials for local corrosion are exceeded (see also Section 8.3). Chlorine may cause corrosion on several other materials as well.

In connection with corrosion experiments involving cathodic polarization in fluids containing chloride, one should be aware that chlorine is developed (in addition to oxygen) on the anode (the counter electrode) by reaction (6.4) in the reverse direction. In order to prevent this affecting the conditions at the working electrode (the test specimen), the counter electrode should rather be positioned in its own compartment, separated from the main compartment of the cell with an ion–conducting material that prevents mixing of the anolyte and the catholyte (e.g. agar or wood containing salt water).

HNO_3 (see Equation (6.5)) is an example of a strong oxidizer, which (contrary to Cl_2) promotes passivation of steel. Because HNO_3 lifts the potential strongly, it may cause heavy corrosion on relatively noble metals and alloys (e.g. copper and its alloys).

References

6.1 Haagenrud SE. Dr. ing. Thesis, Institutt for mekanisk teknologi, NTH, Trondheim, 1972. (In Norwegian.)

6.2 Parsons R. Handbook of Electrochemical Constants. London: Butterworths, 1959.

6.3 Johnsen R, Bardal E. Cathodic properties of different stainless steels in natural seawater. Corrosion: 41, 1985, 296.

6.4 Bardal E, Drugli JM, Gartland PO. The behaviour of corrosion–resistant steels in seawater: A review. Corrosion Science, 35, 1–4, 1993: 257–267.

6.5 Vetter KJ. Electrochemische Kinetik. Berlin–Heidelberg–New York: Springer-Verlag, 1961.

6.6 Tomashov ND. Theory of Corrosion and Protection of Metals. New York: MacMillan, 1966.

6.7 Gartland PO. Effects of flow on cathodic protection of a steel cylinder in seawater. SINTEF–report STF16 F80007, Trondheim, 1980.

6.8 Bardal E. Effects of flow conditions on corrosion. Lecture Series on Two–phase Flow, NTH, Trondheim, 1984.

6.9 Schlichting H. Boundary Layer Theory. New York: McGraw–Hill, 1979.

6.10 Levich VG. Physiochemical Hydrodynamics. Prentice–Hall, 1962.

6.11 Shreir LL. Corrosion, Vol. 1, 2nd Ed. London–Boston: Newnes-Butterworths, 1976.

6.12 Uhlig HH. Corrosion Handbook. New York: John Wiley and Sons, 1948.

6.13 Lotz U, Heitz E. Flow dependent corrosion. I Current understanding of the mechanisms involved. Werkstoffe und Korrosion, 34, 1983: 454–461.

6.14 Lotz U. Velocity effects in flow induced corrosion. Corrosion/90. Paper No 27, National Association of Corrosion Engineers, NACE, Houston, Texas, 1990.
6.15 Kaesche H. Korrosion der Metalle. Berlin–Heidelberg–New York: Springer-Verlag, 1966.
6.16 Metals Handbook, 9th Ed. Vol. 13, Corrosion. Metals Park, Ohio: ASM International, 1987.
6.17 Marcus P. Sulfur-assisted corrosion mechanisms and the role of alloyed elements, in Marcus P, editor. Corrosion Mechanisms in Theory and Practice, 2nd Ed., Marcel Dekker, New York, 2002; 287–310.
6.18 de Waard C, Lotz U. Prediction of CO_2 corrosion of carbon steel. Corrosion/93, Paper No 69, NACE, Houston, Texas, 1993.
6.19 Gray LGS, Anderson BG, Danysh MJ, Tremaine PR. Effect of pH and temperature on the mechanism of carbon steel corrosion by aqueous carbon dioxide. Corrosion/90, Paper No 40, NACE, Houston, Texas, 1990.
6.20 Schmitt G, Rothmann B. Studies on the corrosion mechanism of unalloyed steel etc., Part I. Kinetics of the liberation of hydrogen. Advances in CO_2 Corrosion, Vol. 1, NACE, Houston, 1984; 72.
6.21 Videm K, Dugsstad A. Corrosion of carbon steel in an aqueous carbon dioxide environment. Part 2 Film formation. Materials Performance, April 1989; 46–50.
6.22 Dugstad A. The importance of $FeCO_3$ supersaturation on the CO_2 corrosion of carbon steels. Corrosion/92, Paper No 14, NACE, Houston, Texas, 1992.
6.23 de Waard C, Milliams DE. Carbonic acid corrosion of steel. Corrosion 31, 5, 1975: 177.
6.24 de Waard C, Lotz U, Milliams DE. Predictive modell for CO_2 corrosion engineering in wet natural gas pipelines. Corrosion, 47, 1991: 976–985.
6.25 Srinivasan S, Kane RD. Prediction of corrosivity of CO_2/H_2S Production Environments. Corrosion/96, Paper No 11, NACE, Houston, Texas, 1996.
6.26 API Recommended Practice for Design and Installation of Offshore Production Platform Piping Systems, API RP 14E, API, Dallas, Texas, 1991.
6.27 Salama MM. Erosional velocity limits for water injection systems. Corrosion/93, Paper No 62, NACE, Houston, Texas, 1993.
6.28 Nesic S, Postlethwaite J, Olsen S. An electrochemical model for prediction of corrosion of mild steel in aqueous carbon dioxide solutions. Corrosion/95, Paper No 131, NACE, Houston, Texas, 1995.
6.29 de Waard C, Lotz U, Dugstad A. Influence of liquid flow velocity on CO_2 corrosion: A semi–empirical model. Corrosion/95, Paper No 128, NACE, Houston, Texas, 1995.
6.30 NORSOK Standard M–506. CO_2 corrosion rate calculation model. Oslo: Norwegian Technology Standard Institution, 1998.
6.31 EFC publication No 16. Guidelines on materials requirements for carbon and low alloy steels for H_2S–containing environments in oil and gas production, 2nd Ed., London: European Federation of Corrosion, 2002.
6.32 Neumann FK, Carius W. Archiv für das Eisenhüttewesen, 30, 5, May 1959.

6.33 Enerhaug J. A Study of Localized Corrosion in Super Martensitic Stainless Steel Weldments, Dr. ing. Thesis, NTNU, Trondheim, 2002.
6.34 Gartland PO, Drugli JM. Methods for evaluation and prevention of local and galvanic corrosion in chlorinated seawater pipelines. Corrosion/92, Paper No 408, NACE, Houston, Texas, 1992.

EXERCISES

1. We have two different air–saturated solutions:
   a) diluted sulphuric acid (H_2SO_4, pH = 3).
   b) an NaOH solution (pH = 13).

   Write down the reaction equation for reduction of O_2 as well as Nernst's equation in each of the two cases. Explain the similarity and the difference between corresponding addends in the Nernst's equation in the two cases.
   Indicate the standard equilibrium potential of the reactions and the equilibrium potential as a function of the pH at $pO_2 = 1$ in a pH–potential diagram.

2. a) A steel plate is exposed to seawater in equilibrium with the atmosphere. The partial pressure of oxygen is $pO_2 = 0.2$ atm, the temperature = 20°C, and the pH of the water = 8.
   Determine the equilibrium potential for the O_2 reduction reaction on the steel plate when the standard equilibrium potential for the reaction $\frac{1}{2}O_2 + 2H^+ + 2e^-$ $= H_2O$ is $E_{0a}^0 = 1.23$ V (SHE).

   b) The water is flowing parallel to the surface of the steel plate with a velocity V = 0.5 m/s.
   Calculate the limiting current density of the oxygen reduction reaction when there is no deposit on the steel surface, and:

   the O_2 concentration $C_b = 2.5 \times 10^{-7}$ mol/cm^3,
   the diffusivity of O_2 $D_{O2} = 2.5 \times 10^{-5}$ cm^2/s.

   The thickness of the diffusion boundary layer can in this case be determined by the following formula (when V is inserted in m/s): $\delta = 10^{-4}/V^{0.5}$ (m).

   c) When the steel plate is inspected after long time, a layer of rust and calcium carbonate, 0.5 mm thick, is found. It is assumed that this layer has been intact most of the exposure time. The average corrosion rate, determined by thickness measurement, has been 0.1 mm/year.
   Assume that the corrosion rate has been controlled by oxygen diffusion, and calculate the oxygen permeability of the surface layer.

Atomic weight of Fe: 56.
Density of Fe: 7.8 g/cm^3.
Faraday's number: 96,485 c/mol e^-.

3. Describe and explain how the corrosion rate of steel depends on the temperature in the following systems:

   a) Open system and equilibrium between the water and the atmosphere.
   b) Tight cover on the container.

4. In a pipe with inner diameter 100 mm where water had circulated, an average corrosion rate of 0.3 mm/year was determined. In the extension of this pipe the inner diameter was only 50 mm.

   Assume that the materials in both pipe sections have been active and without surface deposits, and that the corrosion has been diffusion controlled by a turbulent boundary layer, i.e. the corrosion rate is proportional to (the flow velocity) 0.8. Determine the corrosion rate in the narrow pipe section.

5. On the stay-vane in a water turbine drum the maximum corrosion rate was found to occur at approximately 30 m/s. At even higher velocities the corrosion rate was low. Explain!

6. The concrete coating on an offshore platform riser was damaged so that a part of the steel surface in the splash zone was exposed. Explain which factors may cause extraordinarily high corrosion rates in this case.

7. Present data from Figure 6.12 in a (log i_L)–(–log [H^+]) diagram. Explain the shape of the curve (see Equation (4.8), p.42).

8. Use the de Waard and Milliams nomogram in Figure 6.14 and determine a curve that shows the corrosion rate of steel as a function of temperature in the range 20–140°C for a case with constant CO_2 partial pressure = 1 bar. Explain the shape of the curve.

7. Different Forms of Corrosion Classified on the Basis of Appearence

Happy he, who has availed to
read the causes of things.
VIRGIL

7.1 Introduction

In the previous chapters we have mainly (more or less implicitly) assumed that:

1) electrochemical corrosion is the only deterioration mechanism;
2) anodic and cathodic reactions take place all over the electrode surface, but not simultaneously at the same place, i.e. the anodic and cathodic reactions exchange places, constantly or frequently. Closely related to this dynamic behaviour it is assumed that:
3) there are no significant macroscopic concentration differences in the electrolyte along the metal surface, and the metal is fairly homogeneous.

These three assumptions lead to *uniform (general)* corrosion. But this is only one of several corrosion forms that occur under different conditions. The other forms of corrosion depend on the deviations from the mentioned assumptions. Such deviations may be due to:

a) the design (the macro–geometry of the metal surfaces)
b) the combination of metal and environment
c) the state of the surface (particularly cleanliness and roughness)
d) other deterioration mechanisms

These conditions will cause various deviations in the geometry and appearance of the attack compared with uniform corrosion, and it is convenient to classify corrosion just after the appearance of the corroded surface, as done by Fontana and Greene [7.1] (possibly with supplementary information about important service conditions). The advantage of such a classification is that a corrosion failure can be identified as a certain corrosion form by visual inspection, either by the naked eye or possibly by a magnifying glass or microscope. Since each form of corrosion has its characteristic causes, important steps to a complete diagnosis of failure can often be taken after a simple visual inspection. On this basis, the following corrosion forms can be defined:

1. Uniform (general) corrosion
2. Galvanic (two–metal) corrosion
3. Thermogalvanic corrosion
4. Crevice corrosion (including deposit corrosion)
5. Pitting, pitting corrosion
6. Selective attack, selective leaching (de–alloying)
7. Intergranular corrosion (including exfoliation)
8. Erosion corrosion
9. Cavitation corrosion
10. Fretting corrosion
11. Stress corrosion cracking
12. Corrosion fatigue

A simple illustration of the various forms of corrosion is shown in Figure 7.1.

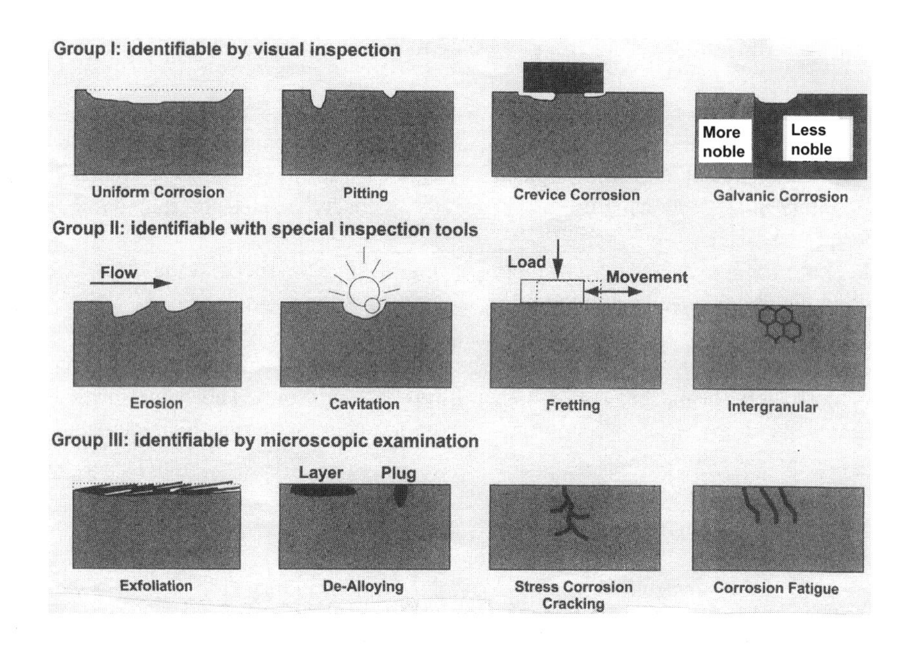

Figure 7.1 Main forms of corrosion grouped by their ease of recognition. (Reproduced from Roberge PR Handbook of Corrosion Engineering. New York: McGraw-Hill, 1977.)

The relative frequency of occurrence of various corrosion forms depends on the type of industry and environment. One example from the chemical industry is shown in Table 7.1. In the heading we notice the high proportion of corrosion failures. The percentages of crevice corrosion and galvanic (!) corrosion are surprisingly low.

Under many other conditions these are much higher. The group named weld corrosion in the table is often a kind of galvanic corrosion. On stainless steel, weld corrosion may in many cases mean crevice corrosion in or at the weld.

Table 7.2 shows another example, namely from Japanese oil refineries and petrochemical industry. Here more than 80% of the material failures are due to corrosion (wet and dry). The high proportion of stress corrosion cracking/corrosion fatigue is particularly noticed.

Table 7.1 Material failures over a two-year period (56.9 % corrosion and 43.1% mechanical) (data from the Du Pont Company) [7.3].

Corrosion failures	%
General corrosion	31.5
Stress corrosion cracking Corrosion fatigue	23.4
Pitting corrosion	15.7
Intergranular corrosion	10.2
Corrosion–erosion Cavitation damage Fretting corrosion	9.0
High–temperature corrosion	2.3
Weld corrosion	2.3
Thermogalvanic corrosion	2.3
Crevice corrosion	1.8
Selective attack	1.1
Hydrogen damage	0.5
Galvanic corrosion	0.0

7.2 Uniform (General) Corrosion

By definition, attacks of this type are quite evenly distributed over the surface, and consequently it leads to a relatively uniform thickness reduction (Figure 7.2). The necessary conditions for uniform corrosion have already been presented (first page of this chapter). Homogeneous materials without a significant passivation tendency in the actual environment are liable to this form of corrosion.

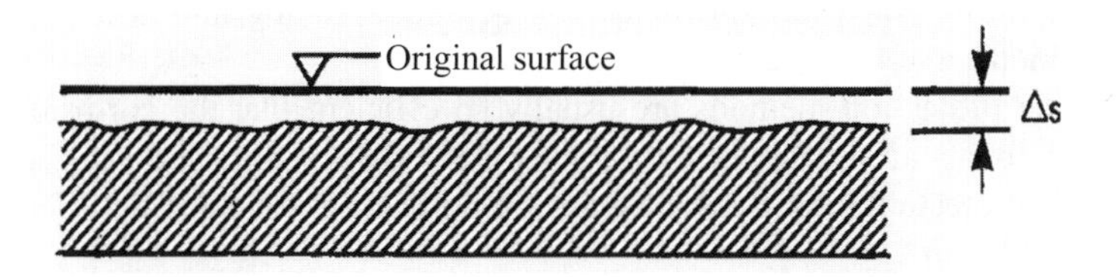

Figure 7.2 Uniform (general) corrosion.

Table 7.2 Materials failures in Japanese oil refineries and petrochemical industry.. (From Yamamoto and Kagawa [7.4]). Reprinted with permission of NACE.

Mode of failure	Percent of total		
Wet environments:	72.1		
1. General corrosion		12.5	
2. Localized corrosion (pitting, dewpoint corrosion, deposit attack, etc.)		15.9	
3. Stress corrosion cracking and corrosion fatigue		39.9	
(a) Austenitic stainless steels			(24.0)
(b) Carbon steels, low-alloy steels and ferritic stainless steels			(10.5)
(c) Cu alloys			(5.4)
4. Miscellaneous		3.8	
Dry environments:	22.5		
1. Corrosion		8.2	
(a) Sulphidation, oxidation, carburization			(4.1)
(b) Vanadium attack			(4.1)
2. Cracking		10.9	
(a) Creep rupture			(5.4)
(b) Mechanical cracking in welds			(4.1)
(c) Cracking by thermal stress			(1.4)
3. Decrease of mechanical properties of materials		1.7	
(a) σ phase and 475°C embrittlement			(1.0)
(b) Hydrogen attack			(0.7)
Materials defects	2.0		
Welding defects	3.4		
Total	100		

Uniform corrosion is assumed to be the most common form of corrosion and particularly responsible for most of the material loss. Traditionally, however, it is not recognized as a dangerous form of corrosion, because:

1. Prediction of thickness reduction rate can be done by means of simple tests. Corresponding corrosion allowance can be added, taking into account strength requirements and lifetime.
2. Available protection methods are usually so efficient that the corrosion rate is reduced to an acceptable level. Actual methods are application of coatings, cathodic protection or possibly change of environment or material.

By use of modern materials with higher and more precisely defined strength, and higher cost, uniform corrosion has required increased attention. This is so because the strength data correspond to thinner and more accurately determined material thickness, and the material cost makes a thinner corrosion allowance attractive.

Description of uniform corrosion by means of polarization and overvoltage curves is relatively simple, since we can consider both the anodic and the cathodic area equal to the total area.

Parallel use of the Pourbaix diagram and overvoltage curves is sometimes useful when evaluating how typical and stable the uniform corrosion is under different conditions. Let us consider a case with iron or unalloyed steel in two different environments:

a) diluted sulphuric acid (pH = 2), and
b) neutral water (pH = 7), and assume the following data:

For the hydrogen reaction H^+/H_2: $i_0 = 10^{-6}$ A/cm², $b_c = 0.12$ V/decade c.d.
For the oxygen reaction: $i_0 = 10^{-8}$ A/cm², $b_c = 0.22$ V/decade c.d., $i_L = 40 \times 10^{-6}$ A/cm²,
For dissolution of iron, Fe/Fe^{2+}: $i_0 = 10^{-6}$ A/cm², $b_a = 0.10$ V/decade c.d.

When evaluating the stability of the corrosion state, the distance to the passivity region in the Pourbaix diagram along the pH axis should be taken into account, because the reactions lead to pH changes close to the metal surface. Figure 7.3 shows the situation in the case of pH = 2. Although H^+-ions are consumed in the cathodic reaction, the high H^+-ion concentration in the solution prevents pH from increasing considerably at the metal surface. The figure shows that the state is far from the passive region, and the corrosion is therefore typically uniform and stable. The same is the case for several other metals such as Zn, Cu, Al and Ni in acids.

Figure 7.4 shows the corresponding picture for iron in neutral water. The O_2-reduction implies production of OH^- and leads to an alkaline layer in the water at the cathodic areas (whether these over time cover the whole surface or parts of it). Depending on the conditions, the pH close to the metal surface can adopt values from 8 to 11 at these areas. In Figure 7.4 it is assumed that some areas are cathodes long enough to give pH = 10. The result would in principle be the same with pH = 9. These areas become passivated. A passivated surface area may be attacked at the boundary, and thereby decrease, but may otherwise be maintained for longer periods. In the figure it is assumed that the passivated areas constitute 50% of the total area, and that the oxygen reduction takes place uniformly over the whole surface (i.e. we disregard any diffusion-limiting deposit on the active parts of the surface), while the anodic reaction takes place on the half of the surface that is not passivated. The real current density on the anode is therefore double the nominal one. A certain potential drop may exist from a passive to an adjacent active area, as shown to the right in the figure. Since only one half of the current is running between the cathodic and anodic regions, the potential drop can be expressed by RI/2.

The situation can be modified when corrosion product deposits are formed on or close to the active regions. These deposits reduce the access of oxygen to the active areas, so that the oxygen reduction may essentially take place on the passivated regions. Then, the potential drop is more correctly expressed by RI than by RI/2.

The area ratio between active and passive regions may approach different values depending on the circumstances. Anyway, under these conditions the form of corrosion is not typically uniform, because the nominal state is so close to the passivity region in the Pourbaix diagram. The result is a transition form between uniform and localized corrosion that often occurs under real conditions.

The data providing the basis for Figures 7.3 and 7.4 are not generally valid, although they are based on tests and partly on calculations. In addition, they are combined in a way that corresponds to the Pourbaix diagram and at the same time gives a corrosion potential and characteristic current densities in reasonable agreement with observed values.

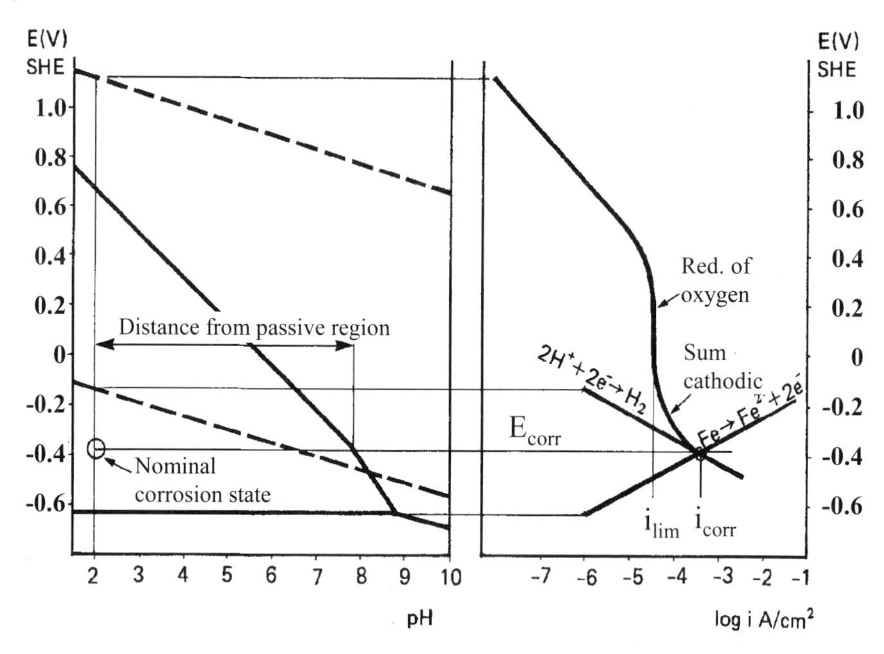

Figure 7.3 Pourbaix diagram and potential – log current density diagram showing the state of iron in diluted acid (pH = 2).

7.3 Galvanic Corrosion

7.3.1 Conditions That Determine Corrosion Rates

When a *metallic* contact is made between a more noble metal and a less noble one, the corrosion rate will increase on the latter and decrease on the former. A necessary condition is that there is also an *electrolytic* connection between the metals, so that a closed circuit is established (Figure 7.5).

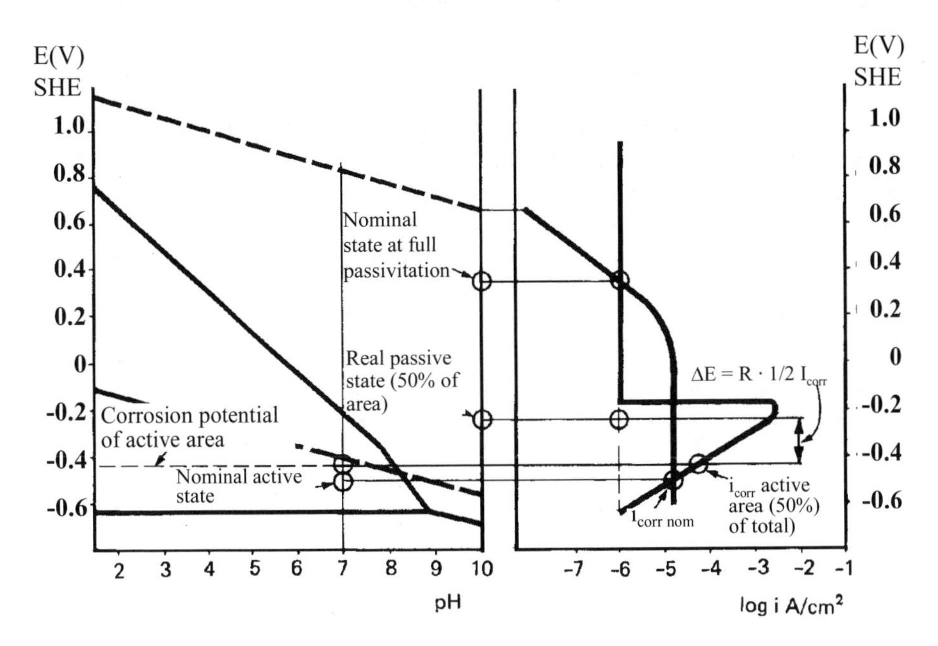

Figure 7.4 As Figure 7.3, but in neutral water (pH = 7).

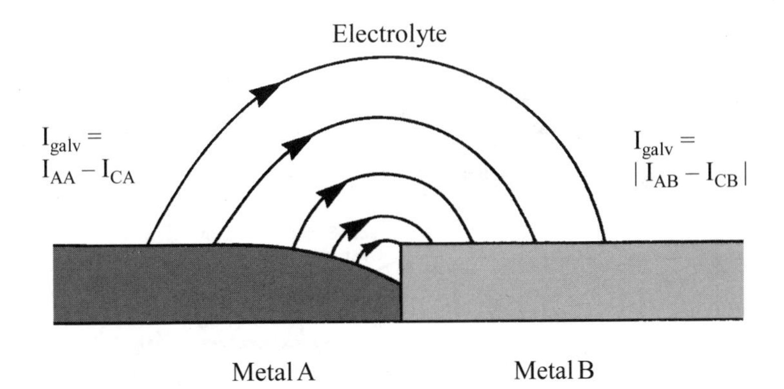

Figure 7.5 Galvanic corrosion.

The series of standard reversible potentials of the various metals (Section 3.6) are now and then used to explain and estimate the risk of galvanic corrosion. This can be very misleading, because 1) these potentials express thermodynamic properties, which do not tell us anything about the reaction rate (e.g. passivation tendencies are not taken into account), and 2) if the potential difference between the two metals in a galvanic couple is large, the more noble metal does not take part in the corrosion process with its own ions. Thus, under this condition, the reversible potential of the more noble metal does not play any role (see explanation on the next page in connection with Figure 7.7).

In addition to the standard reversible potential series we have practical galvanic series that show corrosion potentials of various metals in a given environment.

A new galvanic series of different materials in seawater at 10°C as well as 40°C has been established, see Figure 7.6. It is shown that the corrosion potentials of the most corrosion-resistant materials depend considerably on the temperature. This is due to a microbiological (bacterial) slime layer that is formed at temperatures below about 30°C in northern waters, and below 40°C in warmer waters like that in the Mediterranean. The bacterial activity lifts the potential of these materials (see Sections 6.2.1 and 7.5.4, Figure 7.23). Previously published galvanic series do not show this effect, probably because the exposure periods were too short to imply the potential increase. In addition to the temperature, the flow velocity may have some effect upon the corrosion potential. The data in Figure 7.6 were recorded at 1 m/s.

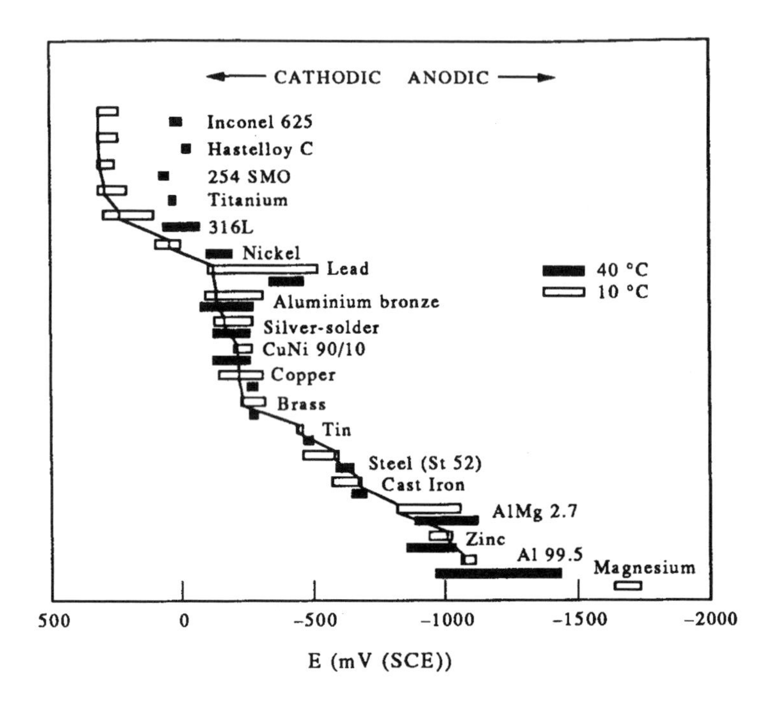

Figure 7.6 Galvanic series of different materials in seawater at 10°C and 40°C [7.5, 7.6].

A relevant series of this type is a more adequate basis for prediction of galvanic corrosion risks than is the standard reversible potential series. However, the practical galvanic series provide only qualitative and incomplete information. To explain what conditions determine the rate of galvanic corrosion, and to predict the rate, we must study the overvoltage curves of the actual reactions that take place (see Figure 7.7).

In order to take the area of each metal into account, it is recommended to draw these curves in the form of a potential–log *current* diagram. If current–density data are given, these data have to be multiplied by the actual area for each reaction. In Figure 7.7, there are two corroding metals, A and B, and a cathodic reaction that can take place on both A and B. On the basis of the potential–log current curve for each reaction (on each metal), sum curves for anodic and cathodic currents, respectively, are drawn. In the first instance, we disregard any potential drop in the solution and in the metal, i.e. the electrode potential is the same on both materials. This common potential E_{corr} is determined as the intersection point of the sum curves, and the current of each reaction is given by the intersection between the line $E = E_{corr}$ and the potential–log current curve for the actual reaction. The corrosion current of metal A increases from I_{corrA} to I'_{corrA}, while that of metal B decreases from I_{corrB} to I'_{corrB} when the two metals are coupled together (metal B is protected). The galvanic current between the two materials is given by the difference between the anodic and cathodic current on the less noble material A, which equals the difference between the cathodic and anodic current on the more noble material B:

$$I_{galv} = I'_{corrA} - I'_{catA} = I'_{catB} - I'_{corrB}. \tag{7.1}$$

Equation (7.1) expresses also that the total corrosion rate of the less noble metal (I'_{corrA}) consists of a contribution of galvanic corrosion (I_{galv}) and a contribution of self–corrosion due to the cathodic reaction on the same metal (I_{catA}). The corrosion current densities of a galvanic couple are finally derived by dividing the corrosion current of each metal by the corresponding area.

It can be seen from Figure 7.7 that the increase in corrosion current of metal A due to the galvanic coupling is not significantly affected by the reaction B/B^+. This will be the case in all situations where the potential distance between the two anodic curves is not particularly small (see below). With this reservation, neither the difference $E_{OB} - E_{OA}$ between the reversible potentials nor the difference $\Delta E_{corr} = E_{corrA} - E_{corrB}$ between the corrosion potentials of the two metals separated can say anything about the intensity of galvanic corrosion. This is determined by the increase in the cathodic reaction available for corrosion of metal A (e.g. oxygen or hydrogen reaction) caused by the galvanic coupling. Thus, the metal property that usually determines the rate of galvanic corrosion of a given less noble metal is the cathodic efficiency of the more noble metal in the couple. The difference in cathodic efficiency between various metals is mainly due to different exchange current densities.

Table 7.3 shows an example of galvanic corrosion rates of aluminium alloys in 3.5% NaCl solution when coupled to different materials. For instance, it is seen that the contact with low-alloy steel gives considerably higher galvanic corrosion rates on aluminium than does contact with the – from a practical point of view – more noble stainless steels as well as Ni- and Ti-based alloys (regarding material descriptions, see Section 10.1). The table reflects the cathodic efficiency of the various materials coupled to aluminium (with the exception of cadmium, zinc and aluminium alloys) in the actual environment.

With real conditions in mind, it should be noticed that natural seawater may give galvanic effects quite different from that of pure NaCl solutions. The microbial slime layer previously mentioned may cause the cathodic efficiency of nobler materials such as stainless steels to increase 1–3 orders of magnitude depending on the conditions. Under unfavourable circumstances galvanic corrosion of somewhat less noble materials (e.g. copper alloys) coupled to stainless steel may increase correspondingly [7.6]. On materials like unalloyed and low-alloy steels the bacterial activity has no significant effect on the oxygen reduction. When these materials are exposed to seawater, deposits of rust and calcium carbonate are formed, with the result that the cathodic current density becomes similar to that indicated for stainless steel in 3.5% NaCl solution in Table 7.3. Contrary to the cases where the materials are immersed in natural seawater, in a marine atmosphere, stainless steels act less efficiently as cathodes than do copper and ordinary structural steels.

The examples dealt with above show for one thing that use of practical galvanic series (E_{corr}) and even more standard equilibrium potential series (E_0^0) is more or less

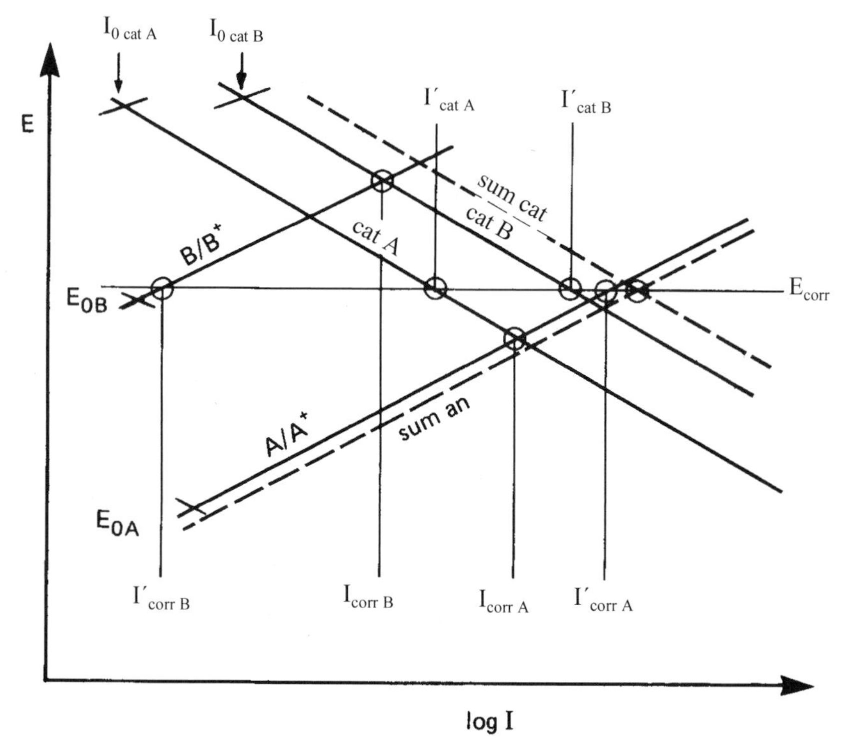

Figure 7.7 Cathodic and anodic overvoltage curves for a galvanic element of the metals A (less noble) and B (more noble).

Table 7.3 Galvanic current densities between Al alloys and various other materials in a 3.5% NaCl solution. Mean values for 24 hours of exposure. Area ratio: 1:1. A corrosion current density of 9.2 μA/cm² is equivalent to a corrosion rate of 0.1 mm/year in average for aluminium. (Adapted from Mansfeld et al. [7.7].)

Material coupled to Al alloy	Galvanic current density i_g, μA/cm²				
	AA1100	AA2024	AA2219	AA6061	AA7075
Silver (99.9%)	54	51	47	55	64
Copper	40	41	36	44	45
Steel (4130)	28	25	20	24	25
Nickel	14	18	11	22	22
PH 13–8 Mo	18	13	8	16	17
A286	14	10	13	15	17
Stainl. steel 304L	12	13	12	11	17
347	12	11	14	14	17
301	17	13	11	12	15
Inco 718	10	9	10	8	12
Haynes 188	6	6	4	5	9
Titanium (Ti–6–4)	8	5	4	5	8
Tin	3.4	2.8	5	1.6	6
Cadmium	1.2	–1.5	–2.1	0.3	6
Zinc	–1.5	–7	–11	–1.5	–3.4
AA2219	3.1	2.8	–	3.1	4.2 (5.88)*
AA2024	1.2	–	–2.3	2.0	2.6 (6.68)*
AA6061	–0.7	2.0	–3.1	–	0.7 (0.71)*
4.2 (5.88)*	–	–1.2	–1.5	0.7	0.3 (0.31)*
AA7075	0.3	–2.6	–3.8	–0.7	(1.21)*

* Corrosion current density for the respective Al alloys separated from other metals.

inadequate when evaluating galvanic corrosion. If, however, overvoltage curves or relevant data like that in Table 7.3 are not available, the practical galvanic series for the actual environment can be applied as follows:

a) If $\Delta E_{corr} < 50$ mV for two metals, galvanic coupling between them will only give small changes in the corrosion rates (if the area ratio between the anode and cathode is not particularly unfavourable, or if the less noble metal is not liable to typical localized corrosion, e.g. pitting).

b) If $\Delta E_{corr} > 50$ mV, there is a risk of increased corrosion rate on the less noble metal after galvanic coupling, but the difference between the corrosion potentials is not a reliable measure for the increase of corrosion rate due to the galvanic contact. In cases where the more noble metal is a very efficient cathode, even equal areas of the two materials may cause relatively rapid failure (compare Table 7.3).

Example 1: Copper accelerates corrosion considerably on unalloyed steel and even more on aluminium alloys in seawater because Cu is a more efficient cathode than steel and particularly aluminium.

Example 2: For copper alloys in contact with stainless steel in seawater, with an area ratio 1:1 between the materials, the galvanic coupling may give an increase in the corrosion rate on copper by a factor of 4–8 [7.8]. (In this case, stainless steel is the most efficient cathode.)

Example 3: Titanium usually causes, in spite of its practical nobleness, little galvanic corrosion in contact with steel when the area ratio is about 1:1, because Ti is not an efficient cathode material (exceptions may occur in seawater with bacterial films). With a relatively large area of titanium, damaging corrosion may take place on steels, copper alloys etc.

c) If the cathodic area is much larger than the anodic area ($A_c >> A_a$), the corrosion rate of the less noble material may increase strongly by galvanic coupling, even with small differences between the corrosion potentials of the separate materials. One example of this is heavy corrosion on bolts of Monel in a seawater pump with housing made of the ferritic–austenitic steel SAF 2205 (UNS S 31503). Failure occurred after 3 months of service. Passive SAF 2205 has about the same corrosion potential as passive AISI 316 in seawater (see Figure 7.6), while the potential of Monel is a little lower. Special precautions should be taken if the less noble metal in the couple is liable to localized corrosion (as is the case for aluminium in aqueous solutions containing chloride).

As already pointed out, *the area ratio* between the metals in a galvanic couple plays a crucial role for the galvanic corrosion rate. The importance is best illustrated in a potential–log current diagram, where the area of the less noble metal conveniently can be set equal to 1 cm^2. In Figure 7.8 this is shown by an example where iron or unalloyed steel in one case is alone (A_c = 1 cm^2) and in other cases is connected to stainless steel so that the total cathodic area is 10 and 100 cm^2, respectively. For the sake of simplicity it has been assumed that the cathodic overvoltage curves are equal on stainless and unalloyed steel. Two different environments are considered:

a) Aerated water, pH = 6, where the corrosion is diffusion controlled. It is seen that the corrosion current in this case is proportional to the cathodic area (the corrosion current density is 30, 300 and 3000 $\mu A/cm^2$, respectively).

b) De-aerated acidic water, pH = 3, where the cathodic reaction is activation controlled. The effect of increasing cathodic area is considerably smaller in this case than in case a) due to the slope of the cathodic curve. It should, however, be noticed that it is the same *parallel shift of the cathodic curve along the current axis* for a certain area increase in both cases, namely one current decade for each decade increase in area.

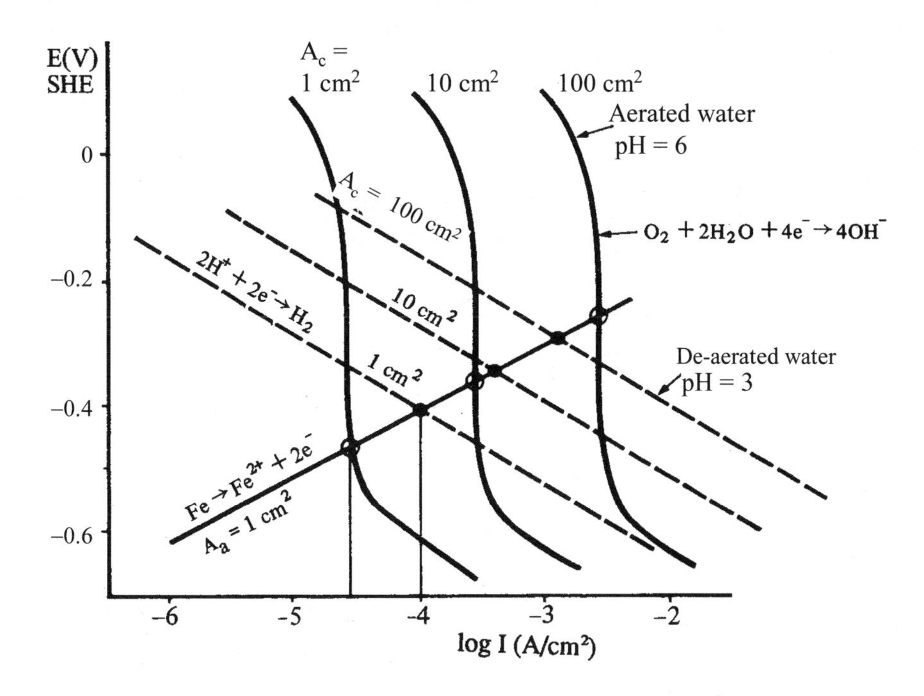

Figure 7.8 Effect of the area ratio between the cathode and the anode in a galvanic couple of carbon steel and stainless steel in two different environments.

Regardless of the details in the procedure for determination of corrosion rate of galvanic couples, we must always stick to the main rule $\Sigma I_a = \Sigma I_c$, which is the basis for derivation of corrosion current I_{corr} and then the corrosion current density $I_{corr} = I_{corr}/A_a$.

It is evident that a large area ratio A_c/A_a must be avoided where galvanic corrosion is possible. If a structure exposed to corrosive conditions is to be painted, it is therefore more important to paint the more noble material than the other parts of the structure. If the more noble material is directly exposed to the environment, small defects in the paint coating on the less noble material will give very high ratios A_c/A_a and intensive corrosion at the coating defects.

In addition to the cathodic efficiency and the area ratio, the *conductivity of the electrolyte* and the *geometry* of the system play important roles because these factors determine if there will be a considerable potential drop in the liquid from one metal to the other. Limitation of galvanic corrosion rate due to a potential drop is shown in a simplified way in Figure 7.9. Here, it is assumed that essentially all the cathodic reaction takes place on the more noble material and essentially all the anodic reaction on the less noble one. (It should, however, be clear from the description above that this assumption is far from being valid in many cases.)

Without any potential drop the corrosion current is I_{corr}. With a potential drop in the liquid there is a difference in the electrode potential between the anodic and the cathodic material, $\Delta E = E_c - E_a$, and the corrosion current is reduced to I'_{corr} (in order to satisfy the condition $I_a = I_c$).

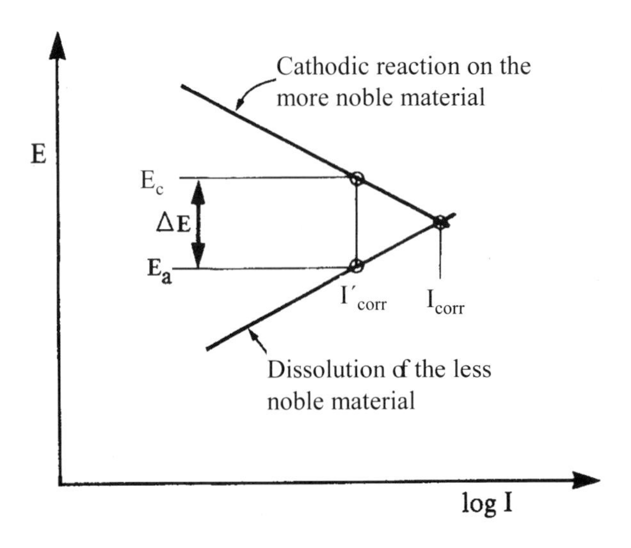

Figure 7.9 Simplified illustration of how an ohmic potential drop in the electrolyte between the metals reduces the rate of galvanic corrosion.

In this simplified picture, it is assumed that E_a and E_c are constant over their respective metal surfaces, but under real conditions these potentials usually vary. This is schematically shown in Figure 7.10, where also some current paths in the electrolyte between the anodic and the cathodic metal are indicated. For sufficiently small area elements of the electrodes, constant potential of each element can be assumed, actually E_{cj} and E_{aj} for the area element A_{cj} on the cathodic and A_{aj} on the anodic metal, respectively. The difference between these two potentials is

$$E_{cj} - E_{aj} = R_j I_j, \text{ where } R_j = \rho \int_j \frac{dl}{A_j(l)} \tag{7.2}$$

R_j is the resistance in the current path j in the liquid. The total galvanic current is given by the sum of the currents in the various paths: $I_{galv} = \Sigma I_j$.

The corrosion attack on the anodic material is reduced with increasing distance from the contact interface between the two materials, in accordance with the falling potential. The magnitude of this *distance effect* is determined by the resistivity of the electrolyte, the geometry of the liquid volume and the polarization properties of the metals. Figure 7.11 shows an example of galvanic corrosion on St.37–2 steel in contact with stainless steel in waters with two different resistivities [7.9]. It should be noticed that – under these and similar conditions – the highest resistivity leads to the highest top value of galvanic corrosion rate (concentrated attack close to the contact interface) in spite of less total galvanic current.

Potential variation and current distribution on galvanic elements can be calculated approximately or accurately as dealt with in Section 10.4, Cathodic Protection.

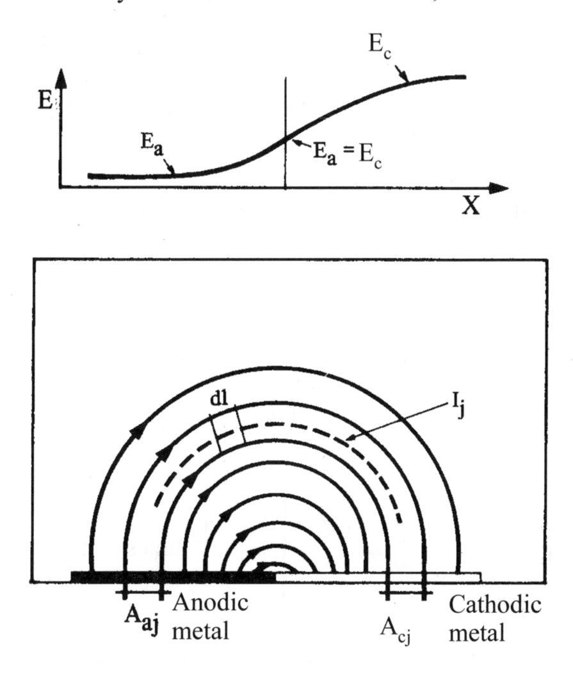

Figure 7.10 Current paths in the electrolyte between two metals arranged in the same plane. Current path j leads from the anodic area element A_{aj} to the cathodic area element A_{cj}. The variation of the electrode potential along the metal surface is shown above.

Corrosion medium and temperature affect galvanic corrosion because the corrosion characteristics of the metals are affected. For example, in the couple Zn/Fe, Fe will usually be protected, but at temperatures above 60–70°C in some types of tap water, Zn is protected by Fe. The explanation of this is that the corrosion products of Zn make the metal more noble. Another example is the relationship between aluminium and structural steel: in waters with chloride content 0.01% (100 ppm) aluminium is cathodic relative to steel at room temperature, but the polarity is reversed when the temperature is raised above 40°C [7.10].

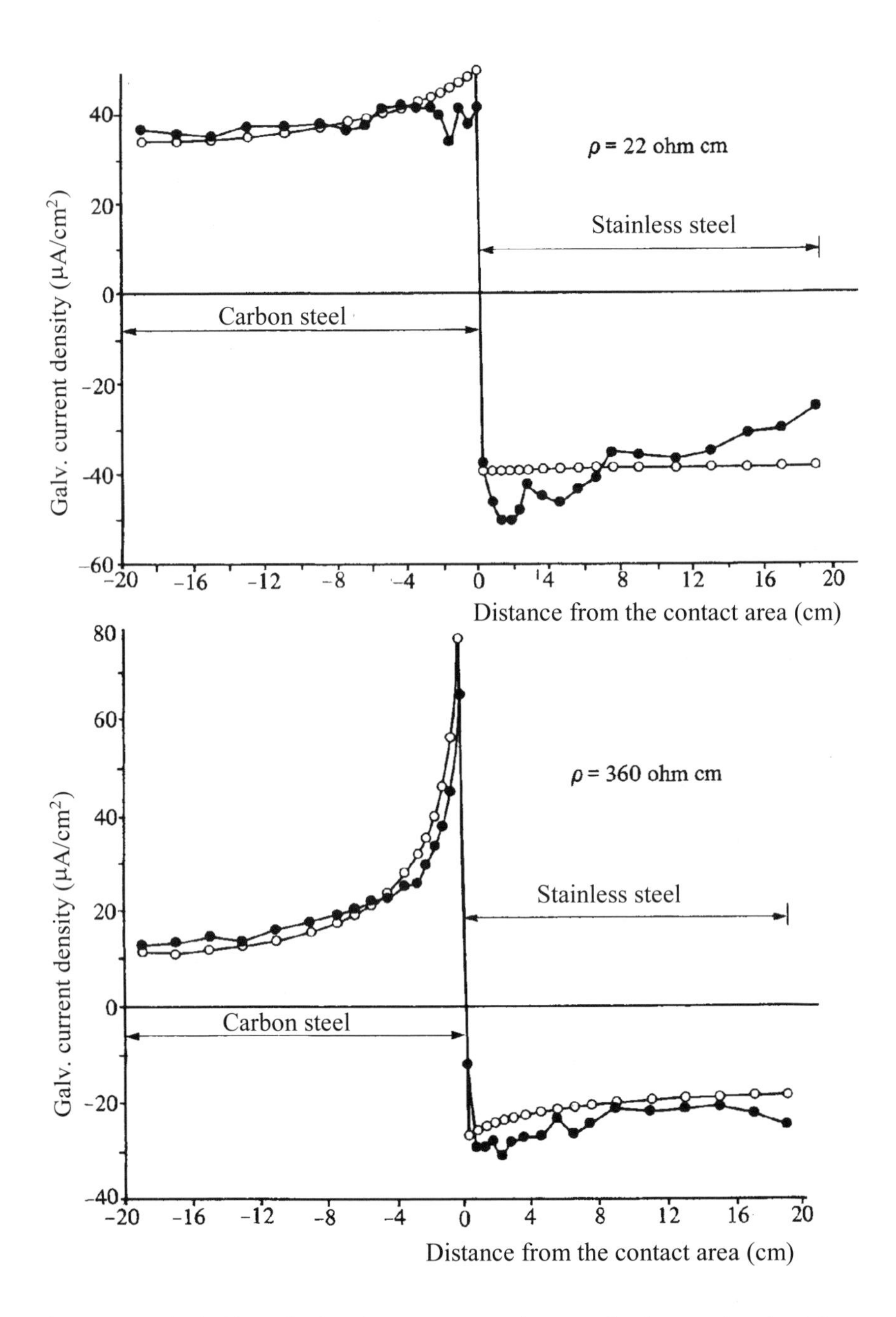

Figure 7.11 Distribution of galvanic current on an element of carbon steel and stainless steel when the resistivity of the water is, respectively, 22 ohm cm (above) and 360 ohm cm (below). (● Experiments; ○ Calculation [7.9].)

As expected from the data in Table 7.3, galvanic corrosion is one of the major practical corrosion problems of aluminium and aluminium alloys. The reason for this is that aluminium is thermodynamically more active (less noble) than most other common structural materials, and that the passive oxide which usually protects aluminium may easily be broken down locally when the potential is raised due to contact with a more noble material. This is particularly the case when aluminium and its alloys are exposed in waters containing chlorides or other aggressive species (see also Section 7.6, Pitting Corrosion).

Galvanic corrosion occurs also in the atmosphere (atmospheric galvanic corrosion). The extent depends on various corrosion–promoting conditions. The atmosphere is far more aggressive near the sea than inland. The reason for this is primarily that the humidity at the sea is combined with salts, which increases the conductivity of the water. Industrial atmosphere with high content of SO_2 is also aggressive. Accumulated corrosion products and deposits of organic or inorganic matter at the connection between the materials are usually hygroscopic and may cause galvanic corrosion in this area (see also Section 8.1).

Galvanic corrosion can usually be prevented by reasonable means (see below). Nevertheless, more or less serious cases of galvanic corrosion occur now and then. A drastic example concerns a safety valve on a foam tank with compressed air. Normally, the valve should open at a pressure somewhat above six bars. But because the valve ball material was low–alloy steel, and the valve housing was made of brass and other parts of martensitic stainless steel, the ball rusted and stuck to the seat. Together with other occasional malfunctions, this caused the tank to explode. Cases mentioned earlier (pp. 99 and 100) also represent examples of failure experienced in real life.

7.3.2 Prevention of Galvanic Corrosion

Galvanic corrosion can be prevented or made harmless in several ways:

1. Selecting adequate materials.
2. Using weld consumables that make the weld metal more noble than the base material.
3. Avoiding the combination of a large area of an old material (which has become more noble due to a surface film) with a small area of new, more active material.
4. Using favourable area ratios between more and less noble materials.
5. Insulating the parts of the couple from each other. Here, the use of "insulation materials" with insufficient resistivity should be warned against. A resistivity $\rho > 10^{10}$ ohm cm and the use of chloroprene rubber have been recommended [7.11]. Design and assembly are specified as well. It should be emphasized that metallic contact must be avoided completely at all parts of the actual components (Figure 7.12a).
6. Applying non–metallic distance pieces (e.g. pipe extensions) so that galvanic corrosion becomes insignificant due to large a voltage drop in the water. (For cases where it is not possible or practical to avoid metallic contact between the

components. This is relevant for pipelines. When the conductivity of the liquid is low, relatively short distance pieces are sufficient.)

7. Using a metallic coating on one of the materials, e.g. thermally sprayed aluminium on steel in contact with aluminium parts.
8. Painting appropriately, i.e. primarily on the more noble material.
9. Applying cathodic protection.
10. Adding inhibitors.
11. Designing to make less noble materials easy to replace or with sufficient corrosion allowance.
12. Avoiding access of humidity to the contact region between the materials, e.g. as shown in Figure 7.12b. A sealer of a polysulphide-based, creep-resistant material may for instance be used [7.11].
13. Avoiding electrolytic deposition of a more noble material on a less noble one. For example, let us look at a joint between an iron pipe and a copper pipe (Figure 7.13). We assume a low conductivity of the liquid and a small pipe diameter so that only a small part of the Cu pipe is effective as a cathode for the couple. Therefore, the normal galvanic corrosion current at the joint becomes quite limited (and it can be calculated, as shown in Section 10.4). Alternatively, the galvanic corrosion can be prevented by insulating the pipes from each other. We will have a potential variation, as shown in Figure 7.13. Both materials are corroding more or less. If the flow direction was opposite to that shown in the figure, copper ions would be carried into the iron pipe where they would be discharged and deposited on the iron surface because of the lower potential in this pipe section (in the immune potential range of copper). Deposition implies the reduction reaction $Cu^{2+} + 2e^{-} \rightarrow Cu$. Iron suffers from increased corrosion where the copper is deposited: firstly, because iron is dissolved correspondingly to the reduction of Cu^{2+} ions, and subsequently because the deposited copper constitutes efficient cathodes finely distributed over the surface. Heavy pitting may be the result. Therefore it must be checked that the materials are oriented appropriately in relation to the flow direction, as shown in Figure 7.13.

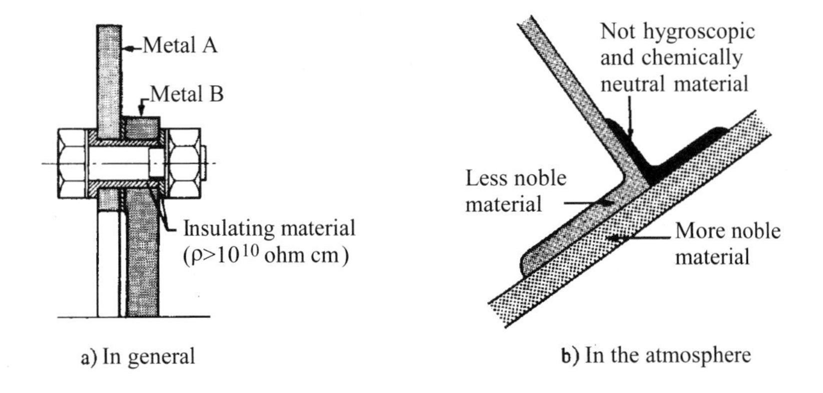

Figure 7.12 Design in order to prevent galvanic corrosion.

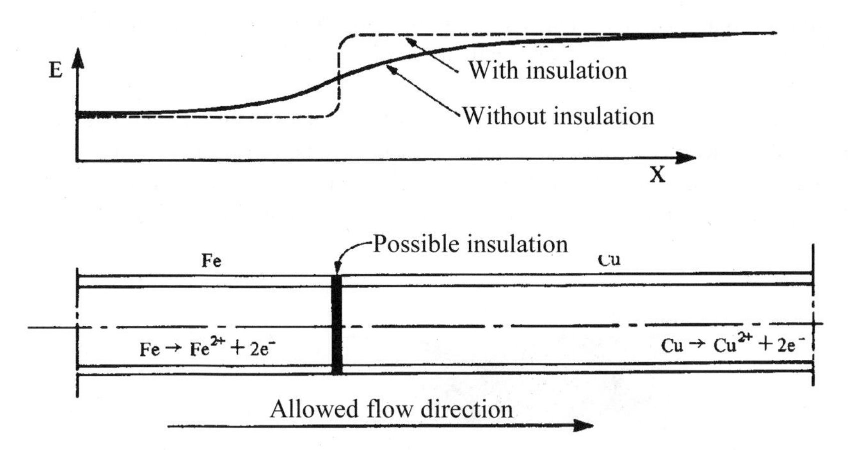

Figure 7.13 Potential variation and indication of appropriate flow direction when a more noble material is coupled to a less noble one.

7.3.3 Application of Galvanic Elements in Corrosion Engineering

The most common example of beneficial application of galvanic elements in corrosion engineering is cathodic protection by use of sacrificial anodes (Section 10.4). A coating of a material less noble than the substrate gives the same effect.

In this connection one should be aware of the possibility to protect critical parts of a system (equipment, plant) by means of less noble and less critical structural components in the same system, i.e. the corrosion is localized to preferred parts of the system. For instance, stainless steel pipes, fittings, pumps and valves in seawater systems can be protected against serious deterioration by use of pipe sections of unalloyed steel or cast iron. These sections must be easy to replace, or they must have relatively large wall thickness and/or large exposed surface area in relation to the more noble materials. In heat exchangers, relatively thick tube plates of steel can be used in combination with tubes of a copper alloy.

7.4 Thermogalvanic Corrosion

When a material in a corrosive environment is subject to a temperature gradient, a galvanic element may arise, causing what we call thermo–galvanic corrosion. Usually, the hot surface forms the anode and the cold one the cathode (Figure 7.14). Partly, this is so because the anodic properties of the material depend on the temperature. In addition, the properties of the environment along the metal surface will also vary due to varying temperature and varying temperature gradient normal to the metal surface. Often this affects the cathodic reaction, which should also be taken into consideration when such a corrosion form is analyzed.

While ordinary galvanic corrosion (two-metal corrosion) can be identified by a visual inspection alone, for thermogalvanic corrosion we need also to know a little about the service conditions.

Thermogalvanic corrosion is prevented by appropriate design and measures to avoid uneven heating/cooling and forming of hot spots. For heat-insulated pipes and equipment is important that the insulation is continuous. The corrosion form may under certain conditions be prevented by cathodic protection or coatings.

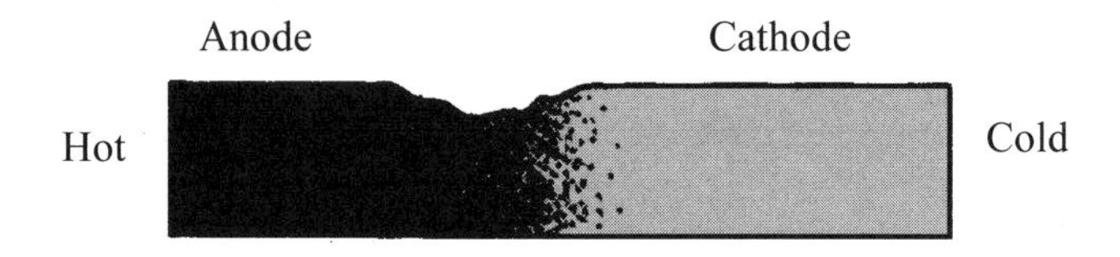

Figure 7.14 Thermogalvanic corrosion.

7.5 Crevice Corrosion

7.5.1 Occurrence, Conditions

This is localized corrosion concentrated in crevices *in which the gap is sufficiently wide for liquid to penetrate into the crevice and sufficiently narrow for the liquid in the crevice to be stagnant.*

Crevice corrosion (CC) occurs beneath flange gaskets, nail and screw heads and paint coating edges, in overlap joints, between tubes and tube plates in heat exchangers etc. The same form of corrosion develops beneath deposits of, e.g. corrosion products, dirt, sand, leaves and marine organisms, hence it is called deposit corrosion in such cases.

The most typical crevice corrosion occurs on materials that are passive beforehand, or materials that can easily be passivated (stainless steels, aluminium, unalloyed or low alloy steels in more or less alkaline environments etc.), when these materials are exposed to aggressive species (e.g. chlorides) that can lead to local breakdown of the surface oxide layer. Materials like conventional stainless steels can be heavily attacked by deposit corrosion in stagnant or slowly flowing seawater. A critical velocity of about 2 m/s has often been assumed, but more recent studies have indicated that crevice corrosion can occur at higher velocities too. The corrosion form is most frequently observed in environments containing chlorides, but can also occur in other salt solutions. A comprehensive survey of the literature on crevice corrosion has been carried out by Ijsseling [7.12]. It comprises mechanisms, modelling, test methods and results, practical experience, protective measures and monitoring. Various materials are dealt with, but much attention is reasonably paid to stainless steels in chloride solutions.

A special form of crevice corrosion that can develop on steel, aluminium and magnesium beneath a protecting film of lacquer, enamel, phosphate or metal is the so–called filiform corrosion, which leads to a characteristic stripe pattern. It has been observed most frequently in cans exposed to the atmosphere.

7.5.2 Mechanism

A review of the mechanisms of crevice corrosion has recently been published by Combrade [7.13]. Reference is also made to the literature survey [7.12].

Figure 7.15 shows two stages in the development of crevice corrosion. A flange joint of a metal M is exposed to neutral water containing oxygen and NaCl. For simplicity, we assume firstly that M is a univalent metal. The development can be divided into the following four stages:

I. In the beginning the metal corrodes at the same rate inside and outside the crevice (Figure 7.15a). The anodic reaction $M \rightarrow M^+ + e^-$ and the cathodic reaction $O_2 + 2H_2O + 4e^- \rightarrow 4OH^-$ occur on area elements distributed between each other all over the surface. If the metal is passive beforehand, the corrosion rate is low (corresponding to the passive current density) and the consumption of oxygen is correspondingly low. In principle, the situation can be illustrated with the potential–log current diagram in Figure 7.16a.

   The oxygen is gradually being consumed within the crevice. This is also the case when the metal surfaces are passive, provided that the crevice is so narrow and deep that oxygen is more slowly transported into the crevice than it is consumed inside it.

II. When the oxygen has been completely consumed inside the crevice, OH^- can no longer be produced there. Conversely, the dissolution of metal inside the crevice continues, supported by the oxygen reduction outside the crevice. The concentration of metal ions within the crevice increases, and with missing OH^- production in the crevice, the charge equilibrium (electrical neutrality) is maintained by migration of Cl^- into the crevice* (Figure 7.15b). In this way an increasing amount of dissociated metal chloride (M^+Cl^-) is produced in the crevice. The metal chloride reacts with water (hydrolyzes):

$$M^+Cl^- + H_2O \rightarrow MOH + H^+Cl^-. \tag{7.3}$$

   Metal hydroxide is deposited and hydrochloric acid is formed in the crevice, which causes a gradual reduction of the pH. This process leads to a critical corrosion state.

III. When the environment has become sufficiently aggressive, the surface oxide film is attacked. The crevice surfaces are transferred to an active state and the corrosion rate increases.

---

* It is mainly Cl^- and not OH^- that migrates into the crevice because the concentration of OH^- is very low except close to the cathodic surface.

IV. The growth (or propagation) phase. Because of increased corrosion rate, the migration of Cl^- increases also, which contributes to a further acceleration of the corrosion process. The process promotes itself, i.e. it is "auto–catalytic". Since the pH has been strongly reduced, the hydrogen reaction $2H^+ + 2e^- \rightarrow H_2$ can also possibly contribute as a second cathodic reaction.

pH inside the crevice can reach values of 0–4 depending on the actual material–environment combination. At the same time, pH may increase to 9–10 on the metal surface outside the crevice, where the oxygen reduction takes place. The conditions at this stage can in principle be described with the overvoltage curves in Figure 7.16b. The corrosion current $I_{corr} = I_o + I_H$, where I_o is the oxygen reduction current outside the crevice and I_H represents the hydrogen reaction $2H^+ + 2e^- \rightarrow H_2$ inside. There is a potential drop $\Delta E = E_o - E_i$ in the electrolyte from the free surface outside to the inside of the crevice. The area of the crevice surface is often small compared with that of the outer surface. This means that a large corrosion current is concentrated on a small area, so that the corrosion current density is very high in unfavourable cases.

The anodic polarization curve for a specimen with an active crevice will be in principle as shown in Figure 7.17. In this case a very small free external surface is assumed, and any internal hydrogen reduction is disregarded. E_{ou} is the potential as measured with the reference electrode positioned outside the crevice. As explained above, the real potential in the crevice, E_{in}, is more negative. The lower limit for corrosion in an active crevice is the protection (or repassivation) potential E_{pr}. However, the critical potential that must be exceeded for initiation of the crevice corrosion process, the crevice corrosion initiation potential, is higher than the protection potential.

For materials that are not passive initially, such as structural steel in a neutral environment, the increased pH on the cathodic area (the free surface outside the crevice) may lead to passivation of this area. Both in this case and in cases with an originally passive surface, the active corrosion will be concentrated within the crevice, which gives the most typical and pronounced crevice corrosion.

Various metals, e.g. Al, Fe, Cr and Ni, may suffer from crevice corrosion. The corresponding hydroxides deposited in the crevice are $Al(OH)_3$, $Fe(OH)_2$, $Cr(OH)_3$ or $Ni(OH)_2$. $Cr(OH)_3$ plays a vital role in crevice corrosion of stainless steel, in which also the content of Mo is important.

The time until the crevice becomes active (stages I–III) is called the incubation (or initiation or induction) period. This may last, e.g. several months (under conditions giving a small passive current density). However, after the incubation period, penetration of plates with thickness of a few millimetres may in extreme cases occur within some days. The length of the incubation time is difficult to predict. Calculation models have been developed, and testing can be carried out, as described in the following sections.

Crevice corrosion is affected by several factors, of a metallurgical, environmental, electrochemical, surface physical, and last but not least, a geometrical nature. One of the most important factors is the crevice gap. This is indicated in Figure 7.18. We can see that the critical gap size, below which crevice corrosion is possible under the

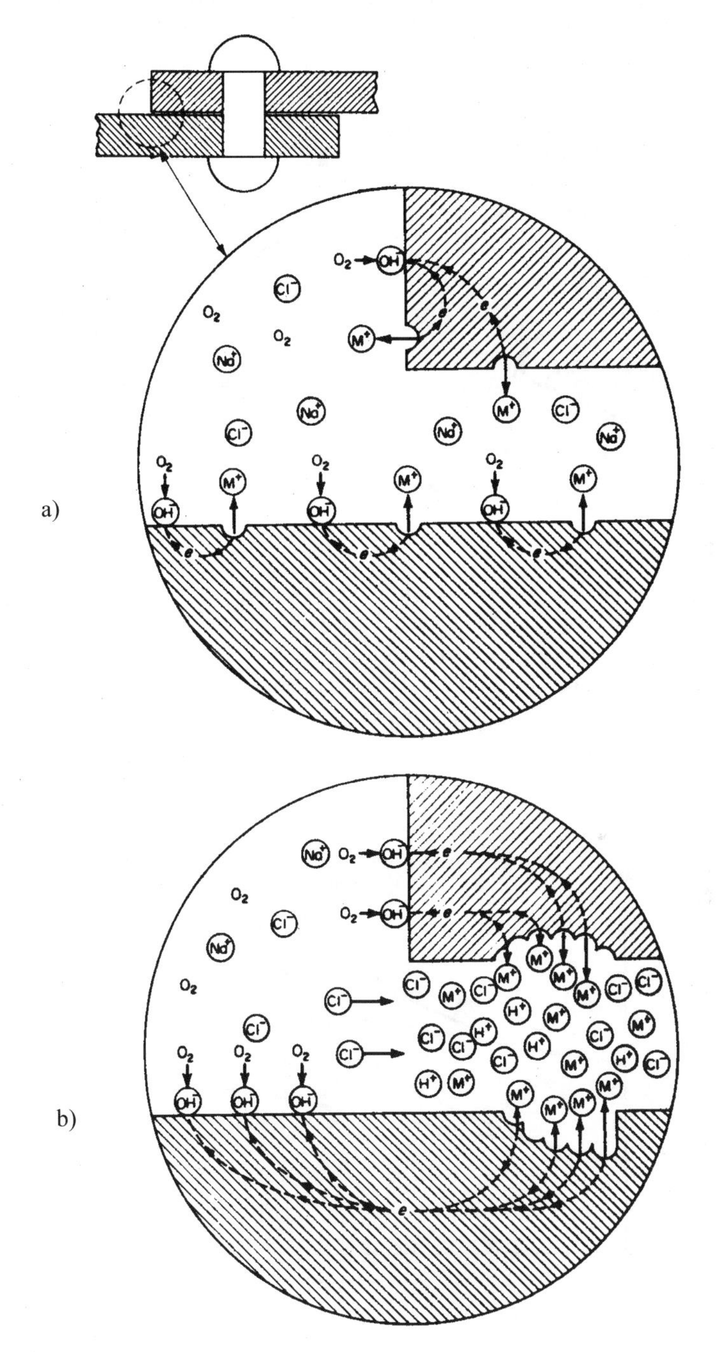

Figure 7.15 Crevice corrosion. a) Initial and b) later stage. (From Fontana and Greene [7.1].) Reproduced with the permission of The McGraw–Hill Companies.

actual conditions, is smaller the more resistant the material is. For description of materials, see Section 10.1.

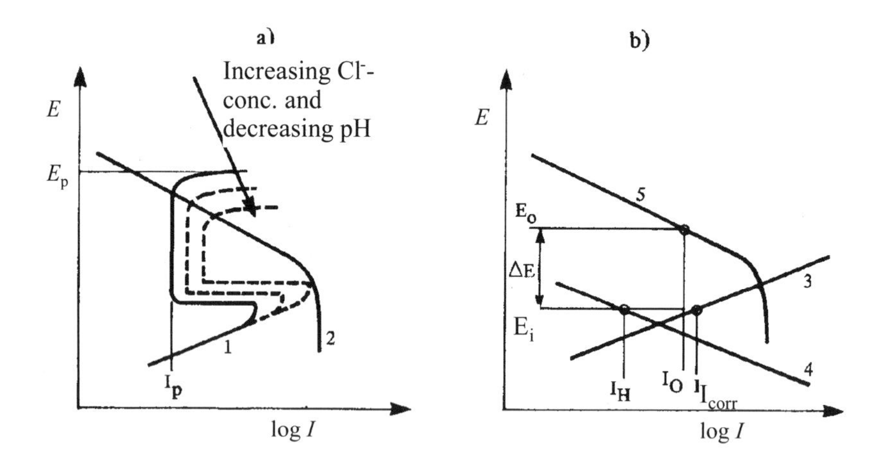

Figure 7.16 Potential–log current diagram for crevice corrosion. a) Initial and b) developed to a late stage. 1) Anodic and 2) cathodic curve (oxygen reduction) in neutral water. 3) Anodic curve in HCl solution, pH = 0–4, corresponding to a crevice solution. 4) Curve for the hydrogen reaction in the crevice (pH = 0–4). 5) Curve for the reduction of oxygen outside the crevice (pH = 8–10).

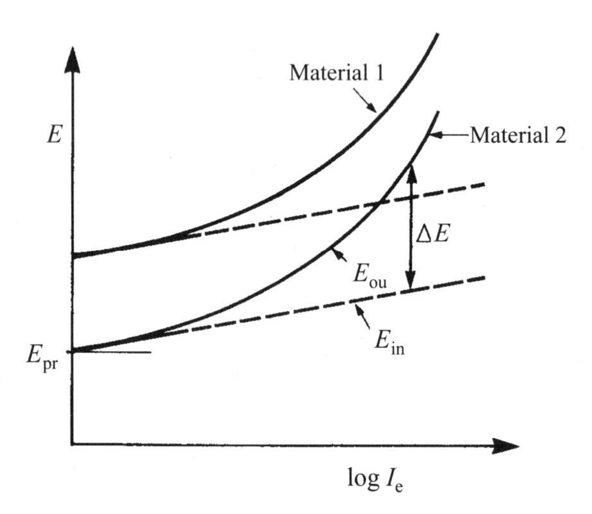

Figure 7.17 Anodic polarization curves for a specimen with an active crevice. E_{ou} is the potential set with the potentiostat (with the reference electrode outside the crevice). E_{in} is the potential inside the crevice. I_E is the external current.

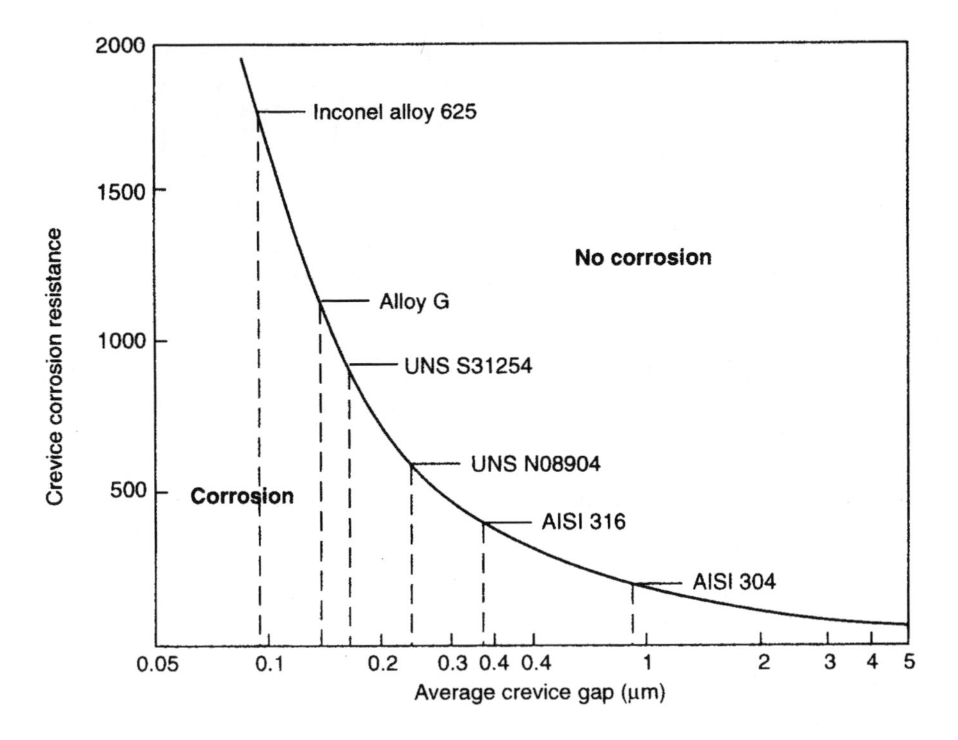

Figure 7.18 Prediction of whether or not crevice corrosion will occur in crevices 5 mm deep in seawater at ambient temperature as a function of crevice gap. (After Oldfield and Todd [7.14].)

7.5.3 Mathematical Models of Crevice Corrosion

Oldfield and Sutton [7.15] proposed in the 1970s a model for calculating the evolution of the first three stages (the incubation period) of crevice corrosion of stainless steel. During the last two decades, several works have been published on CC modelling. Surveys of proposed models have recently been presented in References [7.12] and [7.13]. One of the most advanced models was developed by Gartland during the late 1980s [7.16]. Gartland's model includes also the growth phase, and conditions for repassivation. In addition and in contrast to previous models, it takes into account the potential variation within the crevice as well as effects of possible temperature variations.

The crevice is considered one–dimensional, but the crevice profile may be arbitrary. The crevice and the free surface around it is treated as a galvanic element. The potential variation and current distribution is calculated numerically by the finite difference method (FDM). This is based on the electrolyte in the crevice being divided into finite resistors, as shown in Figure 7.19.

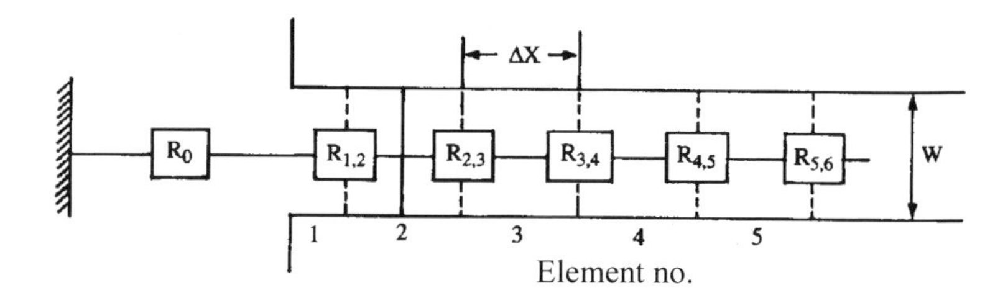

Figure 7.19 Illustration of the crevice as a series of finite resistors [7.16a].

A flow chart of the model is presented in Figure 7.20. As shown at the top, to the left, we start with the alloy composition, the bulk pH and Cl^- concentration of the solution, and the geometry. The additional data basis for the calculation is:

a) Smoothed, experimentally determined anodic and cathodic overvoltage curves recorded when the steel is exposed to simulated crevice solutions (Figure 7.21). These crevice solutions correspond to the successive stages that evolve as crevice corrosion proceeds, with decreasing pH and increasing aggressiveness within the crevice.

b) Chemical equilibria of various corrosion products that are formed. The equilibrium constants form the basis for calculation of ionic strength $I = \frac{1}{2}\sum z_i^2 c_i$, in which z_i and c_i are charge and concentration, respectively of the various species, and for calculation of pH, values of mobility u_i and resistivity ρ.

From the situation at $t = t_o$ as regards potential variation, current distribution, production rate and mass transport of various substances, the conditions at $t = t_o + \Delta t$ are computed. The calculation is finished when a defined time $t = t_c$ has been reached. The output provides the values of the following functions of x (depth in the crevice) at $t = t_c$: resistivity (ρ), ionic strength (I), concentrations of species i (c_i), pH, potential (E), anodic current densities (i_a) and galvanic current densities (i_g).

Figure 7.21a and b show examples of schematic, smoothed anodic overvoltage curves of a high–alloy austenitic stainless steel (20Cr18Ni6Mo) (UNS S 31254), often called 6 Mo steel, exposed at low pH values. The figures show that the passivity of this material is broken down at very low pH values only. Not included in these figures, but shown in Reference [7.16a], is that chlorination affects the overvoltage curve at pH ≥ 0.5, but not in the pH range where the passivity is broken down. This indicates that the effect of chlorination on the crevice corrosion tendency is limited to the consequences of increased corrosion potential (compare Sections 6.7 and 8.3).

Results from the calculation are compared with experimental results in Figure 7.22. The figure, showing crevice gap as a function of depth in the crevice after certain corrosion periods, confirms that the calculation gives realistic results.

The model is also used for computing the effect of temperature. For one thing, it can.be shown that repassivation occurs at a lower temperature than initiation of

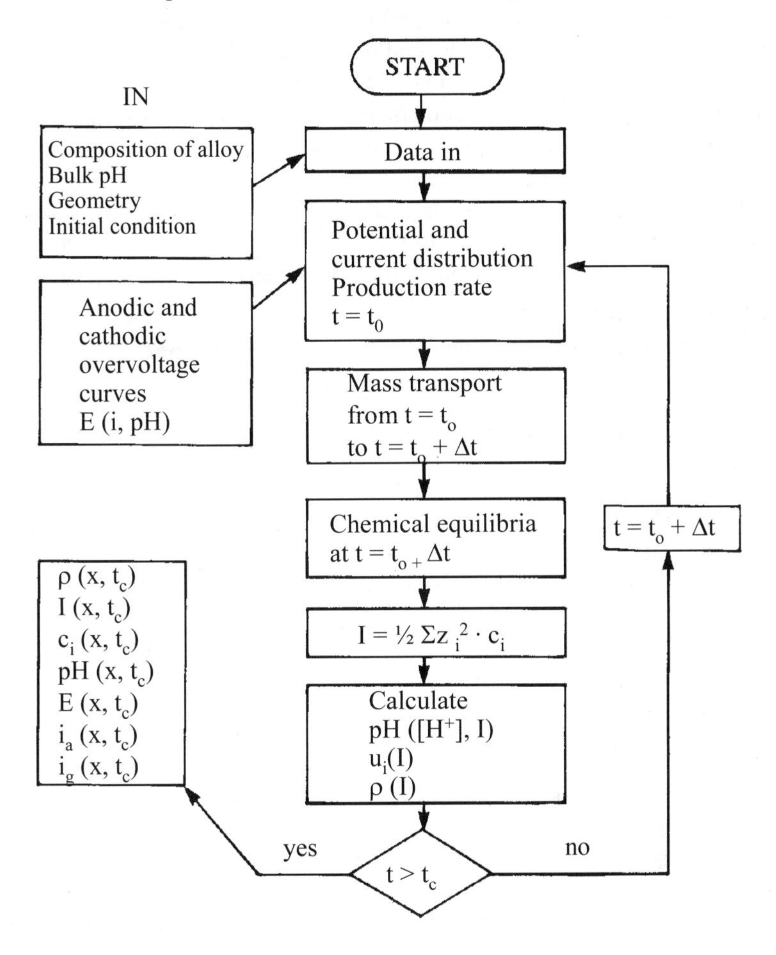

Figure 7.20 Flow chart of the mathematical crevice corrosion model [7.16a].

crevice corrosion, which has also been proved by experiments [7.16b] (see next section).

More recently, Gartland has proposed a simplified model that is based on the computerized model but now described by use of algebraic equations [7.16c].

7.5.4 Crevice Corrosion Testing

As mentioned previously, the corrosion potential increases strongly after a certain time of exposure of stainless steel and other corrosion-resistant materials in natural seawater [7.6, 7.17, 7.18]. This is shown for various stainless steels in Figure 7.23. For conventional stainless steels such as AISI 304 and AISI 316, this potential increase means that more or less well-defined crevice corrosion potentials (critical initiation potentials) are exceeded. The end of the incubation period is recorded by a potential fall, as shown for one of the materials in Figure 7.23.

The figure caption indicates a problem in laboratory testing of such materials: crevice corrosion may occur where the specimen is fixed, which is usually not the intention.

Critical initiation potentials can be defined by means of potentiostatic experiments with crevice specimens, e.g. as shown in Figure 7.24 [7.19, 7.20]. Possible initiation of crevice corrosion (at a given set potential) is indicated by a marked increase in the

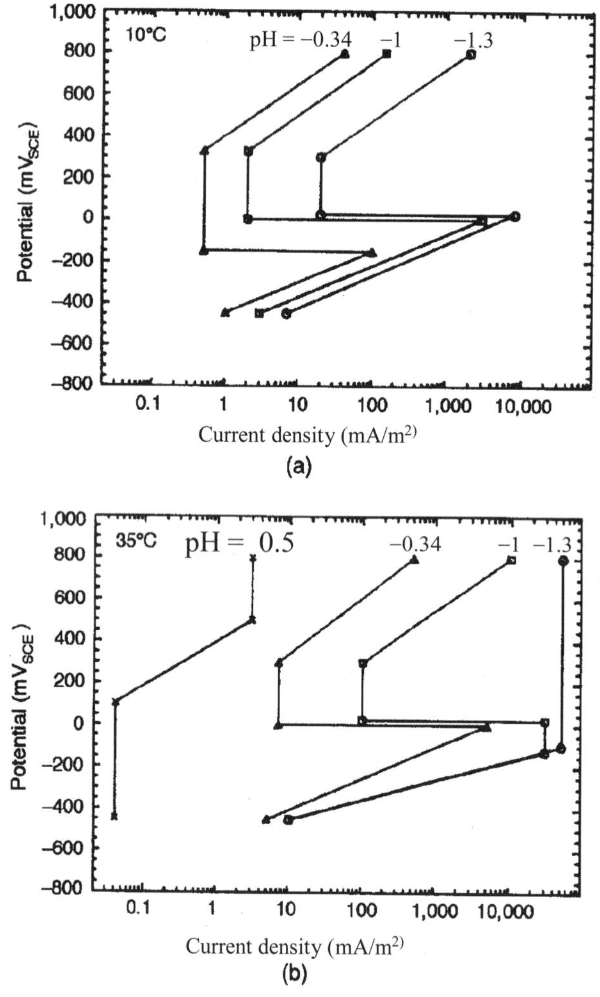

Figure 7.21 Schematic overvoltage curves for steel UNS S 31254 and UNS S 32750 at a) 10 ^{0}C. and b) 35^{0}C [7.16b].

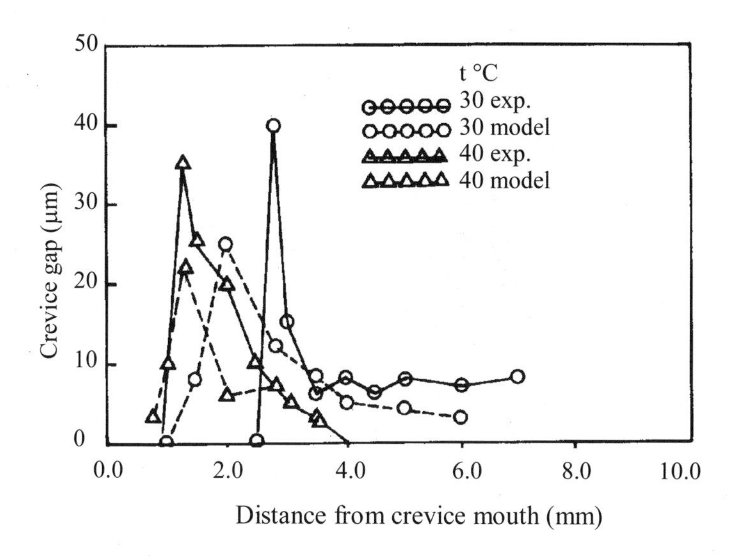

Figure 7.22 Profiles of experimentally determined and calculated crevice attacks on UNS S 31254 steel [7.16].

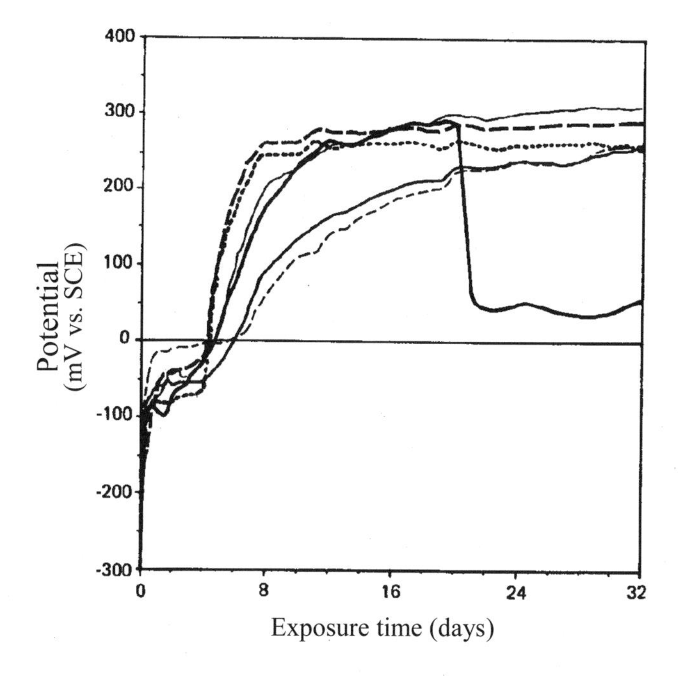

Figure 7.23 Potential as a function of time for various stainless steels in seawater at a flow velocity of 1.2 m/s and temperature 9°C. The potential drop after about 20 days for one of the materials is due to initiation of crevice corrosion.

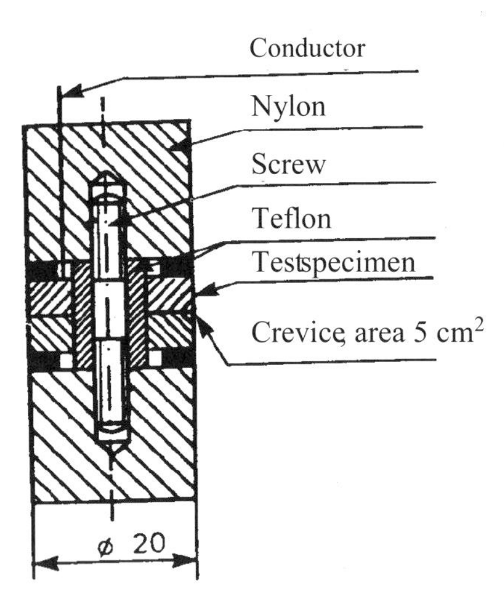

Figure 7.24 Cross-section of crevice specimen assembly.

anodic current. But the time elapsed before this happens (the incubation period) decreases with increasing potential. In order to define a certain initiation potential, a given exposure schedule must be used. That is, the critical initiation potential is the lowest potential leading to active crevice corrosion within a certain time of exposure at constant potential or possibly within a certain procedure and rate of stepwise increase of potential. If by such experiments we determine an initiation potential that is lower than or equal to the stable free potential on passive surfaces (in seawater about + 300 mV References SCE), corrosion can occur in similar crevices in service as well. With practical application in mind it must also be taken into account that different crevice geometry and different gasket materials give different critical initiation potentials.

A procedure has been proposed implying stepwise increase of the potential by means of a potentiostat (with relatively long time at each potential) until corrosion starts. An average potential increase rate of 25 mV/hour has been found sufficient [7.20]. The temperature is kept constant in this case.

For high-alloy stainless steel, it is, however, more suitable to hold the potential of the crevice specimen constant and increase the temperature stepwise until corrosion is initiated. By this a critical crevice temperature (CCT) is determined [7.6, 7.20] (compare determination of a critical temperature for pitting corrosion CPT, Section 7.6). In such testing it is convenient to use a neutral, aerated NaCl solution. A critical crevice corrosion temperature can also be determined by exposure in 6% $FeCl_3$ solution, a method that is standardized in ASTM G48 [7.21]. Both methods are accelerated, and we cannot assume that they give the same critical temperatures as those determined by long-term testing in service environments. As a rule, the short–term tests described above will give useful ranking of various materials.

If the exposed free surface area around a crevice is large compared with the crevice area, the corrosion rate of stainless steel will be high when corrosion has first been initiated. Therefore, it is usually necessary to avoid initiation of crevice corrosion. Consequently, testing the critical conditions for corrosion initiation becomes very important. For cases where the free area is relatively small, it is also of interest to predict the corrosion rate in active crevices, as this may be within the acceptable limits. The prediction can be based on recorded anodic polarization curves for active crevices and cathodic curves for free surfaces, as described in References [7.19].

Because of the importance of this area ratio, crevice test specimens that are not polarized with an external current should have a large area ratio between the free surface and the crevice. Some types of test specimens that are used, even some of the standardized ones, do not satisfy this requirement.

Experiments have shown that if the temperature of an active crevice specimen is lowered from the critical initiation value (CCT), the crevice will not be repassivated before the temperature reaches a considerably lower level (the repassivation temperature) [7.6]. This is very important to know in connection with service conditions that may imply temperatures above the CCT for short periods, sufficient to initiate corrosion. If the normal service temperature is below the CCT but above the repassivation temperature, corrosion will continue after the temperature is back to its normal level.

For the study of experimental details in crevice corrosion testing the reader may consult works by Ijsseling [7.22] and Oldfield [7.14].

7.5.5 Practical Cases of Crevice and Deposit Corrosion

As mentioned, conventional stainless steels are liable to deposit corrosion in seawater, and several cases of very rapid localized corrosion have been experienced in practice. In one case, AISI 316 steel (Section 10.1) was used in pipes carrying seawater to a heat exchanger system. The pipe diameters were 150–400 mm, the flow velocity $\leq$ 1.2 m/s and the water temperature in the range 7–30 °C. Only 2–3 months after installation, penetration of pipe walls of thickness 3–4 mm occurred. The failures were mainly concentrated in and around welds. They were characterized by small, grey–black bulges where the pipe wall could easily be penetrated with a sharp tool. The diameters of the holes varied in the range 1–5 mm. The form of corrosion is considered to be deposit/pitting corrosion. It had occurred because of low flow velocity (allowing biofilm growth) combined with unsatisfactory use of backing gas in the welding process. Insufficient use of backing gas allows the weld zone to oxidize; the Cr content in the surface material and consequently the corrosion resistance is reduced. This is revealed by the colour pattern of the pipe surface at the welds (dark brown–blue regions). For prevention of further corrosion, sacrificial anodes of a zinc alloy were mounted, after previous calculation of ohmic potential drops, in order to find suitable positions for the anodes. If the anodes were replaced after periods of 12–14 months, the plant functioned satisfactorily. An alternative to cathodic protection is to use a higher-alloyed stainless steel (Section 10.1).

A similar case has been experienced in a ship containing tanks of AISI 316L and 317L steels. The tanks were filled with seawater during a maintenance period. Here, attacks had developed to 3–4 mm depth after 1–2 months, also in this case under deposits, mainly in/at welds, slag remnants and small weld defects were considered to play an important role in the corrosion development. Some attacks were also found under remnants of blasting sand and slag outside the welds. The reason for the very high local corrosion rates is that the area ratio between the free surface (cathodes) and the attacked points (anodes) was extremely large.

Rapid corrosion because of deposits has also occurred in other environments, e.g. on boiling tubes for in–evaporation of glue water in a fish–glue production ship. The temperature varied from 40°C to 100°C from stage to stage in the plant. Heavy corrosion attacks were found both on unalloyed and on stainless steel AISI 304. At least partly, the attacks were caused by insufficient cleaning. An example of such an attack on unalloyed steel is shown in Figure 7.25.

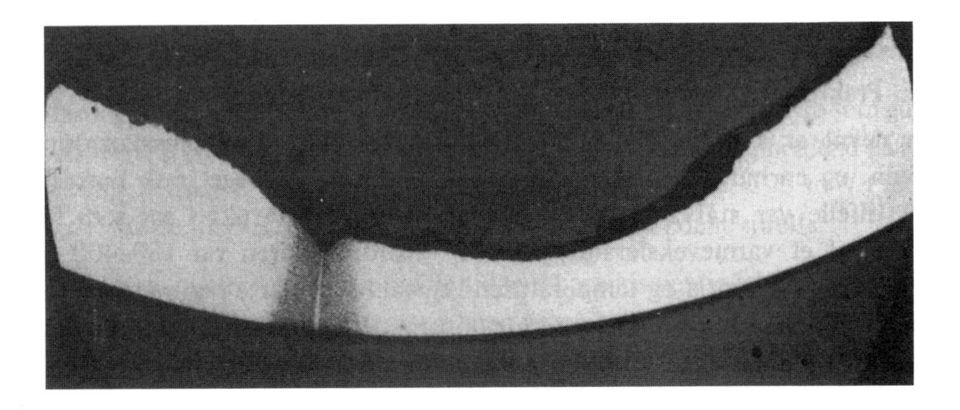

Figure 7.25 Deposit corrosion attack on a boiling tube of St 35 steel in an in-evaporation plant for fish-glue. 5×. (Photo: M. Broli, SINTEF Corrosion Centre.)

7.5.6 Galvanic Effects on Crevice Corrosion

If there is a crevice on a component made of a material liable to crevice corrosion, and this component is connected to a more noble material with free surfaces, crevice corrosion may be intensified strongly. Such a case is a couple of an aluminium component (with a crevice) and a steel plate in water containing some chloride. The corrosion form can be called galvanic crevice corrosion. The crevice corrosion rate will be particularly high if the more noble metal acts as an efficient cathode in the given environment. The explanation is the same as for ordinary galvanic corrosion.

If, on the other hand, a less noble metal with free surface is connected to a more noble component with a crevice, the coupling will counteract crevice corrosion in the latter. Certainly, in many cases of stainless steel exposed to seawater, crevice corrosion is prevented by contact with unalloyed/low-alloy steel or cast iron. An example is stainless steel pumps or valves coupled to ordinary steel or iron pipes.

The reason for the absence of crevice corrosion in such cases is often misunderstood. It is an example of cathodic protection with sacrificial anode.

7.5.7 Prevention of Crevice Corrosion

Crevice corrosion can be prevented or reduced by appropriate:

1. *Selection of material.* As dealt with in previous sections, conventional stainless steels, with martensitic, ferritic, austenitic or ferritic–austenitic (duplex) structure, are sensitive to crevice corrosion (Table 7.4). Newer high-alloy steels with high Mo content show by far better crevice corrosion properties in seawater and other Cl^--containing environments (see Section 10.1).

Table 7.4 Resistance of various materials to crevice corrosion in stagnant seawater (From Fontana and Greene [7.1].) Reprinted with permission of The McGraw-Hill Companies. (For definition/description of materials, see Section 10.1).

Inert	Useful resistance: Best	Useful resistance: Neutral	Useful resistance: Less	Crevices tend to initiate deep pitting
Hastelloy[1] "C"	90–10 copper–nickel 1.5 Fe	Aus. nickel cast iron	Incoloy[2] alloy 825	Type 316 SS
	70–30 copper–nickel 0.5 Fe	Cast iron	Alloy 20	Nickel–chromium alloys
Titanium	Bronze Brass	Carbon steel	Nickel–copper alloy Copper	Type 304 SS Series 400 SS

1) Trademark Union Carbide Corporation.
2) INCO trademark.

2. *Design and production.* Avoid as far as possible crevices and deposition. A couple of examples are given in Figure 7.26a and b. Comments:

   a) Use butt welds instead of overlap joints. Alternatively, overlap joints can be sealed by continuous welds. Be careful about the use of backing gas in welding processes, and check that the penetration is complete. After welding of stainless steels, carry out pickling, brushing, blasting or grinding for removal of less protective and less resistant surface layers.
   b) Design in a way that promotes complete drainage and prevents accumulation of deposits. Make it easy to inspect and clean.

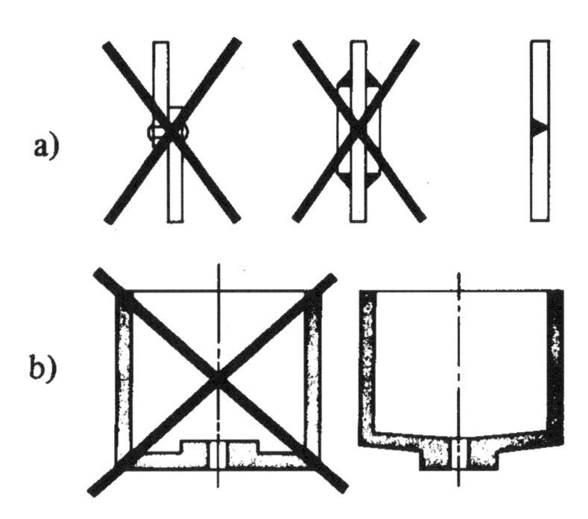

Figure 7.26 Examples of bad and good design.

3. *Cathodic protection.* The potential must be kept below the protection potential. An efficient method for internal protection of pipe joints is the "resistance-controlled cathodic protection" (RCP) (see Section 10.4).4.

4. *Measures for preventing deposition.*

   a) Inspection and cleaning during and between service periods.
   b) Separation of solid material from flowing media in process plants.
   c) Gravel filling around buried piping and structures.

7.6 Pitting Corrosion

7.6.1 Conditions, Characteristic Features and Occurrence

Pitting corrosion (sometimes only called pitting) occurs on more or less passivated metals and alloys in environments containing chloride, bromide, iodide or perchlorate ions when the electrode potential exceeds a critical value, the pitting potential (Figure 7.27), which depends on various conditions. The pitting potential is not a thermodynamically defined potential and depends for one thing upon the rate of potential increase when the polarization curve is recorded.

This form of corrosion is characterized by narrow pits with a radius of the same order of magnitude as, or less than, the depth. The pits may be of different shape, but a common feature is the sharp boundary (Figure 7.28). Pitting is a dangerous form of corrosion since the material in many cases may be penetrated without a clear warning (because the pits often are narrow and covered) and the pit growth is difficult to predict. This is connected to the fact that the extent and the intensity of pitting corrosion is difficult to measure because the number and size of pits (diameter and depth) vary from region to region and within each region. Short-term testing in the laboratory for determination of pit growth is also problematic because, under realistic conditions, it may take long time, e.g. many months, before the pits become visible. Another problem is that the critical size, i.e. the maximum pit depth, increases with increasing surface area.

Aluminium is liable to pitting corrosion in media containing chloride, but the corrosion may not necessarily be of a serious nature. However, if aluminium is

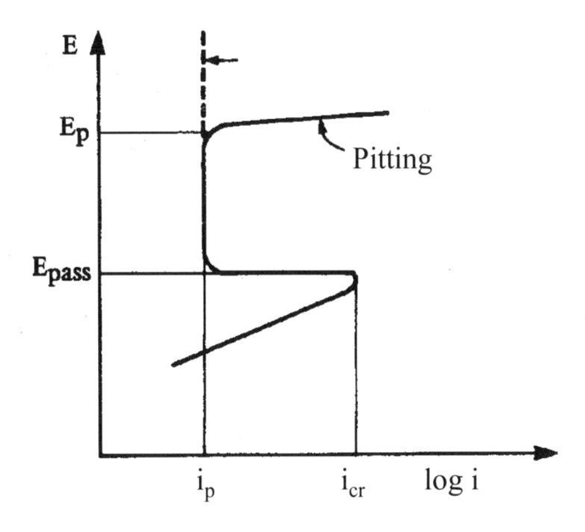

Figure 7.27 Anodic overvoltage curve for an active–passive metal in an environment causing pitting corrosion. E_p = pitting potential, E_{pass} = passivation potential, i_{cr} = critical current density and i_p = passive current density.

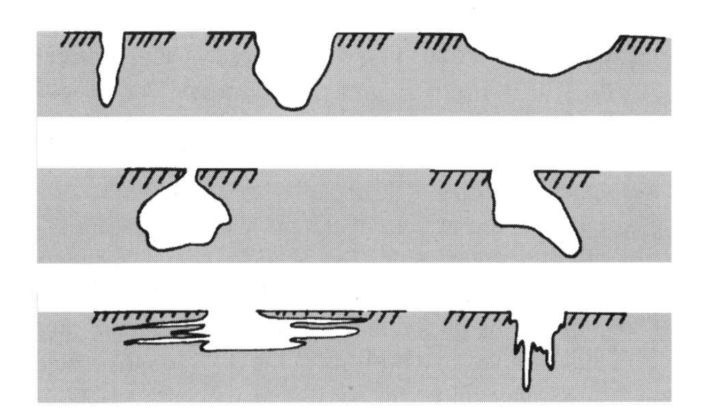

Figure 7.28 Different shapes of corrosion pits. (ASTM [7.21].)

polarized by metallic contact with a more noble material, the former will be heavily attacked. Stainless steels are attacked in strongly oxidizing chloride solutions.

Copper and copper alloys may corrode by pitting in fresh water, particularly at temperatures > 60°C, pH < about 7.4, and where the ratio between sulphate and carbonate contents in the water is relatively high. Growth of organisms also increases the risk of corrosion, and a transition to deposit corrosion may then occur.

7.6.2 Mechanisms

We distinguish roughly between the mechanism of pitting initiation and that of pit growth. For more detailed discussions, however, the pitting process is often considered to consist of the following stages: 1) Local breakdown of passivity (pit nucleation), 2) early pit growth, 3) late (stable) pit growth, and (possibly) 4) repassivation (Strehblow [7.23]). The initial stage is not fully understood, but various theories exist.

On materials that often operate in a less pronounced passive state, such as copper, zinc and tin, pitting corrosion may start in pores in surface layers of corrosion products. For materials that are typically passive initially, e.g. aluminium and stainless steels, it is assumed that pitting is initiated by adsorption of halide ions that penetrate the passive film at certain positions. This happens at weak points of the oxide film, e.g. at irregularities in the oxide structure due to grain boundaries or inclusions in the metal. Absorption of halide ions causes strong increase of the ion conductivity in the oxide film so that metal ions can migrate through the film. In this way localized dissolution occurs, and intrusions are subsequently formed in the metal surface (Kaesche [7.24]). Another theory is that the initial adsorption of aggressive anions at the oxide surface enhances catalytically the transfer of metal cations from the oxide to the electrolyte and thus causes successive local thinning of the oxide film. A third possibility is that the attacks start at fissures in the passive layer [7.23]. Which of the mechanisms is the most effective depends on both material and environment.

After pit nucleation there is a transition (i.e. successive stages of early pit growth) to stable pit growth. The transition, implying development of pits from a few nanometres to the micrometre range, has been studied with scanning tunnelling microscopes (STM) and further with scanning electron microscopes (SEM), as summarized in References [7.23].

The next stage, i.e. the stable growth of the pit, proceeds essentially by the same mechanism as that of crevice corrosion (see Section 7.5.2). Generally, the anodic dissolution occurs inside the pit, often mainly at the bottom, while the cathodic reaction is localized outside the pit, on inclusions or other parts of the surface which are sufficiently efficient cathodes. Hydrolysis in the pit leads to an acidic, aggressive pit solution, which also may allow an additional cathodic reaction here, namely reduction of hydrogen ions. Metal cations from the dissolution reaction migrate and diffuse towards the mouth of the pit where they react with OH^- ions from the cathodic reaction, forming metal hydroxide deposits that may cover the pit more or less. An example of deposits on numerous pits on aluminium is shown in Figure 7.29.

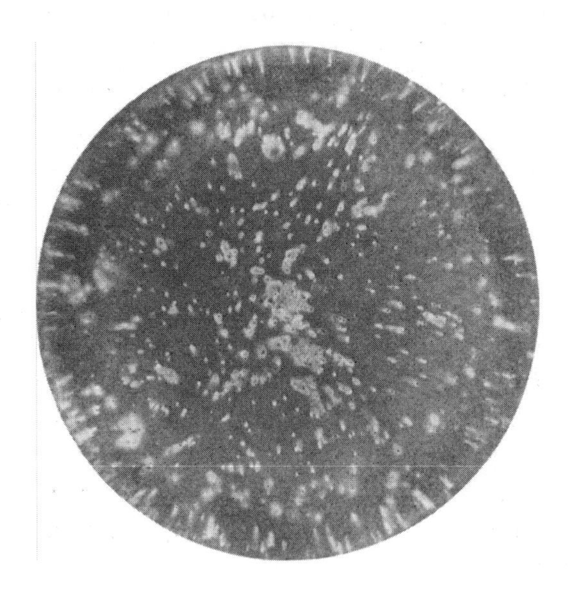

Figure 7.29 Reaction products on an aluminium surface after 150 h exposure in neutral 3% NaCl solution. 3 × .

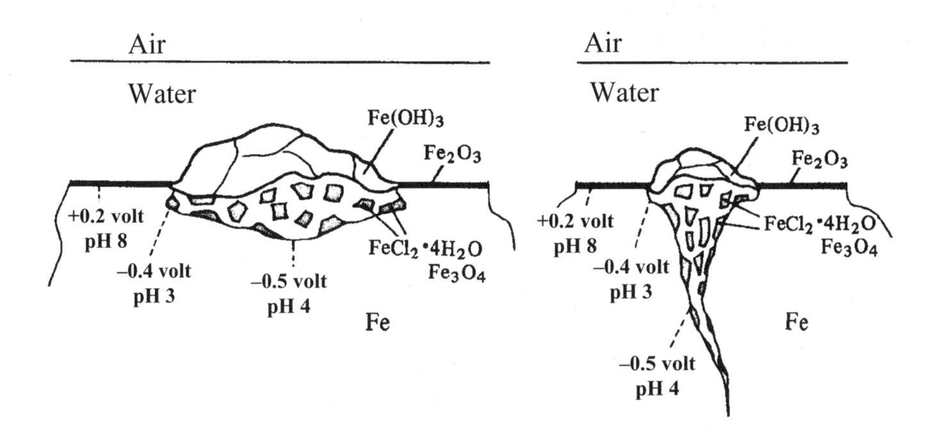

Figure 7.30 Tentative sketch of a corrosion pit and a crack in iron. Potential referred to SHE. (After Pourbaix [7.25].)

Corrosion products covering the pits facilitate further corrosion because they prevent exchange of the electrolyte in the pits, which becomes very acidic and aggressive.

In addition to the pH difference between the outside and the inside of the pit, there is also a potential difference, as shown for pitting of iron in Figure 7.30. The figure illustrates environmental similarities between pitting and stress corrosion cracking [7.25].

7.6.3 Influencing Factors

Factors of general significance for pitting are:

a) pH and chloride concentration. The pitting potential and pitting resistance normally increase with increasing pH and decreasing chloride concentration.
b) Flow velocity. When pitting is initiated, an aggressive environment is established within the pit, and this is combined with an increased pH at the adjacent free surface. The increased pH gives an increased resistance to initiation of new pits around the first one formed (this is the case, e.g. for unalloyed steel, but it is more uncertain with respect to aluminium). If there is no movement of the liquid, these conditions will be conserved, and the result is large and few pits. Liquid flow gives a higher probability for washing away the aggressive environment in the pit, and at the same time it increases the transport of oxygen to the active area, so that the pit may be passivated before it gets the chance to grow to a considerable size. The alkaline layer around an active pit is also washed away to a higher extent, and pit initiation in the neighbourhood of an active pit occurs more easily. This means that increased flow velocity results in smaller but more numerous pits, which makes pitting corrosion less serious.
c) The gravity force. Horizontal top surfaces are often heavily pitted, while underside surfaces are hardly, or not at all, attacked. Vertical surfaces are intermediate as to the extent of pitting. The reason is that the aggressive environment in the pits has a higher density than pure water.
d) Cu^{2+} and Fe^{3+} ions favour pit initiation and accelerate pit growth because they lead to increased potential (they are oxidizers, compare Section 6.1). The Cu ions have a double effect because Cu is precipitated on the material surface and forms efficient cathodes.
e) Metallurgical properties. Impurities and inclusions are important, e.g. AlFe secondary phase, which contributes to the localization of pits on aluminium because 1) the oxide film on top of and around the inclusion is weak, thin and stressed, 2) such inclusions are efficient cathodes. Pitting of this type will usually last only until the pits are so large that the inclusions are falling out.
f) The insulating ability of the oxide. If the oxide insulates efficiently, the surface is inactive as a cathode. This is to a certain extent the case for aluminium in seawater; which is the direct reason for the relative slow growth of pits in this case.
g) Surface roughness. The main trend is that smooth surfaces get few, large pits while rough surfaces get numerous smaller pits.
h) Temperature. Increasing temperature gives usually decreasing pitting potential and increasing liability to pitting corrosion (see Section 7.6.5).
i) Galvanic contact with a more noble material increases the tendency to and the rate of pitting corrosion (the corrosion potential is lifted).

7.6.4 The Time Dependence of Pitting

A material such as stainless steel has a relatively high resistance to initiation of pitting. Therefore, rather few pits are formed. But when a pit has been formed, this may grow very fast. The high corrosion rate in the pit is promoted by large cathodic areas and a thin oxide film that has considerable electrical conductance, i.e. the cathodic current for each pit is allowed to be large. For such a material, it is important to avoid pit initiation completely.

Conversely, for several aluminium alloys, pit initiation can be accepted under many circumstances. This is so because numerous pits are usually formed, and the oxide is insulating and has therefore low cathodic ability, so that the corrosion rate is under cathodic control. However, if the cathodic reaction can occur on a different metal because of a galvanic connection or for instance deposition of Cu on the aluminium surface, the pitting rate may be very high. Since we in other respects can accept pit initiation, the time dependence of pit growth and pit depths is important, and we shall consider this more quantitatively.

Figure 7.31 shows an example of distribution of pit depth at different exposure times for commercially pure aluminium in tap water (after Aziz [7.26]). Aziz considers each curve as a sum of a stationary curve with the shape of a reversed J, and a bell-shaped curve that moves to the right during the first few weeks. The stationary part represents the pits that are formed and successively passivated after short time. The bell-shaped part represents the pits that grow further. After two months of exposure, the mode of the bell-shaped curve becomes stationary, and relatively few pits continue growing.

In service, different aluminium alloys have in many cases been found to acquire a maximum pit depth d according to the following equation:

$$d = C\ t^{1/3}, \tag{7.4}$$

where C is a constant that depends on the alloy, environment and surface area, and t is the exposure time.

According to Godard et al. [7.27] this relation has been found valid in laboratory testing of 200 different fresh waters, furthermore for a water pipeline (of AlMg) 600–700 m long in industrial service during a 13-year period, and also in seawater (with an example for an AlMgSi alloy shown by Godard). AlMg, AlMgMn and AlMn alloys are the best ones in seawater, with a typical largest pit depth of 1.3 mm after 5 years.

The cubic root relation in Equation (7.4) is not always valid. There is considerable scatter, the exponent may deviate from 1/3, and other deviations may exist too, particularly during the first few time intervals. Figure 7.32 shows an example of pit depth development expressed by functions of the form $d = a + b \log t$. The results deviate little from the cubic root relationship. Altogether, Equation (7.4) can be considered as reasonably representative, at least for the more corrosion-resistant aluminium alloys (those represented in Figure 7.32 as well as the AlMn alloys).

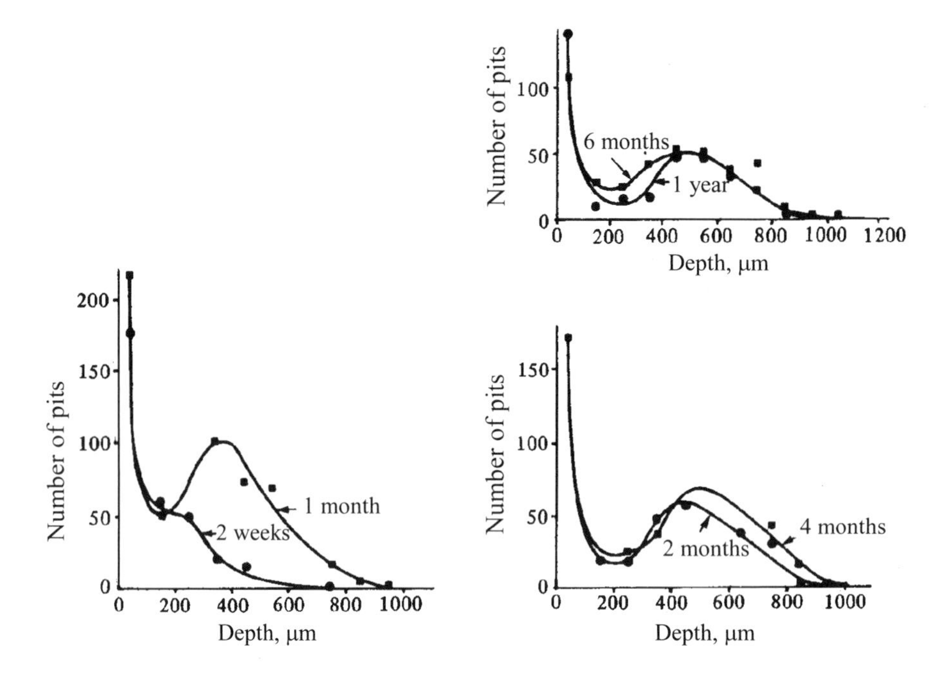

Figure 7.31. Distribution of pit depths on commercially pure aluminium in tap water at various exposure times [7.26].

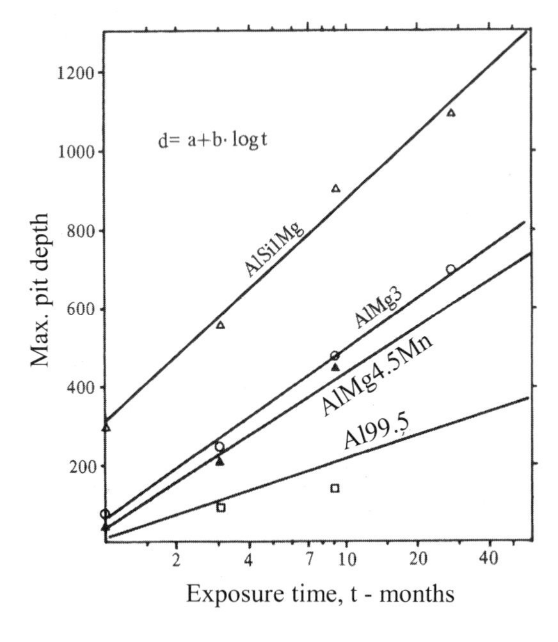

Figure 7.32 Maximum pit depth (98% significance) on a surface area of ½ m^2 of various aluminium alloys in artificial seawater [7.28].

Equation (7.4) can be explained theoretically by means of a simple model (Figure 7.33). The model is based on three assumptions:

1. There is a constant number of growing pits.
2. The total cathodic current does not change with time. With a constant distribution of current on the various growing pits, it follows that the cathodic current for each pit, including the deepest one, is independent of time.
3. The pits are hemispherical.

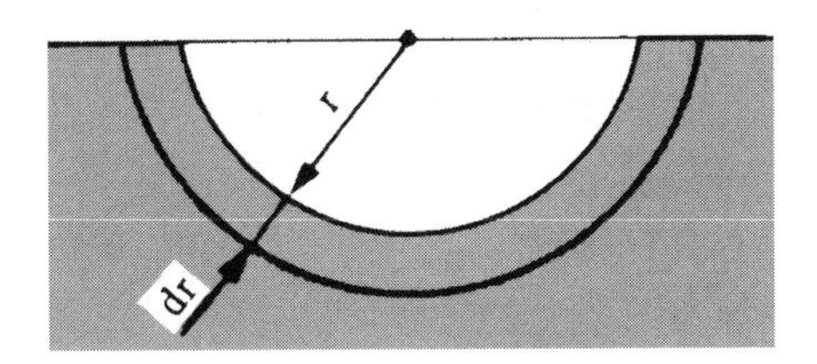

Figure 7.33 Hemispherical model pit.

With these conditions we will have for the largest pit: anodic current = cathodic current, i.e.:

$$I_a = 2\pi r^2 \, i_a = I_{cat} = C_1$$

$$i_a = \frac{C_1}{2\pi r^2} \tag{7.5}$$

where i_c is the corrosion current density in the pit and C_1 is a constant.

Depth increment

$$dr = C_2 \, i_a \, dt \tag{7.6}$$

From Equations (7.5) and (7.6):

$$dt = \frac{dr}{C_2 \, i_a} = \frac{dr}{C_2 \dfrac{C_1}{2\pi r^2}} = C_3 \, r^2 dr$$

$$t = \int_0^t dt = C_3 \int_0^r r^2 dr = C_4 \, r^3$$

The depth at time t: $d = r = C_1 t^{1/3}$.

Discussion of the assumptions for the model.

Assumption 1 means that we neglect the effects of the many initiated pits that passivate after short exposure. Figure 7.31 indicates that this is a reasonable assumption after a few months of exposure (there is no change of the left part of the curve).

Regarding assumption 2, experiments have shown that the average corrosion rate of aluminium decreases with the time from the first few weeks to 1–2 years, i.e. the average cathodic current density decreases. The assumption for the model is not in agreement with this experience. However, the reduction in average cathodic current density is assumed to correspond partly to the passivation of pits during the first 2–3 months, and partly to some reduction of the number of pits growing after that time. In other words, the errors in assumptions 1 and 2 will more or less compensate each other. Altogether, these conditions indicate that the total cathodic efficiency decreases somewhat with the time, which contributes to the very favourable relationship between pit depth and time.

Assumption 3 about hemispherical pits is a reasonable approach in some cases, but not always.

7.6.5 Pitting Corrosion Testing

Relationships like Equation (7.4) and that shown in Figure 7.32 are useful in connection with *testing*. The testing time must last some months so that the constants in the equations can be determined with sufficient accuracy. From this we can predict maximum pit depths after many years of exposure.

Pitting testing can also be carried out with electrochemical methods, e.g. recording of polarization curves for determination of the pitting potential E_p, and for comparison, the free corrosion potential E_{corr}. As mentioned before, one must have $E_{corr} < E_p$ in order to avoid pitting. Pitting tendency of stainless steels is also determined by measuring the critical pitting temperature, CPT. Except that specimens without crevices are used, the test is carried out in the same way as determination of the critical crevice corrosion temperature, CCT (see Section 7.5.4). By potentiostatic determination of the CPT, the same ranking order of various stainless steels has been obtained as that found by calculation of the pitting resistance equivalent PRE_N by the following equation:

$$PRE_N = \%\,Cr + 3.3\,(\%\,Mo + 0.5\%\,W) + x\,\%\,N, \tag{7.7}$$

where x = 16 for ferritic–austenitic (duplex) steel and x = 30 for austenitic steel.

ASTM has standardized methods for investigation/evaluation of pitting attack and pitting testing in 6% $FeCl_3$ (G46 and G48, respectively) [7.21].

7.6.6 Prevention of Pitting Corrosion

The best way to prevent pitting is *appropriate materials selection.* Among aluminium alloys, those alloyed with magnesium and/or manganese and the commercially pure grades are the best ones. These perform quite well in seawater. Aluminium with magnesium (e.g. AlMg 4.5 Mn) is used in hulls of high-speed vessels and small boats, in deck structures on ships and boats and in helicopter decks on oil and gas platforms. AlMgSi alloys will normally acquire somewhat larger pits (see Figure 7.32), but they are used in profiles, e.g. in marine atmospheres. Aluminium alloys with Cu and Zn are generally less corrosion resistant.

In stainless steels, increased content of Cr as well as N and Mo gives higher resistance against pitting, while Si, S and C have the opposite effect. Modern high-alloy steels with higher content of Mo (6–7%) have very good pitting resistance. Some Ni alloys and particularly titanium are even better (Section 10.1).

If general corrosion can be accepted, pitting corrosion can be avoided by selecting materials that are sufficiently active in the environment in question.

Cathodic protection can also be applied to prevent pitting. Regarding aluminium, strong cathodic polarization should be avoided because this can lead to a large increase of pH close to the metal surface, which can cause so-called alkaline corrosion (compare with the Pourbaix diagram for aluminium in Figure 3.11, Section 3.8). Use of sacrificial anodes of Zn or Al alloys is therefore safer than impressed current.

Pitting can also be counteracted by *change of environment.* For instance, pitting of copper and copper alloys in hot water can be prevented by water treatment, implying pH > 8.0 and $[HCO_3^-]/[SO_4^{2-}] > 1$ [7.29].

7.6 Intergranular Corrosion

7.7.1 General Characteristics, Causes and Occurrence

Intergranular corrosion is localized attack on or at grain boundaries with insignificant corrosion on other parts of the surface. The attacks propagate into the material. This is a dangerous form of corrosion because the cohesive forces between the grains may be too small to withstand tensile stresses; the toughness of the material is seriously reduced at a relatively early stage, and fracture can occur without warning. Grains may fall out, leaving pits or grooves, but this may not be particularly important.

The general cause of intergranular corrosion is the presence of galvanic elements due to differences in concentration of impurities or alloying elements between the material in or at the grain boundaries and the interior of the grains:

a) Impurities segregated to the grain boundaries (causing, e.g. the AlFe secondary phase in aluminium).

b) Larger amount of a dissolved alloying element at the grain boundaries (e.g. Zn in brass).
c) Smaller amount of a dissolved alloying element at the grain boundaries (e.g. Cr in stainless steel).

In most cases there is a zone of less noble material in/at the grain boundaries, which acts as an anode, while the other parts of the surface form the cathode. The area ratio between the cathode and the anode is very large, and the corrosion intensity can therefore be high.

In some cases, precipitates at the grain boundaries may be more noble than the bulk material; these precipitates will stimulate grain boundary attacks by acting as efficient local cathodes (e.g. $CuAl_2$ in aluminium alloys).

Intergranular corrosion occurs in stainless steels and alloys based on nickel, aluminium, magnesium, copper and cast zinc. In the following sections we shall look at the three former groups in some detail.

7.7.2 Austenitic Stainless Steels

The most familiar example of intergranular corrosion is connected with certain austenitic stainless steels, particularly 18–8 CrNi steels with 0.06–0.08% C. After cooling the steel from high temperature, Cr and C are in solid solution in the austenite. At room temperature the austenitic phase is supersaturated with these elements, and when the material is heated again to 500–800°C, a chromium carbide, $Cr_{23}C_6$ is precipitated at the grain boundaries. Cr diffuses relatively slowly, and therefore the amount of Cr forming carbide is taken from the very nearest region. Here, the Cr concentration gradient becomes large, as shown in Figure 7.34a. The carbon content in the steel is more than two orders of magnitude less than the chromium content, but the carbide precipitation is possible because C diffuses much more easily than Cr and is therefore taken from larger parts of the grains.

The corrosion properties depend in principle on the Cr content, as illustrated in Figure 7.34b. The curves can be understood in two ways: 1) As pure overvoltage curves, with the simplification that the passive current density on regions other than the grain boundaries is disregarded. 2) The cathodic curve is understood as the sum of the cathodic polarization curve of the cathodic areas on the surface and the overvoltage curve of a possible cathodic reaction (hydrogen evolution reaction) within the grooves or cracks between the grains.

The important relationship is that the critical current density (maximum current density on the anodic curve) increases with decreasing Cr content. When this content decreases to a critical value, C_{cr}, corrosion is initiated. At the grain boundaries a region with active material is produced on both sides of the carbide precipitate film, as illustrated in Figure 7.34a.

As shown in Figure 7.34b, the critical concentration C_{Cr} depends on the cathodic curve, i.e. on the surface state of the alloy, the concentration of oxygen and other oxidizers as well as the pH of the environment, the temperature and possibly the flow conditions (compare with Section 6.1). Serious material failure may occur after different times of exposure, depending on the environment. In seawater this may happen after a few weeks or months [7.30].

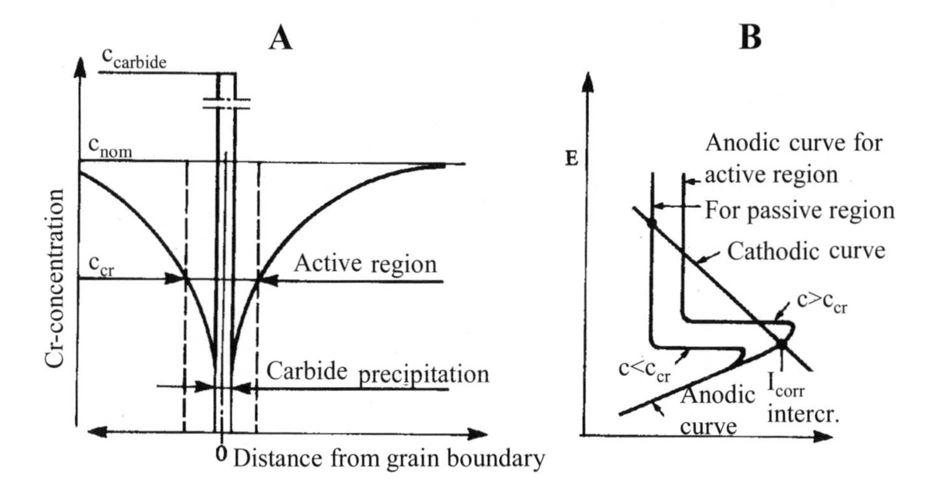

Figure 7.34 a) Schematic illustration of chromium depletion at a grain boundary due to carbide precipitation during annealing of a CrNi steel at 500–800°C.
b) Potential–log current diagram for the galvanic element constituted by the grain boundary region and the other parts of the surface, depending on the concentration of Cr.

A well-known problem is intergranular corrosion of welded stainless steel. The reason for this problem is that welding takes the temperature up to the range where carbide precipitation occurs. The area that is most liable to intergranular corrosion is usually located some millimetres from the fusion zone of the weld, where the temperature is kept within the critical range for the longest time intervals.

For thin plates (thickness < 3 mm) electrically welded with a single pass, the problem is avoided because of the short time at critical temperatures. This makes electric arc welding favourable compared with gas welding [7.1].

Intergranular corrosion of stainless steel does not occur in all environments. Thus inferior corrosion resistance at the grain boundaries does not necessarily lead to practical problems. However, a reasonable aim is to obtain corrosion properties at the welds equal to those elsewhere on the steel surface, in order to utilize the material as much as possible.

The risk of intergranular corrosion of austenitic stainless steel can usually be eliminated by:

a) Annealing at about 1100°C (at which the carbide is dissolved) with subsequent rapid cooling.
b) Alloying with Nb or Ti, which have a stronger affinity than Cr to C.
c) Reduction of carbon content (%C < 0.03) (Figure 7.35).

Figure 7.35 Elimination of intergranular corrosion by using AISI 304L steel (%C < 0.03) instead of AISI 304 (%C = 0.06–0.08). Vertical grooves are due to a weld on the back side of the plate. (From Fontana and Greene [7.1].) Reproduced with permission of The McGraw-Hill Companies.

A special form of intergranular corrosion is knife-line attack, which can occur in Nb/Ti-stabilized austenitic stainless steels after incorrect heat treatment [7.1].

ASTM A-262 [7.21] has standardized five different methods for testing intergranular corrosion of austenitic stainless steels, after Streicher (two), Huey, Warren and Strauss. The Warren test, using exposure to nitric acid–fluoric acid, is particularly suited for molybdenium-alloyed steels.

7.7.3 Ferritic Stainless Steels [7.30, 7.31]

These steels can also be liable to intergranular corrosion. In addition to Cr carbides, Cr nitrides may be precipitated and thus contribute to the depletion of Cr at the grain boundaries. The mechanism is in principle the same as for austenitic steels, but in the ferritic steels the attacks can occur for a wider spectrum of environments and lower contents of C + N. The critical range of temperatures that makes the metal sensitive is higher, namely above 925°C, and the attack occurs therefore close to the weld or in the weld metal itself. The material becomes resistant again after annealing for 10–60 minutes at 650–815°C. It is noted that the effects of the different temperature ranges are opposite to what is the case for the austenitic steels.

Prevention of intergranular corrosion of ferritic stainless steels is done by the same methods as for austenitic steels, with the differences in annealing temperature and maximum content of C + N as mentioned above. The necessary content of Nb or Ti for stabilization of ferritic steels is determined by the content of C + N.

7.7.4 Ni-based Alloys [7.32]

Nickel alloys may be attacked by intergranular corrosion in certain very aggressive environments after incorrect heat treatment. In NiCr alloys, chromium carbide is precipitated in the same temperature range as for the austenitic stainless steels. The NiCr alloys are primarily attacked by strong oxidizers such as hot nitric acid. The prevention measures are mainly the same as for the austenitic stainless steels.

7.7.5 Aluminium Alloys [7.27]

In some aluminium alloys, anodic phases may be precipitated at the grain boundaries, which may lead to intergranular corrosion. The actual phases are Mg_5Al_8 in AlMg alloys containing more than 3% Mg and $MgZn_2$ in AlZnMg alloys. A well-known alloy is AlMg4.5Mn (with about 4.5% Mg and 0.7% Mn). Provided that there is appropriate *production and forming* and application at ambient temperature, this and similar alloys are resistant in natural environments, and they are very suitable under marine conditions. An AlZnMgCu alloy like AA-7075 is liable to intergranular corrosion under such conditions.

Cathodic precipitates are formed in Cu-alloyed aluminium, which will corrode intergranularly in polluted industrial atmospheres, in severe marine atmospheres, and in seawater.

AlMgSi alloys, where the Mg and Si contents are balanced to a ratio giving Mg_2Si (e.g. approximately 1% Mg and 0.5% Si) are less susceptible to intergranular corrosion. Their normal corrosion form may be called pitting with an intergranular tendency. Excessive Si content gives increased liability to intergranular attack.

A special form of corrosion in aluminium alloys is exfoliation corrosion, which usually propagates in grain boundaries parallel to the surface. Exfoliation corrosion is most common in AlCuMg alloys, but it has also been observed in AlMg, AlZnMg and AlMgSi alloys. Standard testing is carried out according to ASTM G-34 and G-66 [7.21]. ASTM G-34 seems to distinguish most clearly between the various alloys with respect to the tendency to different corrosion forms. Both exfoliation corrosion and other types of intergranular corrosion are efficiently prevented with a coating of a more resistant aluminium alloy or commercially pure aluminium (cladding, alclad). Such coatings are most often applied by rolling, but thermal spraying has also been used.

7.8 Selective Corrosion (Selective Leaching)

This form of corrosion is observed in alloys in which one element is clearly less noble than the other(s). The corrosion mechanism implies that the less noble element is removed from the material. A porous material with very low strength and ductility is the result. Regions that are selectively corroded are sometimes covered with corrosion products or other deposits, and since the component keeps exactly the original shape, the attacks may be difficult to discover. Serious material failure may therefore occur without warning.

The most common example of selective corrosion is dezincification of brass, in which zinc is removed from the alloy and copper remains. After cleaning the surface, dezincification is easy to demonstrate because the Zn–depleted regions have a characteristically red copper colour in contrast to the original yellow brass. Dezincification occurs in two forms (see Figure 7.36):

a) Uniform (layer) dezincification, where the front of the selective attack moves more or less uniformly through the material. This form is favoured by high Zn content and acidic solutions.

b) Localized (plug-type) dezincification, where the localization and form of attack are determined more or less arbitrarily. This type occurs more often in alloys with somewhat lower Zn content and in neutral, alkaline and slightly acidic solutions.

The liability to dezincification increases with increasing Zn content and temperature (Figure 7.37) and is promoted by a stagnant solution, porous surface layers and oxygen in the corrosive medium (but oxygen is not strictly necessary).

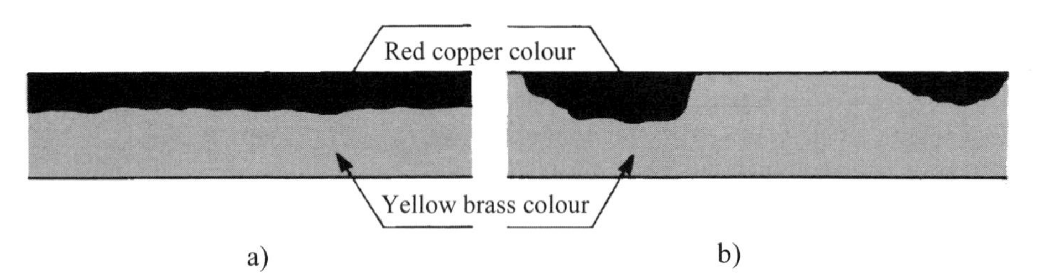

Figure 7.36 a) Uniform (layer) dezincification and b) localized (plug–type) dezincification of brass.

There has been some doubt about the mechanism, but the following is commonly accepted: brass dissolves, subsequently zinc ions stay in solution while copper is deposited electrolytically and forms a porous material with poor strength and ductility.

Dezincification can be reduced or prevented by:

a) Removing oxygen from the solution.
b) Cathodic protection.
(For economical reasons the use of a) and b) has been limited.)
c) Appropriate selection of alloy.

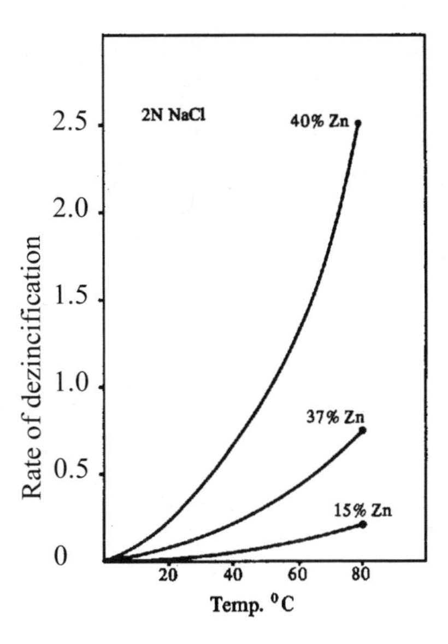

Figure 7.37 Dezincification rate as a function of temperature for three different brasses. (From Reference [7.1].)

Brass with 15% Zn is almost immune in many environments. In addition, alloys that are stabilized with about 1% Sn or of the order of 0.01% of As, Sb, or Pb exist [7.30]. Cu–alloys with better resistance to selective corrosion can also be chosen (copper–nickel, (nickel)–aluminium bronze, or other bronzes).

Regarding aluminium bronzes, it should be noticed that these as well may suffer from selective corrosion. Dealuminization may occur in acids and strongly polluted seawater. In seawater, rapid selective corrosion can occur in aluminium bronzes and nickel–aluminium bronzes, particularly under conditions where hydrogen sulphide is formed. One example of this is found in a water jet engine made of nickel–aluminium bronze (10% Al, 5% Ni, 4% Fe, 1% Mn, balance Cu), which was used in a high-speed vessel [7.33]. After about one year in the sea, for most of the time at the quay, many areas of attack were discovered. Visual inspection showed a clear difference between the surface areas that had been immersed only when the vessel was in service, where no attack was found, and the regions that were immersed also when the boat was at the quay, on which larger or smaller attacked areas were detected. Therefore it was concluded that the areas of attack were developed when the vessel was stationary at the quay, and they were explained by the presence of strongly polluted seawater. Another example of selective corrosion in nickel–aluminium bronze is shown in Figure 7.38.

Selective corrosion occurs also in silicon bronze, by which silicon is removed from the material.

An example of selective attack in a different group of material is the graphitization of grey cast iron, where the iron is corroded away and graphite is left on the surface.

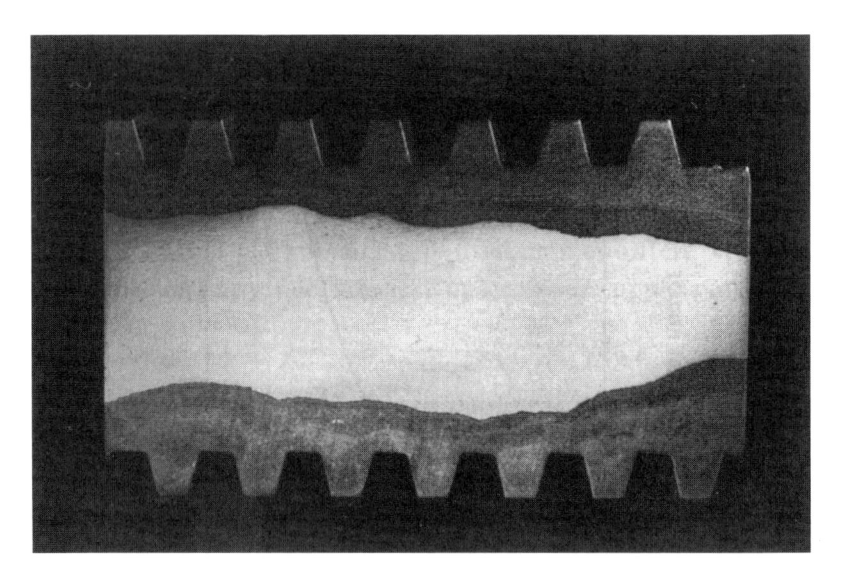

Figure 7.38 Uniform selective corrosion in a spindle made of Ni–Al bronze used in a valve carrying seawater.

7.9 Erosion and Abrasion Corrosion

7.9.1 Characteristic Features and Occurrence

When there is a relative movement between a corrosive fluid and a metallic material immersed in it, the material surface is in many cases exposed to mechanical wear effects leading to increased corrosion, which we usually call erosion corrosion.

The mechanism is that deposits of corrosion products, or salts precipitated because of the corrosion process, are worn off, dissolved or prevented from being formed, so that the material surface becomes metallically clean and therefore more active. In extreme cases, erosion corrosion may be accompanied by pure mechanical erosion, by which solid particles in the fluid may tear out particles from the material itself and cause plastic deformation, which may make the metal even more active.

The results of erosion corrosion are grooves or pits with a pattern determined by the flow direction and the local flow conditions (Figures 7.39 a,b and 7.44).

Reasonably, the corrosion form is typical at relatively high velocities between the material surface and the fluid, and it is particularly intensive in cases of two-phase or multiphase flow, i.e. liquid–gas and liquid–solid particle flow. Components often liable to erosion corrosion are propellers, pumps, turbine parts, valves, heat exchanger tubes, nozzles, bends, and equipment exposed to liquid sputter or jets. Most sensitive materials are those normally protected by corrosion products with inferior strength and adhesion to the substrate, e.g. lead, copper and its alloys, steel, and under some conditions aluminium/aluminium alloys. Stainless steel, titanium and nickel alloys are much more resistant because of a passive surface film with high strength and adhesion.

In some cases, the wear effects are more abrasive than erosive. Erosion is characterized by impingement, which implies a finite angle of impact relative to the material surface. If, on the other hand, wearing particles are mowed in contact with and parallel to the material surface, we have a case of abrasion, and a more correct name of the corresponding corrosion form would be abrasion corrosion. This form will reasonably also comprise cases where corrosion products between two components are removed by relative movement of the components.

7.9.2 Types and Mechanisms

Leaving cavitation corrosion and fretting aside as separate corrosion forms (see Sections 7.10 and 7.11), erosion and abrasion corrosion can be divided into three types, a), b) and c), as described below. The first two types are erosion corrosion, while type c) is to be considered as abrasion corrosion. The three types may overlap each other and partly occur simultaneously in the same system.

a) Impingement corrosion, often occurring in systems with two-phase or multiphase flow, particularly where the flow is forced to change direction. Numerous impacts from liquid drops in a gas stream, or particles or gas bubbles in a liquid flow lead to pits with a direction pattern as shown in Figure 7.39a. In cases with solid particles, the situation can be illustrated as in Figure 7.41, where corrosion products are removed and the surface locally activated.

b) Turbulence corrosion, which occurs in areas with particularly strong turbulence such as the inlet end of heat exchanger tubes (Figure 7.39b).

c) Increased corrosion due to removal of corrosion products by wear due to particles moving along and in contact with the corroding surface, or by wear between components in moving contacts with each other.

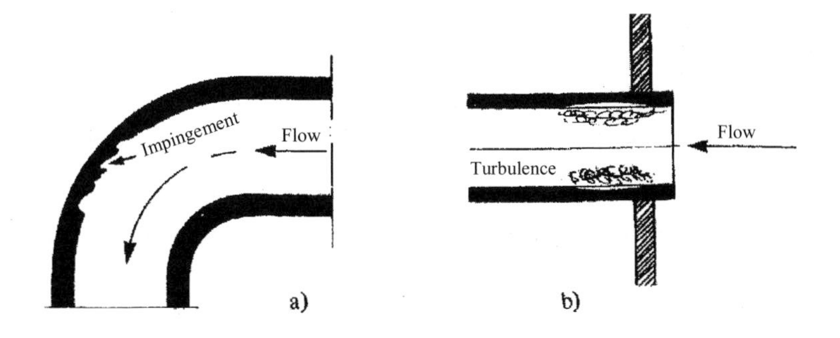

Figure 7.39 a) Impingement, and b) turbulence corrosion.

In addition to the general mechanisms of erosion corrosion described in this section, there are also some special mechanisms occurring under certain conditions of materials and environments (see next section).

As mentioned, the removal of corrosion products may occur by impacts, where we have a force component normal to the material surface. In some cases deposits may also be removed by high shear stresses (force components parallel to the surface). The shear stresses may vary considerably as a consequence of flow fluctuations or repeated impacts. Therefore, it is possible that deposits of corrosion products are destroyed either by extreme values of shear stress or by fatigue.

The effects of high flow velocity and erosion on the reaction kinetics are in principle described in Figure 7.40. Both the effect on the cathodic curve (increased transport of oxygen to the surface) and on the anodic curve (increased activation of the metal in the corrosion range of potential) are shown. The indicated passivation (case d) is possible only under certain conditions, like for instance on steel in fresh water without particles and at very high velocity, as frequently occurs in sections of water turbines.

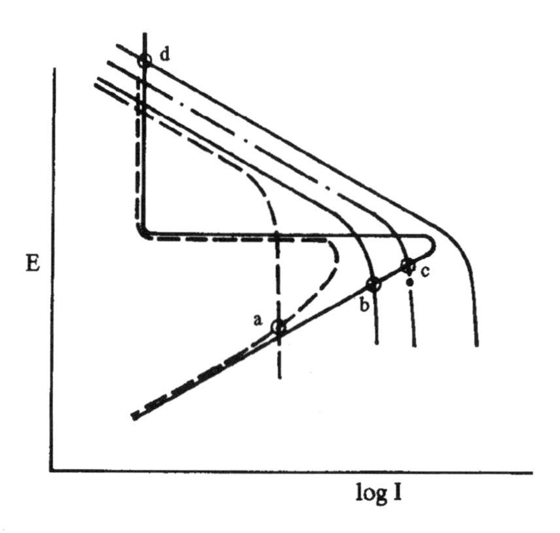

Figure 7.40 Overvoltage curves at a) low velocity and formation of a layer of corrosion product on the surface (----), b) high flow velocity and removed corrosion product, c) increased cathodic reaction as a result of galvanic contact with a more noble metal, and d) flow velocity high enough to cause passivity.

7.9.3 Erosion and Erosion Corrosion in Liquid Flow with Solid Particles

As mentioned previously, erosion corrosion will be particularly intensive when the flowing medium contains solid particles. At relatively low velocities, we may have the situation illustrated in Figure 7.41. The particles hit the surface with a velocity v at an angle β. Corrosion products are removed from the surface, which is activated, and the corrosion rate on the activated spots may increase strongly. If only corrosion products are removed, all the material loss can (under stable conditions) be measured as electrochemical corrosion, because the electrochemical reaction is necessary for making new corrosion products to compensate for the removed ones.

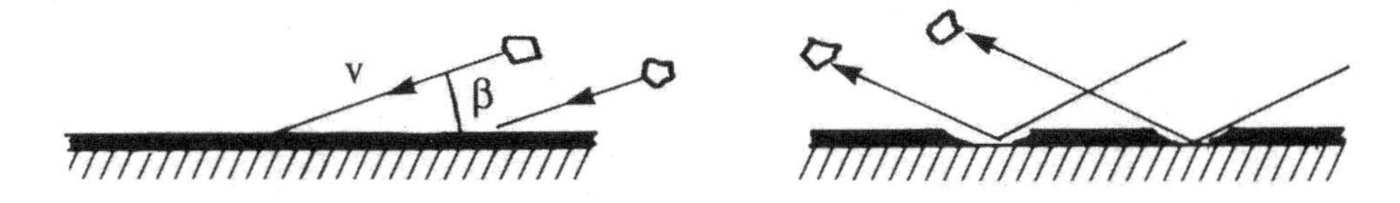

Figure 7.41 Impacts from solid particles in a liquid flow causing removal of corrosion products from the surface (erosion corrosion).

With increasing velocity the impact energy increases as well, and above a certain level not only the corrosion products, but also small particles of the material itself will be removed. This contribution to the material deterioration is a pure mechanical process, i.e. pure erosion. Mechanisms and factors that influence pure erosion are described in, e.g., References [7.34]. In strongly erosive environments (high particle concentration and/or high flow velocity), erosion is usually the dominating deterioration process, while the corrosion contribution forms a smaller proportion (although the absolute corrosion rate may be high). The pure erosion rate can be expressed by the following formula:

$$W\ (\text{mm/year}) = K_{mat} \times K_{env} \times c \times v^n \times f(\beta), \qquad (7.8)$$

where K_{mat} is a material factor depending in a complex manner on (among other properties) hardness and ductility of the substrate; K_{env} is an environmental factor that includes the effects of size, shape (sharpness), density and hardness of the particles; c is the concentration of particles, n is the so–called velocity exponent, v is the particle velocity, and β the impact angle shown in Figure 7.41.

For a certain geometrical element it is convenient to replace $f(\beta)$ by a geometrical function, which, e.g. for pipe bends includes pipe diameter and bend radius. Such functions as well as empirical values of the constants in Equation (7.8) have been determined for fluids containing silica sand [7.35].

Equation (7.8) is very useful in connection with erosion testing in the laboratory. The proportionality with the particle concentration gives a unique possibility for realistic acceleration of the tests by using a larger sand concentration than that existing under service conditions. It also makes it easy to transfer the results to real sand concentrations. In this way, the testing time can be strongly reduced.

For pure particle erosion, a value of n of the order of 3 is frequently found. This can be explained by a consideration of the kinetic energy of the particles hitting the surface. The total impact energy per area unit and time unit is $\frac{1}{2} mv^2$. m is the sum of the particle mass per area and time unit. At constant concentration c, m is proportional to v, and hence, the impact energy is proportional to v^3. The effect of the impact angle β is very strong, but quite different for ductile and brittle base materials, as shown schematically in Figure 7.42.

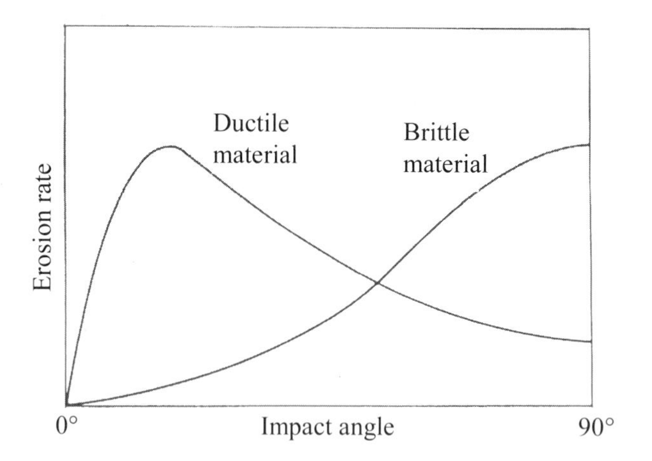

Figure 7.42 Relative erosion rate as a function of impact angle, for ductile and brittle materials, respectively (airborne particles). (After Ives and Ruff [7.36].)

The figure is based on experiments with an air jet carrying sharp particles. For *liquid* flow with sharp particles, the top of the curve for ductile materials is less marked and usually located at a somewhat higher angle (e.g. 30–40°). Otherwise, the difference between ductile and brittle materials is similar for the latter case also. The curves can be explained by different erosion mechanisms for ductile and brittle materials. The large difference between materials as to the effect of the impact angle is important for appropriate materials selection under different flow and geometrical conditions.

Equation (7.8) is also of interest when erosion is combined with corrosion. It can be used in modified form for limited ranges of the involved parameters. The limitation is particularly caused by the fact that the velocity exponent for erosion corrosion is much less than that for pure erosion ($n_{corr} \leq 1.5$, usually $n_{corr} \leq 1.0$) [7.37]. In addition, erosion corrosion increases less than proportionally with particle concentration. Together these relationships mean that for combined erosion and corrosion (corrosive–erosive wear) $W \propto c^x v^y$, where $x < 1$ and $1 < y < 3$, and both x and y depend on the relative contribution of erosion and corrosion (they increase with increasing erosivity and decreasing corrosivity).

Because of these relationships, we need to be very careful when using Equation (7.8) on combined erosion and corrosion. To make such calculations reasonably accurate they must be combined with and based upon several experiments under relevant conditions.

More thorough analyses and mechanism studies show that there often is a considerable synergy effect of erosion and corrosion. Generally, the total material loss rate W_T for such material deterioration can be expressed by

$$W_T = W_E + W_C + W_{EC} + W_{CE}. \tag{7.9}$$

W_E is the pure erosion rate, i.e. mass loss rate when corrosion is eliminated, and W_c is the corrosion rate in the absence of sand erosion. W_{EC} and W_{CE} are both synergy effects: W_{EC} is the increase in erosion rate due to corrosion and W_{CE} is the increase in corrosion rate due to erosion.

It is possible to determine the four contributions to the total material loss rate by the following experimental principles: the total material loss rate W_T is determined by weighing the specimen before and after exposure under combined erosive and corrosive conditions. The sum of W_C and W_{CE} (the corrosion components) can be measured by electrochemical methods during the same exposure (the methods described in Section 9.2 can also be used under erosive conditions). W_E is determined by weighing the specimen before and after exposure in special tests where corrosion is eliminated by cathodic protection (or possibly by other means) but otherwise under the same conditions as in the former experiments. W_C can be measured electrochemically in tests like the original ones but with all solid particles excluded. Finally, the synergy components, W_{CE} and W_{EC}, can be derived from Equation (7.9) and the mentioned experiments.

In several cases, materials for combined erosive and corrosive conditions have been evaluated on the basis of separate erosion and corrosion studies and data, with the consequence that the synergistic effects are left out of the evaluation. Since one or the other of these effects may be large, the conclusions may be quite wrong. For materials that usually are passive due to a dense oxide film, such as stainless steels, W_C is by definition very low. But since sand erosion more or less destroys the passive film, the corrosion rate increases strongly and may reach very high values, i.e. the contribution of W_{CE} may be particularly high for these materials. The other synergy effect, W_{EC}, is most pronounced for ceramic–metallic materials in which the metallic phase has inferior corrosion resistance, e.g. for a cemented carbide with a metallic phase of cobalt (WC–Co).

The deterioration mechanism for WC–Co (and similar materials) is assumed to be as follows: the binder phase of metal around the carbide particles corrodes away so that the WC particles are more easily removed by erosion. The transition from the unexposed state to a corroded and eroded state is schematically illustrated in Figure 7.43. It should be noticed that under extremely erosive conditions or if the metal phase is highly corrosion resistant in the actual environment, erosion is dominating and the effect of the shown mechanism may be insignificant.

These relationships and mechanisms of combined erosion and corrosion on different types of material under different conditions are illustrated by experimental results available in various publications, e.g. in References [7.38, 7.39], which also describe a method implying electrochemical determination of corrosion rates simultaneously on a number of specimens.

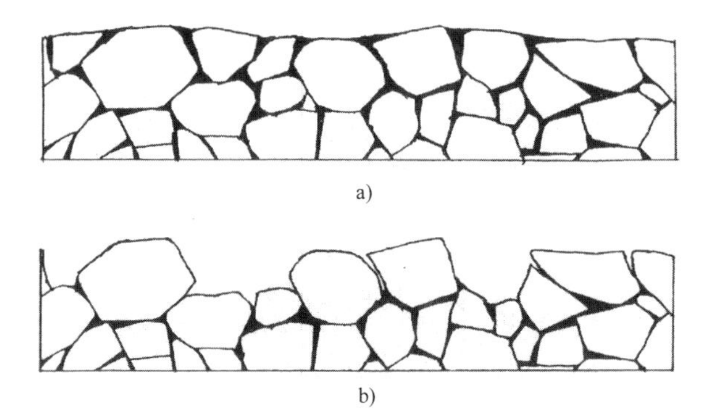

Figure 7.43 Schematic view of a cemented carbide material a) before testing, and b) after exposure under corrosive and low-erosive conditions.

7.9.4 Influencing Factors and Conditions in Liquids and Liquid–Gas Mixtures

Some of the factors discussed in Section 7.9.3 are also important in cases without solid particles present. Local *geometry* and *irregularities* on the surface affect the local flow pattern and are therefore of great significance for erosion corrosion. *Flow velocity* affects corrosion in different ways, e.g. as shown in Table 7.5 (see further discussion below). In some cases, increased flow velocities may reduce the corrosion rate by removing deposits and improving transport of oxygen, thereby promoting passivation, or by improving transport of inhibitors to the metal surface. But in most cases, increased flow velocity will increase the corrosion rate as described in Sections 6.2.4, 7.9.1, 7.9.2 and 7.9.3.

The *materials properties* most important in relation to erosion corrosion are thermodynamic nobleness, the ability to form mechanically stable and protecting surface films, and the liability to rapid passivation at the start of exposure as well as rapid repassivation after removal of surface films. The materials listed in Table 7.5 can roughly be divided into three groups based on their behaviour at the three velocities represented in the table:

a) Carbon steel and iron are active in the whole actual range of flow rates. The corrosion rate increases steadily with increasing flow velocity, mainly due to more efficient supply of oxygen. There are some corrosion products (rust mixed with calcium carbonate) intact on the steel surface at all three velocities, which limit the corrosion rate to some extent. The critical velocity for complete removal of deposits on steel depends for one thing on the relative amount of rust, as shown in some experiments in connection with cathodic protection.
b) All the copper alloys are protected reasonably well by surface oxides and hydroxides at the two lower velocities. When the velocity exceeds a critical value, the protective layers are dissolved. The critical velocity for erosion

corrosion of most copper alloys is in the range 1–5 m/s (see next section). Special grades of copper–nickel and nickel–aluminium bronze can withstand higher flow rates.

c) Stainless steel, nickel alloys (Monel and Hastelloy C) and titanium are satisfactorily passive at all the three velocities. (It should however be emphasized that most stainless steels are liable to crevice corrosion in seawater.)

Table 7.5 Corrosion rates in seawater at different flow velocities (data from International Nickel Co). (Adapted from Fontana and Greene [7.1].)

	Tyical corrosion rates mg/dm^2/day (mdd[1)])		
Material	0.3 m/s [2)]	1.2 m/s [3)]	8.2 m/s [4)]
Carbon steel	34	72	254
Cast iron	45	–	270
Silicon bronze	1	2	343
Admiralty brass	2	20	170
Hydraulic bronze	4	1	339
G bronze	7	2	280
Al bronze (10% Al)	5	–	236
Aluminium brass	2	–	105
9–10 CuNi (0.8% Fe)	5	–	99
70–30 CuNi (0.05% Fe)	2	–	199
70–30 CuNi (0.5% Fe)	<1	<1	39
Monel	<1	<1	4
Stainless steel type 316	1	0	<1
Hastelloy C	<1	–	3
Titanium	0	–	0

[1]For the given materials except for titanium, 1 mdd corresponds to 4–5 × 10^{-3} mm/y, i.e. the values about 200–300 mdd in the right column equal about 1–1.5 mm/y.
[2]Immersed in tidal current.
[3]Immersed in seawater flume.
[4]Attached to immersed rotating disk.

A special example is described by Fontana and Greene [7.1]: The erosion corrosion rate on carbon steel in water may be higher at pH = 7–9 than at pH = 6 as well as at pH = 10. The reason is assumed to be that the oxide layer at pH =7–9 consists of granular Fe_3O_4, which is less erosion resistant than the $Fe(OH)_2$ or $Fe(OH)_3$ dominating at the lower and higher pH, respectively. Increased *content of oxidizers* may also decrease or increase erosion corrosion rate depending on whether passivation is obtained or not. The *temperature* is another important factor. It may affect the properties of the deposits in addition to the shear stress and mass transport (see next section and Section 6.2.2).

Galvanic coupling to a more noble material may give an increased corrosion rate, as illustrated in Figure 7.40.

Unfortunate effects of *two-phase or multiphase mixtures* have been mentioned. In two-phase flow of gas and liquid, slugs may be formed moving at high velocity through the pipeline. These slugs have strong erosion effects on corrosion products, and may cause pronounced erosion corrosion, e.g. in flow lines for oil and gas with precipitated water. Without erosion, protecting films, to a large extent consisting of $FeCO_3$, are formed at higher temperatures (see Section 6.5). When these are destroyed by erosion, the corrosion rate increases strongly.

An example of corrosion in a 22-inch (56 cm) flow line for oil and gas at a field in the North Sea is shown in Figure 7.44. During a service period of three years attacks as deep as 6.5 mm have developed. The temperature and the pressure (at least in periods) have been 75–80°C and 60–80 bar. Inhibitor treatment was carried out during the last half year before the actual pipe section was removed. The flow velocity has been 4–5 m/s most of the time. The flow conditions have not been extreme, but the shape and position of the attacks show that they are flow-affected. It is assumed that corrosion can be prevented if the temperature on the inner pipe wall is kept above the dewpoint of the water vapour in the gas, which may possibly be achieved by arrangements for reduction of the heat loss to the surroundings or change of the internal pressure and temperature. An alternative measure is to reduce the flow rate combined with appropriate addition of inhibitor.

Figure 7.45a and b shows an example of corrosion in a spiral casing in a water turbine. The flow rate was very high, about 20 m/s at the inlet to the the spiral casing. The casing had been painted, but the internal paint coating was thinner than specified, so that corrosion occurred and the paint film partly disappeared. In the area encompassed in the figure (a and b), the corrosion rate has been up to 3–4 mm/year. Conversely, on paint–free (rust–coloured) areas on the sides of the stay–vane there have not been similar large attacks. It is assumed that the steel in these regions has been passivated because of even higher flow velocities and more turbulence. This is a demonstration of the relationships shown in Figures 7.40, 6.9 and 6.10.

7.9.5 Critical Velocities

Table 7.5 does not give any clear information about critical velocities, but it indicates that such thresholds exist for the copper alloys in the velocity range represented in the table (1.2–8.2 m/s). More specifically, both Figure 7.46 and Table 7.6 show examples of critical velocities for erosion corrosion. The values are not absolute; they depend on the composition of the environment, the temperature, geometrical conditions, the exposure history, the exact composition and treatment of the material etc. In connection with Figure 7.46 it can be mentioned that austenitic stainless steels show excellent resistance to erosion corrosion in pure liquid flow at high velocities, while some ferritic [7.42] and ferritic–austenitic steels are attacked less than the austenitic ones if the liquid carries solid particles. The data in Table 7.6 originate from work by Efird [7.43], who interpret his results as follows: for each alloy in a certain environment, there exists a critical shear stress between the liquid and the material surface. When this shear stress is exceeded, surface films are removed and the corrosion rate increases markedly.

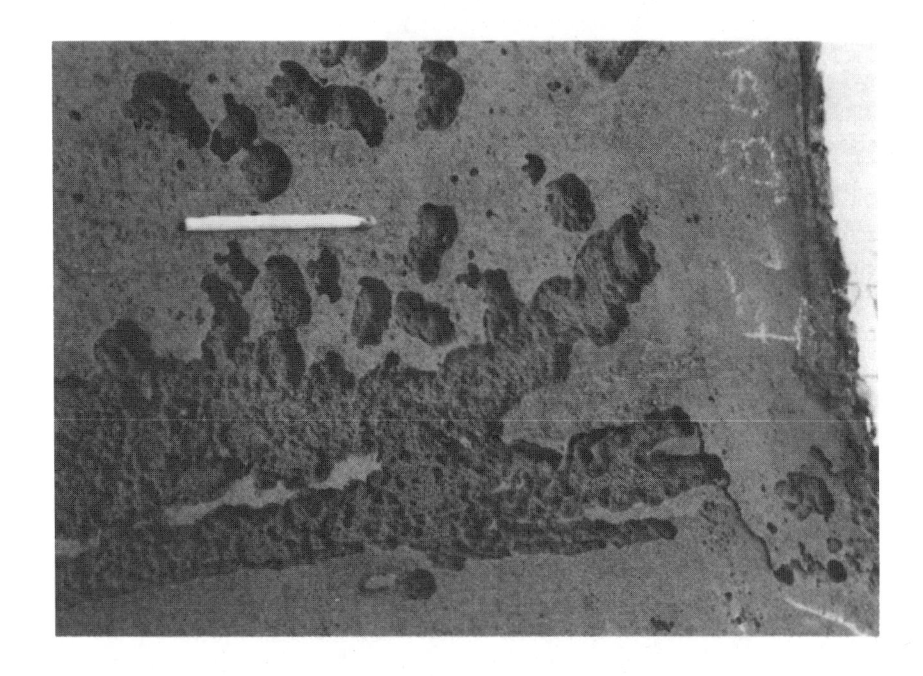

a)

b)

Figure 7.44 Flow-affected corrosion in a flowline for oil and gas in a field in the North Sea. a) A view of the pipe bottom. b) Cross-section of typical pits (2.5 ×). (Photo: J.M. Drugli, SINTEF Corrosion Centre.)

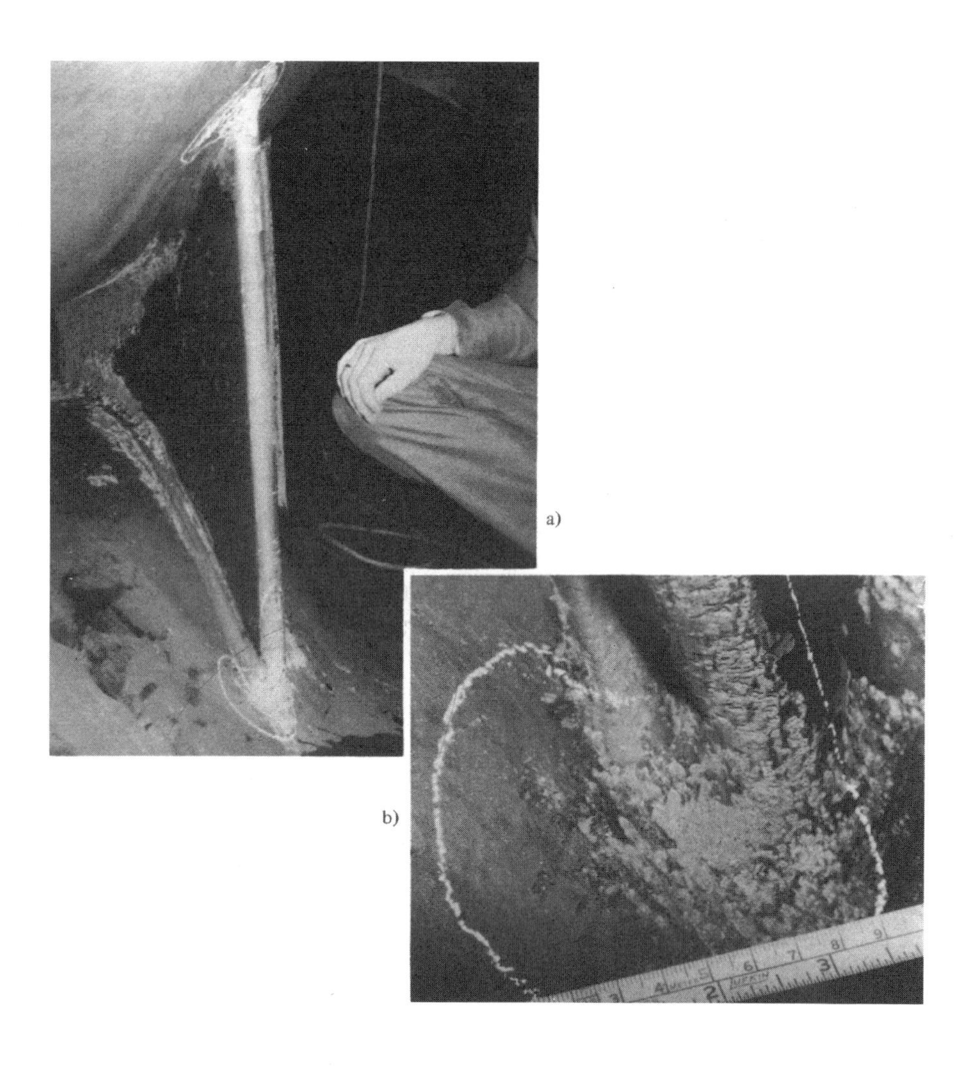

Figure 7.45 Corrosion attacks and partly damaged coating in a hydroturbine drum.
a) Overview. b) Close–up of marked area. (Photo: P.O. Gartland, SINTEF Corrosion Centre.)

Other researchers have stated that erosion corrosion of copper alloys is controlled by mass transport, i.e. diffusion of copper ions away from the metal surface [7.44]. Their opinion is that the shear stresses at the actual flow conditions are too low for tearing off particles from the surface films. A combination of chemical–electrochemical and mechanical effects has also been proposed.

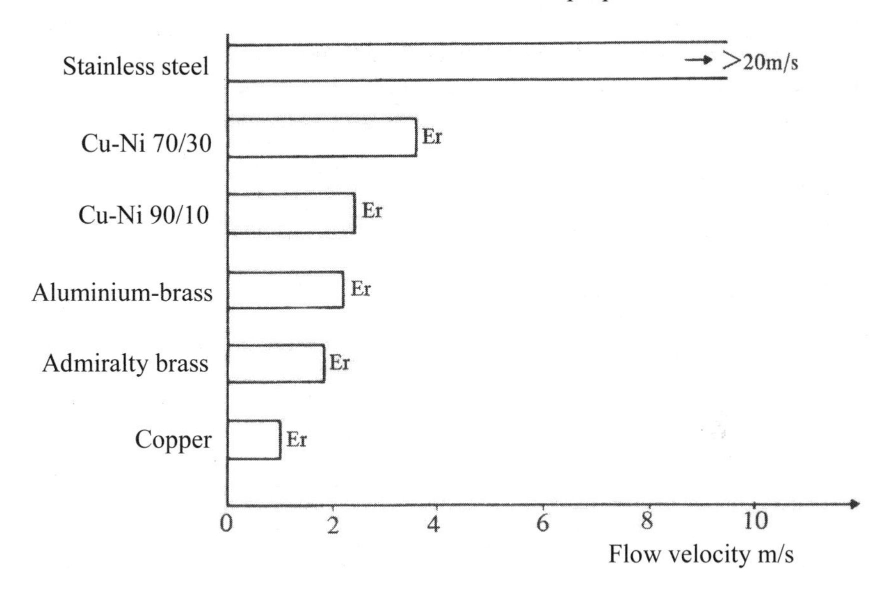

Figure 7.46 Critical velocities for erosion corrosion of different materials in seawater. (From Bernhardson et al. [7.41].)

Table 7.6 Critical velocities for copper and copper alloys in seawater. (Adapted from Efird [7.43].)

Metal/alloy	Critical flow velocity, m/s	Temperature, C°
99.9% Cu, deoxidized with P	1.3	17
Aluminium brass CuZn20Al2	2.2	12
Copper–nickel CuNi10Fe1Mn	4.5	27
Copper–nickel CuNi30Mn1Fe	4.1	12
Copper–nickel CuNi16Fe1Cr	12.0	27

7.9.6 Abrasion and Other Wear Processes Combined with Corrosion

Combination of abrasion and corrosion is also called *corrosive–abrasive wear.* Particle abrasion with low pressure between the particles and the material surface is closely related to particle erosion at low impact angles. Corrosive–abrasive wear has therefore several features in common with combined erosion and corrosion (corrosive–erosive wear). The corrosion mechanism and activation is based upon the abrasive removal of corrosion products, and this may cause large synergy effects. This is particularly the case at slow and moderate wear combined with rapid corrosion. On the other hand, rapid abrasion is not significantly accelerated by moderate or slow corrosion. In the latter case, the corrosion is slower than the wear of the base material, thus no corrosion product film is established. In agreement with these relationships we have the following rough guidelines for materials selection: for cases where slow abrasion is combined with corrosion, high-molecular polymers or elastomers may give lower wear rates than hard and corrodible materials. Conversely, strongly abrasive conditions should be met with hard materials with a corrosion resistance matching the corrosivity of the environment. It should be noticed that abrasion resistance is more closely related to hardness than is erosion resistance. These guidelines apply also to selection of coatings, for which good adhesion to the base material and sufficient thickness are other important properties. A method for studying corrosive-abrasive wear, applying electrochemical methods in a "pin-on-disc" test, is described in Reference [7.45].

Other types of *corrosive wear* can take place e.g. between machine parts or other structural components which are in sliding or rolling contact with each other and exposed to a corrosive environment [7.46]. The most common basic mechanism is, also in such cases, that corrosion products are worn off so that fresh and active metal surface is exposed and rapid corrosion takes place.

7.9.7 Preventive Measures

The most important measures to avoid or reduce the extent of erosion and abrasion corrosion are:

a) *Suitable selection of materials* (see Sections 7.9.1–7.9.6).
b) *Appropriate design.* Numerous examples of good design of flow systems and other equipment are given by Pludek [7.3]. Pipe dimensions and flow cross-sections are to be chosen with the aim of keeping the flow velocity below the critical level. It should be noticed that disturbing local geometrical elements still may cause local turbulence and subsequent local attack. Therefore, such elements should be avoided as far as possible, and any change of cross section or flow direction should be sufficiently smooth and streamlined (Figure 7.47a). Turbulence should be avoided,. particularly in the most critical parts of the system, e.g. by positioning bends sufficiently far from turbulence-producing elements (Figure 7.47b). If possible, regions where corrosion is difficult to avoid should be positioned such that the attack does not cause any harm, e.g. by extending outlet pipe ends into the tank, as shown in Figure 7.47c.
c) *Change of environment,* e.g. as shown in Figure 7.48: filtration or precipitation of solid particles (a), precipitation of water droplets from air, gases or steam (b), and removal of air or other gases from liquid flow (c). In all these cases it is important to have the separation as early in the process as possible.

Upstream to the separation point, the pipe diameter should be large (low velocity) or resistant/replaceable parts should be used. Furthermore, it is important to 1) reduce the temperature, and 2) keep the pressure high to avoid gas in liquid systems. Use of inhibitors and removal of air may be effective but not always economical methods [7.1].

d) *Application of corrosion-resistant coatings,* e.g. metallic coatings by welding, metallic, metallic–ceramic and ceramic coatings by thermal spraying, concrete or rubber/polymer coatings.

e) *Cathodic protection.* A problem with this is that very high current density is usually needed because the cathodic reaction occurs very efficiently under conditions of high flow velocities. Sacrificial anodes in the form of zinc or steel plates have been used to protect the inlet part of heat exchanger tubes. Sacrificial anodes may also be applied as plugs in pump housings and other components [7.1].

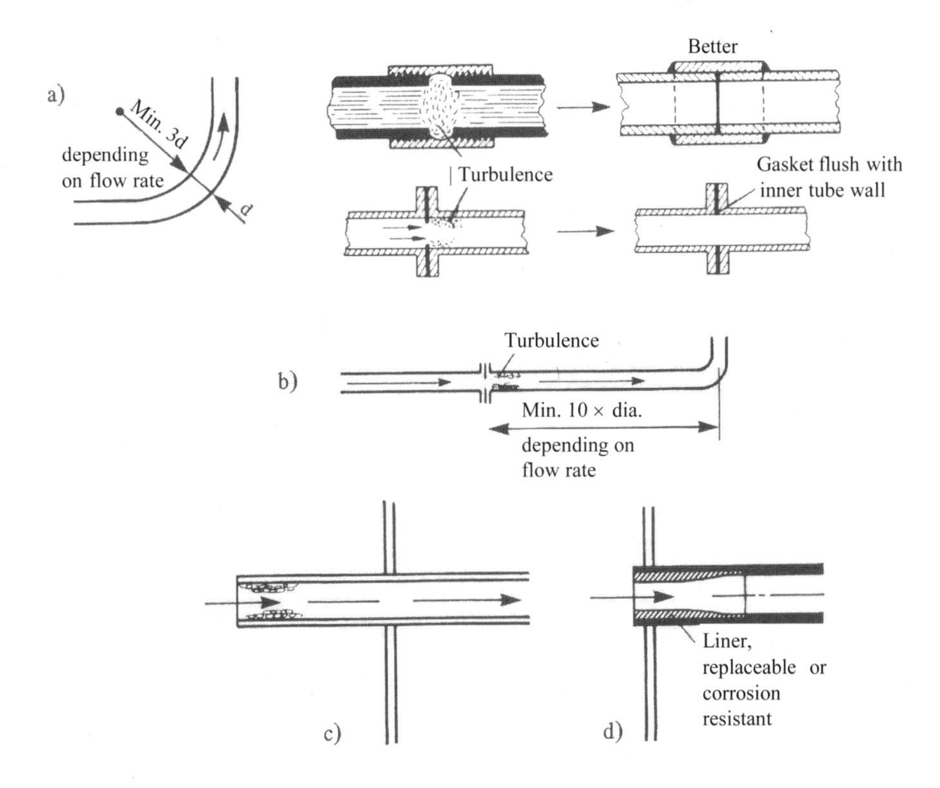

Figure 7.47 Prevention of erosion corrosion by design. (Partly reproduced from Pludek [7.3] with permission of Palgrave MacMillan.)

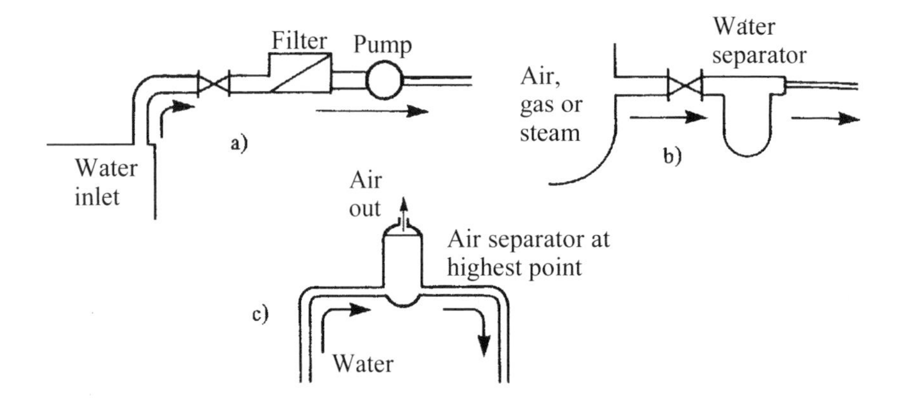

Figure 7.48 Prevention of erosion corrosion by separation of different phases.

7.10 Cavitation Corrosion

This corrosion form is closely related to erosion corrosion, but the appearance of the attack (Figures 7.49 and 7.50) differs from the erosion corrosion attacks described in the last section. While the latter has a pattern reflecting the flow direction, cavitation attacks are deep pits grown perpendicularly to the surface. The pits are often localized close to each other or grown together over smaller or larger areas, making a rough, spongy surface.

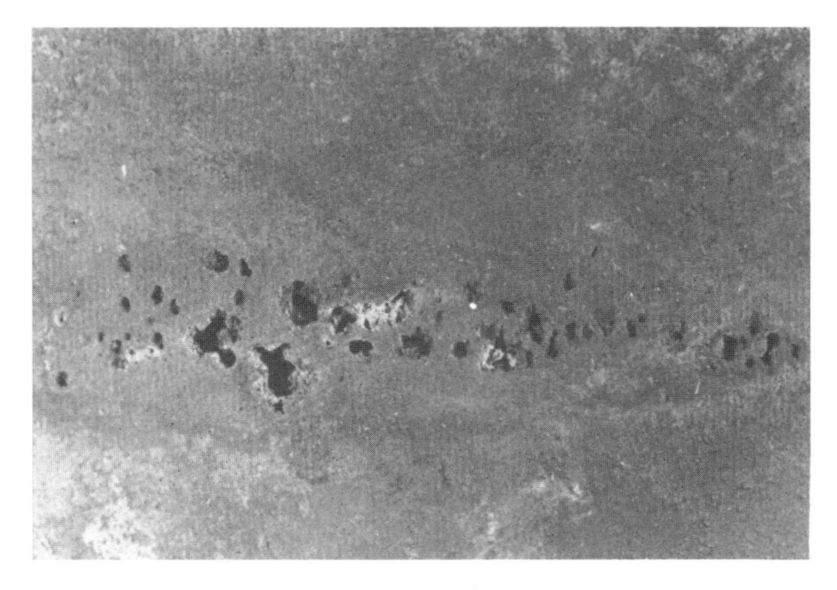

Figure 7.49 External cavitation corrosion on a cast iron cylinder lining in a diesel engine. 2×. (Photo: E. Abusland, Sintef Corrosion Centre.)

Cavitation corrosion has a special mechanism, which is also one of the reasons that it is considered as a separate corrosion form in the present book. In some other books it is considered as a subgroup of erosion corrosion.

Cavitation corrosion occurs at high flow velocities and fluid dynamic conditions causing large pressure variations, as often is the case for water turbines, propellers, pump rotors and the external surface of wet cylinder linings in diesel engines. Vapour bubbles formed in low-pressure zones, or at moments of low pressure at the actual positions, collapse very rapidly when they suddenly enter a high-pressure zone or high-pressure moment. When this happens close to the metal surface, the bubble collapse causes a concentrated and intense impact against the metal, with the induction of high local stress and possibly local plastic deformation of the material. Parts of any protecting film are removed. Repeated impacts may lead to microscopic fatigue and crack formation, and subsequent removal of particles from the material itself. During this process the material is strongly activated at the attacked points, and high local corrosion rates are possible. In some cases, however, the corrosion contribution is small and the deterioration form is more correctly called cavitation erosion.

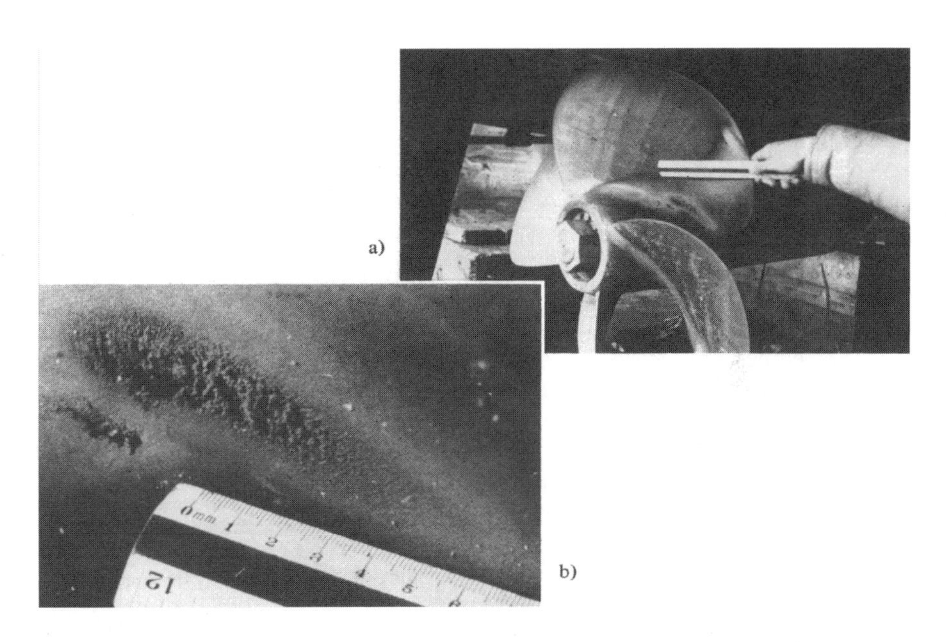

Figure 7.50 a) Cavitation corrosion on the propeller of a high-speed passenger ship. b) Close up of the damages. (The material is nickel–aluminium bronze.) (Photos: T.E. Hammervold and O Sætre.)

A standardized method for cavitation corrosion testing is described in ASTM G32 [7.21].

Both high corrosion resistance and high hardness improve the resistance to cavitation attacks. Relative resistance of various steels, stainless steels, irons, copper alloys and nickel alloys are reported in References [7.1]. Cavitation attacks may also be prevented by increasing the fluid pressure so that gas bubbles are not formed, by reducing surface roughness or by improving design and service conditions. Coatings of corrosion-resistant inorganic materials with high strength, or for some applications rubbers or polymers, may be used. Good adhesion of the coating to the base material is important. Cathodic polarization may counteract cavitation corrosion, for one thing by formation of hydrogen gas giving a cushion effect when the bubbles collapse, and thus reducing the impact pressure. A similar effect may be obtained by injecting air bubbles into the liquid. For cooling systems in engines, inhibitors, e.g. 2000 ppm sodium chromate, may be effective [7.30].

Suitable design is also important, e.g. to avoid vibration. Transfer of vibration can be reduced by vibration isolation, vibration absorbers, stiffeners etc. [7.3].

7.11 Fretting Corrosion (Fretting Oxidation) 7.30, 7.47, 7.48]

Fretting wear occurs at the interface between two closely fitting components when they are subject to repeated slight relative motion (slip). The relative motion may vary from less than a nanometre to several micrometres in amplitude. Vulnerable objects are shrink fits, press fits, bolted joints, and other assemblies where the interface is under load. A critical position is at the end of a collar, as shown in Figure 7.51. Sometimes the attack is serious, particularly because it may lead to macroscopic motion between the parts, or fatigue cracks may develop in the shaft.

The mechanical contribution to fretting damage may include elements of adhesive wear, microscopic fatigue crack development and delamination that result in removal of small particles from the metal lattice. The particles form a debris, which may partly adhere to the fretting surfaces and be trapped between these, and may partly escape from the fretting area.

The exact mechanism of combined fretting and oxidation is not fully understood. The following successive steps in the process have been proposed: the relative motion between the parts may promote oxidation of the surface, the oxide film is partly worn off, the fresh metal surface is highly active and oxidizes again, and this circular process is repeated. Another explanation is that, firstly, metal particles are released from the crystal structure by adhesive wear, microfatigue and delamination, then the particles oxidize, forming a debris consisting of brittle and friable oxide particles. This debris will to a considerable extent move out from the interfacial area between the components, and thus allow the deterioration process to continue at a high rate. Possible effects in this process are that the microfatigue contribution is accelerated by the environment, and that sharp oxide particles, which are much harder than the metal particles, cause some abrasive wear in spite of the small size of particles. The acceleration effect of the environment depends on the access of oxygen but not necessarily of moisture. The friction coefficient is affected by the kind of debris. On steel parts this is composed of α-Fe_2O_3 particles mixed with smaller amounts of iron powder. The visual results of fretting oxidation on steel interfaces are (in addition to the red–brown debris) pits or intrusions, which may lead to fatigue cracks. The attacked surfaces may also be discoloured.

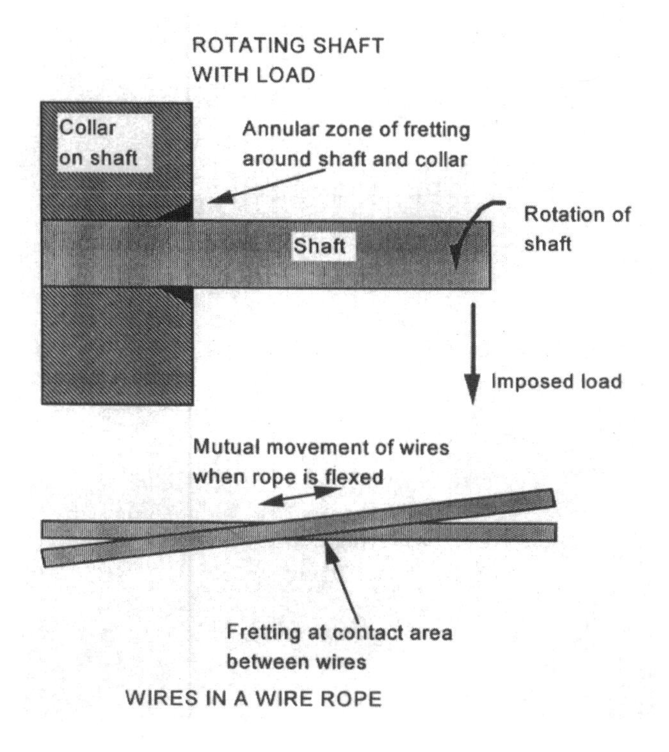

Figure 7.51 Location of fretting wear in some common engineering components. (From Batchelor et al. [7.46].)

According to References [7.30], the intensity of combined fretting and oxidation is increased by reduced (!) temperature and air humidity, and by increased pressure and slip. The material loss rate has been expressed by the equation

$$W = (k_0 L^{1/2} - k_1 L)\, C/f + k_2 l L C, \qquad (7.10)$$

where L is the load, C is the number of cycles, f is the frequency, l is the slip, and k_0, k_1, k_2 are constants. The two first addends on the right-hand side express the chemical contribution and the third one the mechanical effect.

Fretting oxidation is reduced or prevented by use of lubricants (e.g. low-iscosity oil, molybdenum sulphide), which hinder the access of oxygen and at the same time suppress adhesive wear. Other protection methods are to prevent oxygen access by use of gaskets or sealing materials, to change the mechanical parameters affecting the factors in Equation (7.10), and to use suitable materials in the components. One can use hard materials in both parts, alternatively a hard material in one part and a soft one in the other. The soft material may be a coating (e.g. of Sn, Pb, Ag) when both components are made of steel.

7.12 Stress Corrosion Cracking (SCC)

7.12.1 Characteristic Features and Occurrence [7.49]

Stress corrosion cracking can be defined as crack formation due to simultaneous effects of static tensile stresses and corrosion. The tensile stresses may originate from external load, centrifugal forces or temperature changes, or they may be internal stresses induced by cold working, welding or heat treatment. The cracks are mainly formed in planes normal to the tensile stresses, and propagate intergranularly or transgranularly, more or less branched. If they are not detected in time, they will cause fast, unstable fracture.

Macroscopically, the crack surfaces may look brittle and discoloured, dull or darkened by oxide layers. The initiation site may be discovered as a pit (formed by pitting or deposit corrosion) or, e.g. as a defect caused by forming or machining. On parts in service it may, however, be difficult to discover cracks, because they are narrow and filled or covered by corrosion products. As shown in the examples in Tables 7.1 and 7.2, SCC is a common and widespread corrosion form in the process industry.

Since SCC was first discovered and systematically recorded during the First World War, several surveys of risky combinations of alloy and environment have been published. Nevertheless, unexpected cases of SCC still occur under service conditions. The reason for this is partly the incomplete knowledge of designers and service engineers, and partly that higher design stresses and more high-strength materials are used for various fields of applications. Table 7.7 shows a list of materials and environments where SCC has occurred [7.49]. Earlier, SCC was always associated with alloys, but during the last few decades it has also been detected in pure metals, e.g. copper. It should be noticed that SCC has occurred in low-alloy steel with medium or low strength, for instance in clean, saturated water vapour.

SCC materials data have often been presented in the form of diagrams showing time to fracture as a function of nominal stress. The diagrams show that a minimum (threshold stress) is necessary to cause SCC. However, a better measure of critical stress is the critical stress intensity factor (see Section 7.12.3).

Table 7.7 Material–environment combinations that have caused stress corrosion cracking. (Adapted from Lees [7.49].)

Alloy	Environment
Mild steel (ferritic)	OH^{-*}, NO_3^{-*}, CN^-, $NH_{4(an)}^+$, H_2O, moist CO/CO_2 – gas CO_3^{--}/HCO_3^{-*}, molybdates, salts from acetic acid *, phosphates, saturated H_2O vapour, acid SO_4^{--*}, SO_4^{--} + H_2S^*.
Austenitic stainless steels conventional)	Cl^{-*}, OH^{-*}, $H_2O(O_2)^{*□}$, SO_4^{--}/Cl^-, $Na_{(liq)}$, H_2O contam in ated by Pb^*, SO_4^-, saturated H_2O vapour, $F^{-□}$, seawater□.
High strength steels	H_2, $H_2O_{(liq)}$, CN^-, Cl^-, $HCl_{(g)}$, $HBr_{(g)}$, $H_2S_{(g)}$, $HCl_{(liq)}$, $H_2 + O_2$, saturated H_2O vapour, $NH_{3(g)}$, acidic SO_4^{--}, $NH_3/O_2/CO_2$, $Cl_{2(g)}$.
Copper alloys	NH_4^+, salts from citric acid, $FeCl_3$, $H_2O_{(g)}$, moist $SO_{2(g)}$, moist NO_x, moist NH_3 vapour, $H_2O_{(liq)}$.
Aluminium alloys	Cl^-, Br^{-*}, I^-, $H_2O_{(liq)}$, organic substances containing H_2O, moist H_2, $H_2O_{(g)}$.
Titanium alloys	Cl^-, HNO_3^*, molten salts, $H_2O(O_2)$, CH_3OH, CCl_4, $Br_{2(g)}$, $H_{2(g)}$, $N_2O_{4(liq)}$, $H_2O_{(g)}$, distilled water.
Nickel alloys	$H_2O^*{}_{(liq,g)}$, OH^{-*}, HF–vapour.

* Boiling or hot (> 50°C)
□ Only in sensitized condition.

7.12.2 Mechanisms

The factors determining the mechanisms and the course of development can be sorted into three main groups:

1. Environmental and electrochemical factors.
2. Metallurgical factors.
3. Mechanical stress and strain.

The study of SCC is typically multidisciplinary, i.e. the description of the three groups of factors and their effects belong to three different technological disciplines, namely electrochemistry/corrosion, physical metallurgy and fracture mechanics. The relative importance of electrochemical, metallurgical and mechanical factors varies strongly from one material–environment system to another.

Firstly, let us consider *initiation of SCC* on parts without original surface defects. Under such conditions it may take up to several thousand hours before a growing crack is formed. According to Lees [7.49], during this time one or more of the following processes take place: i) Metallurgical changes, e.g. accumulated deformation by logarithmic creep until the stage at which a crack is formed in the passivating surface film and pure metal is exposed. ii) Development of geometrical defects and stress concentrations due to localized corrosion on the surface, i.e. pitting, crevice/deposit corrosion, intergranular attack, fretting or possibly selective corrosion. iii) Production of a more aggressive environment as regards SCC.

However, defects of some kind often exist before exposure. They may be metallurgical, such as non-metallic inclusions, segregation zones at the grain boundaries and hard spots due to heat treatment. These defects may form mechanically weak spots from the beginning or they promote rapid local corrosion and thereby formation of geometrical defects and stress concentrations. Inclusions and segregation zones may also affect the local environment so that it becomes more aggressive. Original geometrical defects may be the results of forming, machining, pickling or other surface treatment.

Regardless of the original state, after shorter or longer incubation time the further development of cracks is based on the following conditions: i) Fracture of the protecting oxide film. ii) Geometrical defects giving high local stress and strain. iii) Local environment favouring SCC.

The mechanism of *further development and growth* depends on the combination of material and environment. The cracks may be transgranular, intergranular or a mixture of these. In particular, two mechanisms seem to dominate *transgranular crack development and growth*: 1. Accelerated anodic dissolution at the crack tip where some material is subject to continual plastic deformation. 2. Hydrogen-induced or hydrogen-assisted crack formation (often denoted hydrogen embrittlement). In addition, a mechanism involving stepwise crack growth due to fracture of a very thin corrosion product film has been proposed.

For some material–environment combinations it has been shown that accelerated anodic dissolution of yielding metal is the significant mechanism. This is the case for austenitic stainless steels in acidic chloride solutions. In these steels, plastic deformation is characterized by a dislocation pattern giving wide slip steps on the surface. For such systems, Scully [7.50] has proposed a model for initiation and development of stress corrosion cracks, which has been supported by other scientists [7.51]. The model in its simplest form is illustrated in Figure 7.52. A necessary condition is that the surface from the beginning is covered by a passivating film (A).

Plastic deformation gives a slip step with a surface that is metallically clean and very active (B). The further development depends on the *repassivation rate:* a) In an intermediate range of this rate, most of the new surface is passivated before significant corrosion occurs, and the attack is concentrated on a narrow region where the dislocation density is highest (C). This gives initiation and growth of cracks. b) If the repassivation rate is higher, the critical region is passivated before significant corrosion occurs. c) If the repassivation rate is low, corrosion is spread over a larger part of the new surface, which will result in a kind of pit [7.51].

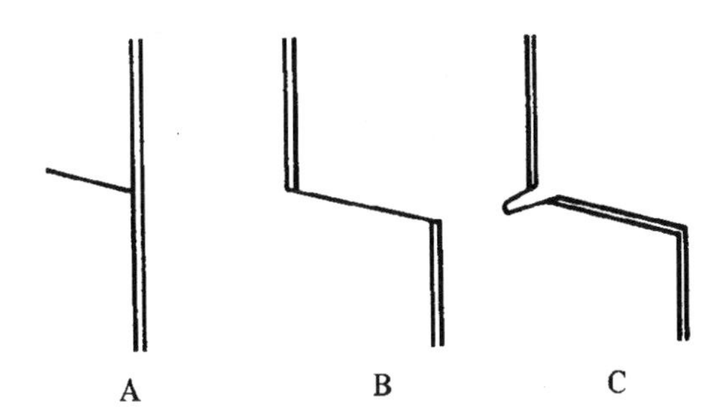

Figure 7.52 Schematic illustration of the mechanism of initiation and evolution of stress corrosion cracking [7.50].

Scully has also thrown light on the mechanism with schematic overvoltage curves for different parts of the crack surface close to the crack front (Figure 7.53a and b). Here, the regions 1 and 2 have been passivated more or less; on region 3 there are conditions causing corrosion (and since the area is limited it takes the form of pitting), and in region 4 the material is extremely active. It should be noticed here that the shown cathodic curve gives a highly simplified picture. Really, the cathodic current density varies over the crack surface and the free surface, furthermore the cathodic area is larger than the anodic area, and finally, the potential drop inside the crack affects the corrosion potential at the respective depths in the crack. The picture illustrates only qualitatively the variation of passivity/activity over the crack surface. Whether one will get pitting or SCC is determined by the difference between the curves 4 and 3 (Figure 7.53a). If this difference is small, the attack is extended relatively little in depth, and the result is pitting. If the difference is large, the corrosion will be relatively faster along the slip plane at point 4, resulting in SCC. Another important feature suggested by Figure 7.53a is that we find the critical potential range around an active–passive transition potential. This is also illustrated in Figure 7.54, which shows a nominal anodic overvoltage curve for non–deformed material (overvoltage curve where the potential is referred to the free surface). Compare also with Figure 7.58.

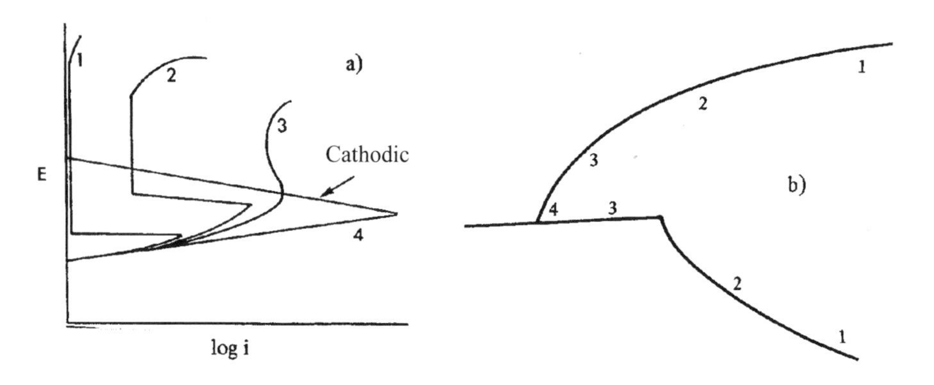

Figure 7.53 a) Schematic overvoltage curves for b) different parts of the crack surface near the crack front [7.50].

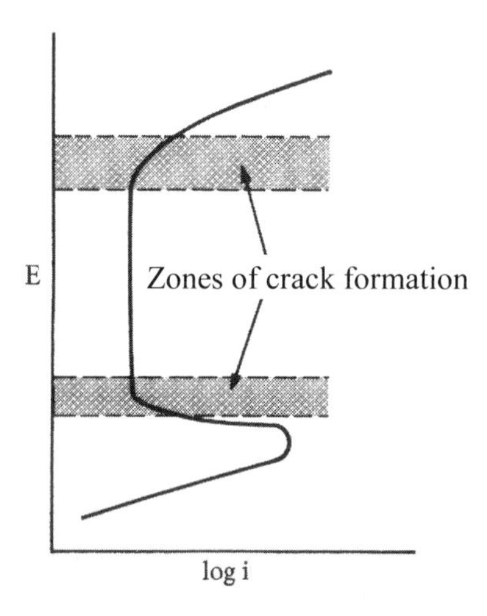

Figure 7.54 Anodic overvoltage curve for non–deformed material and indicated potential ranges where stress corrosion cracking can occur, i.e. ranges where the passive film is unstable [7.51].

In materials with more or less elongated inclusions a combination of different mechanisms may exist. One example is C-Mn steels (and possibly other low-alloy steels) with inclusions of iron/manganese sulphides exposed to acidic solutions [7.49]. The matrix along the inclusions corrodes, and some sulphide dissolves at the same time. Crack growth normal to the rolling direction may occur by ductile tearing of the matrix between adjacent parallel inclusions, while crack growth parallel with the rolling direction occurs in the corrosion direction along inclusions. The dissolution of sulphides leads to formation of H_2S which a) may affect a possible passivation of crack surfaces near the crack front by formation of a film of FeS (by this we might imagine a crack growth mechanism similar to the one described for stainless steels above), and b) has been shown to increase the transfer of hydrogen atoms to the crack tip. The same effects are possible if H_2S – instead of originating from sulphide inclusions – is originally dissolved in the solution and/or possibly formed by sulphate reducing bacteria (compare with Sections 6.4 and 6.6).

Under item b) in the last paragraph we have touched on the second main mechanism of transgranular stress corrosion crack growth, i.e. hydrogen-induced cracking, also called hydrogen-assisted crack growth or simply hydrogen embrittlement. A simple description of the mechanism is shown in Figure 7.55. The origin is a more or less narrow pit or groove, where an acidic environment has been developed, and which has grown to a crack–like defect. Hydrogen atoms from the cathodic reaction $2H^+ + 2e^- \rightarrow 2H$ are only partly combined to hydrogen gas, H_2. The other H atoms diffuse into the material, particularly to the region where the tensile stress is highest, i.e. in front of the crack tip. The hydrogen makes the material more brittle, it obstructs the plastic deformation and thereby increases the stress concentrations ahead of the crack tip, which leads to a small brittle crack extending towards a more ductile material region. In turn, this region is charged with hydrogen so that a new crack growth step is promoted.

Brown and co-workers [7.52] have argued that the mechanism of SCC in high-strength steels in chloride solutions is hydrogen-assisted crack formation. As a part of the work, they have shown by means of potential and pH measurements at the crack front that the conditions for hydrogen reduction are present in growing cracks, both under free corrosion (with local pH = 4) and under cathodic as well as anodic polarization. The point is that the potential is below the equilibrium potential for the hydrogen reaction $2H^+ + 2e^- = H_2$, as shown for free corrosion in Figure 7.56.

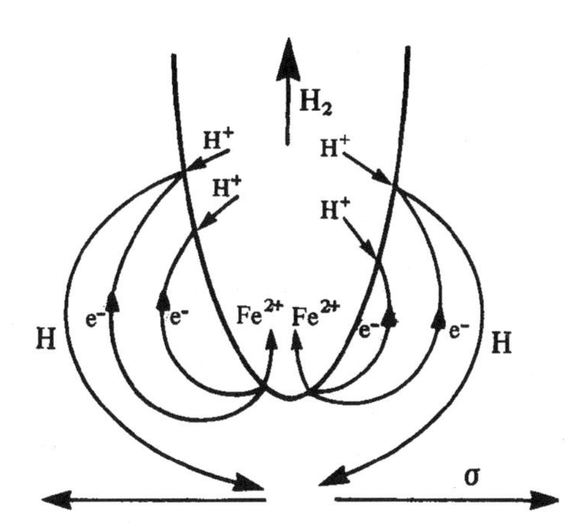

Figure 7.55 Process at the crack front promoting hydrogen-induced (or hydrogen-assisted) crack formation.

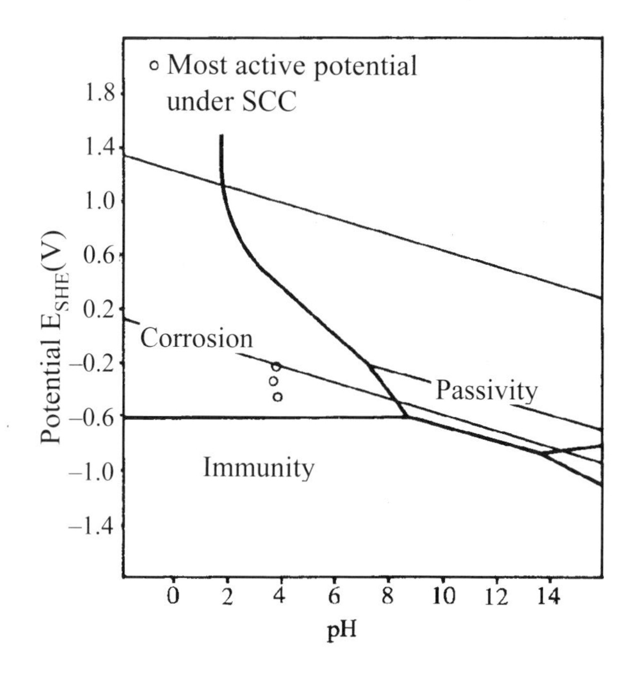

Figure 7.56 Potential and pH at the crack front for some alloy steels under free corrosion conditions [7.52].

Some authors distinguish between crack growth due to hydrogen embrittlement and SCC. Then SCC should only comprise cracks growing by concentrated anodic dissolution at the crack tip and by fracture of corrosion product film. The description above, indicating that combined mechanisms exist, and that the hydrogen may be produced by corrosion at the crack tip, is an argument for a wide definition of SCC such as given in the beginning of the present section. On the other hand, it does not seem reasonable to consider crack formation due to hydrogen developed by complete cathodic protection as a kind of SCC, since in this case there is no corrosion of the material.

In *intergranular stress corrosion crack growth* the grain boundaries act as a preferred active path for anodic dissolution. This is due to chemical heterogeneity caused by segregation or precipitation of a secondary phase. This form of SCC occurs also in materials that are liable to intergranular corrosion in the stress-free state. However, most material–environment combinations liable to intergranular SCC undergo limited attack under stress-free conditions.

The mechanisms of intergranular SCC may also comprise anodic dissolution in slip bands after fracture in passivating films and/or hydrogen-induced cracking, as the case is for transgranular crack formation.

7.12.3 Fracture Mechanics Quantities

The quantity determining whether a brittle fracture will propagate or not is not the nominal stress in the material, but the stress intensity at the crack tip. This quantity can be expressed by the stress intensity factor K:

$$K = Y\sigma \sqrt{\pi a}, \tag{7.11}$$

where σ is the nominal stress, a is the crack length and Y is a geometrical function that depends on the shape of the component and on the orientation of the load relative to the crack.

When the stress intensity increases it will finally reach a critical value K_C above which the crack propagates. If the load and crack orientation is like that shown in Figure 7.57, the actual and the critical stress intensity factor are denoted by K_I and K_{IC}, respectively. K_{IC} is also called the fracture toughness of the material. A necessary condition for the applicability of linear elastic fracture mechanics is that the plastic zone ahead of the crack tip is small compared with the thickness of the component and with the crack length. To satisfy this condition we must have

$$B \geq 2.5\,(K_{IC}/\sigma_y)^2, \tag{7.12a}$$

$$a \geq 2.5\,(K_{IC}/\sigma_y)^2, \tag{7.12b}$$

where B is the thickness of the sample and σ_y is the yield strength of the material (see Figure 7.57).

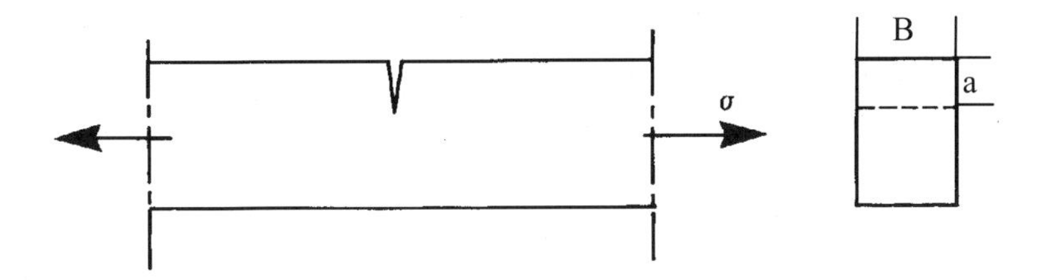

Figure 7.57 Geometrical orientation and size of crack

Under conditions of SCC the crack will grow at lower stress intensities than the critical value under non-corrosive conditions, i.e. we have a lower critical stress intensity factor, which we denote by K_{ISCC}. When Equations (7.12) (with K_{ISCC} instead of K_{IC}) are satisfied, K_{ISCC} is a very useful quantity, much more generally relevant than a threshold value of nominal stress.

With a given size of a crack-like surface flaw (a) and a known geometrical function Y, the values of K_{IC} and K_{ISCC} can be used to calculate the permissible stress by Equation (7.11), for applications in air and the actual corrosive environment, respectively. For corrosive conditions one must take into account the deepest local corrosion defect that may develop during the planned lifetime.

7.12.4 Cracking Course and Data for Some SCC Conditions

Transgranular stress corrosion cracks are known [7.49] from i) austenitic steels in acidic chloride solutions, ii) low-strength ferritic steels in acidic media, iii) ferritic steels in phosphate solutions, iv) carbon steel in water saturated with CO_2 and CO, v) α-brass in ammonia solutions that do not cause surface films, vi) aluminium alloys in $NaCl/K_2CrO_4$ solutions and vii) magnesium alloys in diluted fluoride solutions. For further study of fracture surface appearance, see, e.g. Lees [7.49] and Scully [7.53].

Examples of intergranular cracking [7.49] comprise: i) carbon steel in caustic, nitrate, acetate and carbonate/bicarbonate solutions, ii) low-alloy steels in pure water, iii) stainless steels that are liable to ordinary intergranular corrosion in oxygen-containing water, iv) α-brass in ammonia solutions that cause surface films, v) aluminium alloys in water vapour and humid hydrogen gas, vi) β-titanium alloys in metanol solutions, vii) tempered martensitic stainless steels in chloride solutions and viii) nickel alloys in very pure water and alkaline solutions.

In scanning electron microscopy (SEM), intergranular SCC is often characterized by clearly defined facets in the crack surfaces, each facet with a series of parallel slip bands.

In many cases can be observed a transition from one crack type to another due to the change of stress, environmental composition or electrode potential. For instance, a transition from intergranular to transgranular cracking is often found when the crack intensity factor, K_I, increases. Low-strength (mild) steel is liable to SCC in phosphate, nitrate, carbonate and hydroxide solutions in the pH–potential regions shown in Figure 7.58 [7.54]. The steel is hardly sensitive to hydrogen and tolerates cathodic protection relatively well.

Conversely, high-strength steels react with hydrogen, and the critical stress intensity factor K_{ISCC} therefore depends on the potential, as shown by the examples in Figure 7.59. As is seen, cathodic polarization is beneficial down to about –750 mV referred to the Ag/AgCl electrode, but K_{ISCC} is reduced with further lowering of the potential. In Figure 7.60 is shown how K_{IC} in air and K_{ISCC} in seawater vary with the strength of the steel represented in Figure 7.58. For yield stresses < 1000 MPa the effect of seawater is relatively low. For comparison, results from References [7.55] for a steel with yield stress about 900 MPa in a 5% NaCl solution containing H_2S is shown in Figure 7.61. The grey zone indicates some scatter in the results. The quantity $K_{applied}$ in the figure is not identical to K_I because the conditions for plane strain are not satisfied at the lower H_2S contents. Correction for this indicates that the critical K_I value at 0% H_2S is at least 20% lower than the corresponding critical $K_{applied}$ in Figure 7.61. However, the effect of H_2S concentration is clear. The actual steel has an R_c hardness of 32, while NACE has stated a maximum R_c hardness of 22 for production tubing of carbon steels for "sour" oil/gas wells. In solutions containing H_2S, SCC can also occur in martensitic stainless steels, at high H_2S

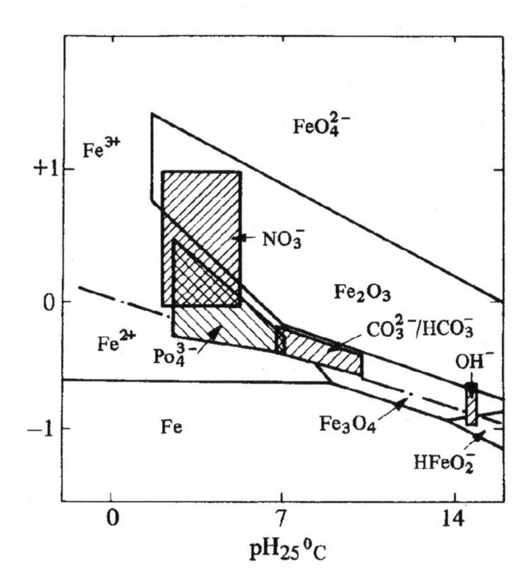

Figure 7.58 pH–potential regions in which mild steel is liable to SCC in different environments. Note that there is a strong tendency to SCC in the regions where a protecting film is unstable (i.e. if the film is damaged locally, corrosion can occur). (After Ford [7.54].)

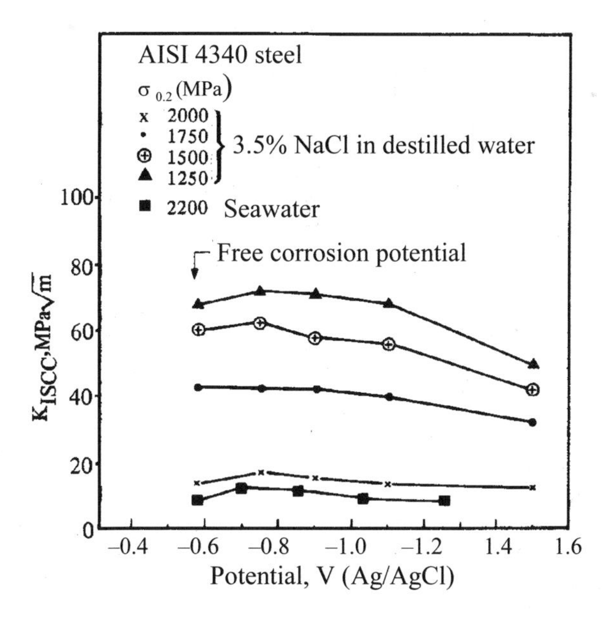

Figure 7.59 Effect of potential on K_{ISCC} for steels with different strength. (After Sandoz et al. [7.52].)

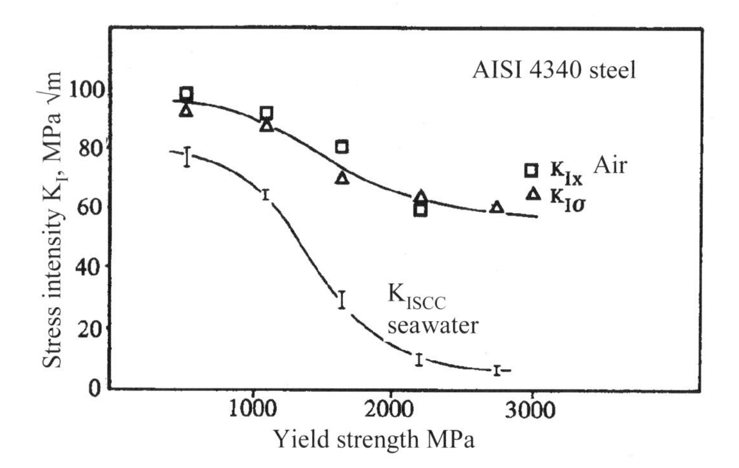

Figure 7.60 Effect of yield strength on resistance to SCC. K_{Ix} and $K_{I\sigma}$ correspond approximately to fracture toughness [7.52].

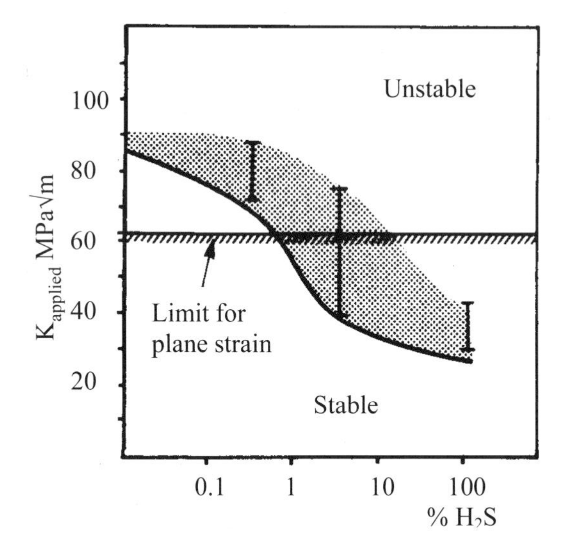

Figure 7.61 Regions of stress intensity factor and H_2S concentration which may or may not lead to crack formation in a casing steel with yield strength σ_y = 900 MPa. Steel thickness: 9 mm. Environment: 5% NaCl solution at 25–80 °C (further explanation is given in the text). (After Edwards [7.55].)

concentration and high temperature in ferritic–austenitic (duplex) stainless steels as well.

SCC data have traditionally often been presented in stress–log time diagrams, as shown for three stainless steels in a chloride solution in Figure 7.62. SCC is avoided when the stress is below a certain threshold value that depends on the steel type and environment. The standardized test environment, 42% boiling magnesium chloride solution, is very aggressive and hardly realistic for most applications.

Both the oxygen and the chloride content are very important (Figure 7.63). It is seen that SCC of austenitic stainless steel can be avoided by removing one of these. The temperature is also a critical factor for stainless steel–chloride combinations. The example in Figure 7.64 indicates a critical temperature of roughly 100°C, but under more unfavourable conditions, these steels are liable to SCC at lower temperatures, in the worst cases down to 60°C. Newer high-alloy austenitic stainless steels, e.g. with 6 or 7% Mo (see Table 10.6 p. 245), have much higher resistance to SCC. Laboratory tests indicate that these are resistant at, and probably considerably above, 100°C.

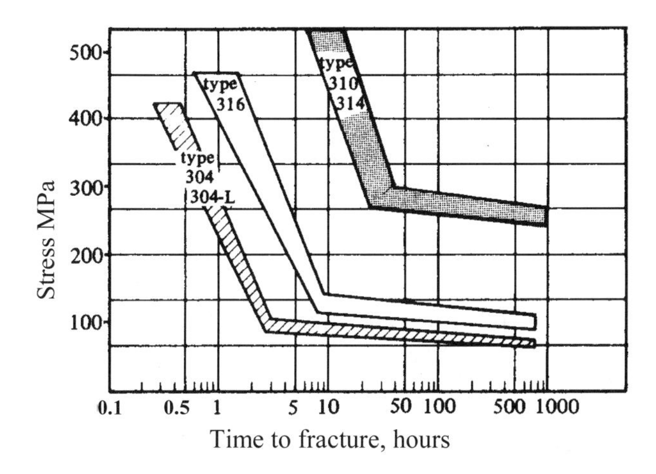

Figure 7.62 Dependence of time to fracture on stress for different stainless steels in 42% boiling $MgCl_2$ solution. (Adapted from Fontana and Greene [7.1].)

In recent decades, also relatively high alloy ferritic–austenitic steels (duplex steels) have been developed. Earlier, these steels were considered immune to chloride SCC. However, both laboratory experiments and experience of fracture on installations in the North Sea have shown that SCC can occur on duplex steels under unfavourable conditions developed beneath the thermal insulation on hot steel surfaces. The insulation is used to reduce heat loss and to protect people. If seawater penetrates the insulation and reaches the steel surface, the water will evaporate and this causes chloride enrichment. Maximum unfavourable conditions can be established: high chloride content, efficient oxygen access (there is a thin water film at the most critical sites), and temperatures above 100°C. Ferritic–austenitic stainless steels should not be used under these conditions.

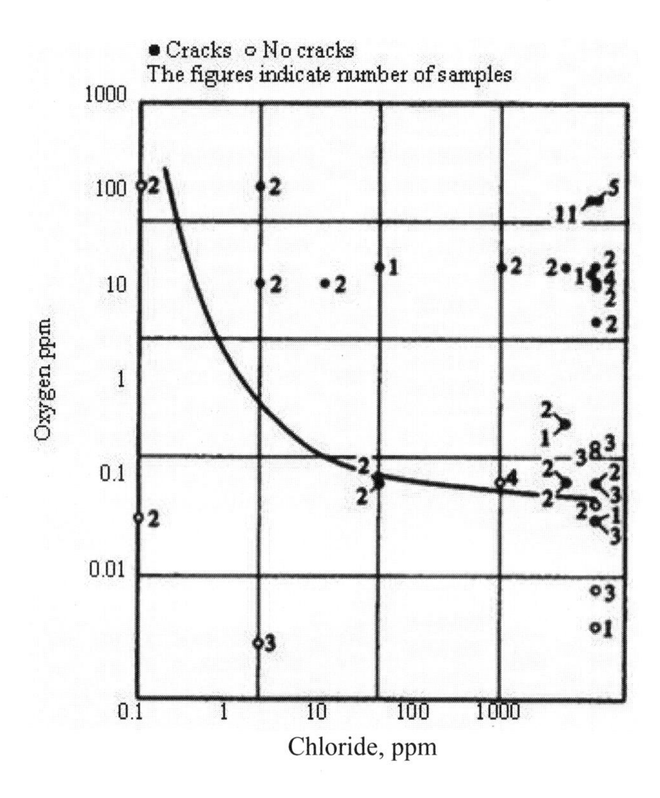

Figure 7.63 Significance of oxygen and chloride content for SCC of an austenitic stainless steel in steam from alkaline–phosphate treated boiler water with intermittent wetting. (From [7.1] after Lee Williams.)

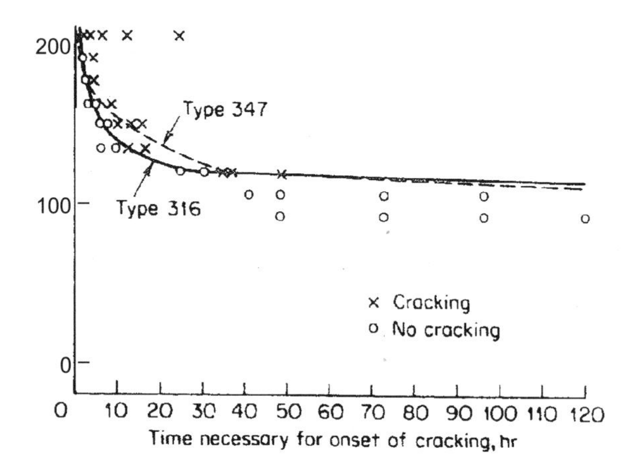

Figure 7.64 Effect of temperature on initiation time of SCC for two stainless steels in water with 875 ppm NaCl. (Adapted from [7.1], after Kirk et al.)

Figure 7.65 shows stress corrosion cracks in a pipe of stainless steel AISI 304 that has carried glucose containing about 16% water, 0.02–0.03% NaCl, with a pH of 5.2. The cracks have developed from small corrosion pits in the weld (associated with small weld defects) on pipe sections heated by a heating cable.

Figure 7.65 Stress corrosion cracks in tubes of AISI 304 steel. 35 × . Photo: J.M. Drugli, SINTEF Corrosion Centre.)

In another case, SCC has occurred in a pipe of AISI 304 steel carrying fresh water of maximum temperature 90°C, pH = 6.5, maximum chloride content 12.5 ppm, and with water velocity 1.2 m/s. The chloride content was so low that SCC was completely unexpected. The explanation is that the attacks have started by crevice/deposit corrosion at the welds due to insufficient gas protection of the weld root. Then the chloride content has increased due to the corrosion process, thus leading to conditions where SCC can take place. Measures to avoid this are dealt with in Section 7.5.7.

Among aluminium alloys, particularly those with high strength within the AlZnMg and AlCu series have been liable to SCC. In heavily cold-worked AlMg alloys with Mg content > 5% and in cast AlCuSi alloys, SCC may also occur. Finally, it may be developed in AlMgSi alloys after poor heat treatment [7.27].

The SCC properties are typically anisotropic properties. Most cases of SCC in high-strength aluminium alloys have been connected with stresses in the short transverse direction. The cracks in Al alloys may be branched on a macro or micro scale depending on the load level. SCC in aluminium alloys may be prevented by cathodic protection.

Standard testing of SCC is described in ASTM G30 (U-bend testing), G35–G39, G41, G44, G47 (different combinations of material and environment), G49, G58 (welds) and G64 (classification of Al alloys).

7.12.5 Prevention of SCC [7.1, 7.30]

Stress corrosion cracking can be prevented by affecting one or more of the determining metallurgical, mechanical, environmental or electrochemical factors dealt with in the previous sections:

a) Selecting the right material. In chloride solutions it may, for instance be appropriate to use carbon steel or ferritic stainless steel (or duplex steel if the temperature is not too high) instead of a conventional austenitic stainless steel. Where high general corrosion resistance is required, high-alloy austenitic stainless steels, Ni alloys or Ti alloys may be a good solution.
b) Reducing the stress and stress intensity to below the threshold values. Annealing to remove residual stresses due to welding or cold working.
c) Making the environment less aggressive by removal of oxygen, distillation or ion exchange.
d) Using cathodic protection by sacrificial anodes (in cases where the material is not liable to hydrogen embrittlement).
e) Supplying inhibitors, particularly in less aggressive corrosion media. To avoid caustic embrittlement in common structural steels in boilers, sodium nitrate in concentrations of 20–40% of the NaOH concentration has been used.

7.13 Corrosion Fatigue

7.13.1 Definition, Characteristic Features and Occurrence

Corrosion fatigue (CF) is crack formation due to varying stresses combined with corrosion. Alternatively, it may be defined as fatigue stimulated and accelerated by corrosion. The problem is caused by tensile stresses, just as it is for SCC. The external difference between these two forms of deterioration is only that CF develops under varying stresses and SCC under static stresses.

Typical fatigue fractures that have occurred in non-corrosive environments show a large smooth crack surface area where the crack has grown by fatigue and an (often smaller) area with a rough and crystalline surface formed by fast fracture when the maximum stress reached the ultimate strength. Characteristic beach marks on the smooth part show the form and position of the crack front at different stages. Fatigue fractures of cathodically protected steel in seawater are also characterized by such beach marks. An example is shown in Figure 7.66.

Usually, CF under free corrosion gives beach marks too, but sometimes they may be less visible because of corrosion products. Corrosion fatigue fractures are normally transgranular, and contrary to "dry" fatigue often branched.

Corrosion fatigue occurs for all material–environment combinations where general or localized surface corrosion is developed. Fatigue and CF became subject to much attention in connection with platforms in the North Sea, because the load spectrum caused by the waves made design against fatigue more relevant in the North Sea than in oil fields developed earlier, e.g. in the Mexican Gulf.

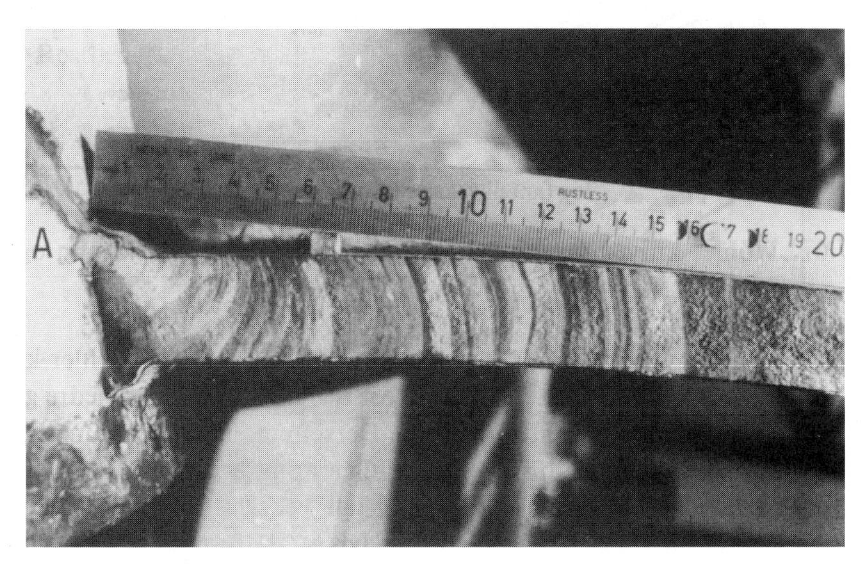

Figure 7.66 Crack surface after fatigue of steel in seawater under cathodic protection. (From the hotel platform "Alexander Kielland". Photo: B. Lian, Statoil.)

7.13.2 Influencing Factors and Mechanisms

In order to understand the mechanisms of CF, it seems reasonable to start with fatigue under noncorrosive conditions, for simplicity here called dry fatigue, and then to look at changes due to a corrosive environment.

The most traditional way to display fatigue data is by Wöhler curves, often called S-N curves, schematically shown for steel in Figure 7.67. For dry fatigue of structural steels, there is a lower limit for stresses causing fatigue, namely the fatigue limit. If we have a notch on the steel surface or a weld, which imply smaller or larger geometrical defects and thereby stress concentrations, the number of cycles to fracture is less and the fatigue limit lower, as shown in the figure. If a corrosive environment is added, in particular two effects are introduced: a) the number of cycles to fracture decreases, and the fatigue limit is eliminated, or at least drastically reduced, and b) the number of cycles to fracture becomes dependent on load frequency. By means of a normal degree of cathodic protection one can retain the same fatigue limit and lifetimes as in dry fatigue for ordinary structural steels. Particularly for high-strength steels and high loads, a strong degree of cathodic protection with subsequent hydrogen development will give reduced lifetime (an effect of hydrogen embrittlement).

A typical example of relationship between static strength, environment, and fatigue strength of various steels is shown in Figure 7.68. While the fatigue strength

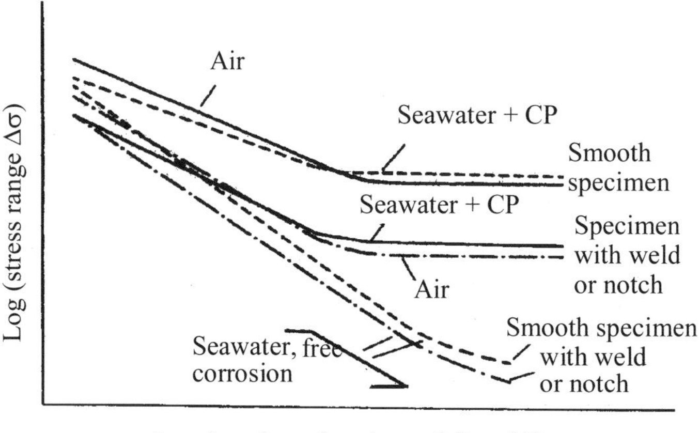

Figure 7.67 S–N curves for steel in air, and in seawater under free corrosion and cathodic protection (CP).

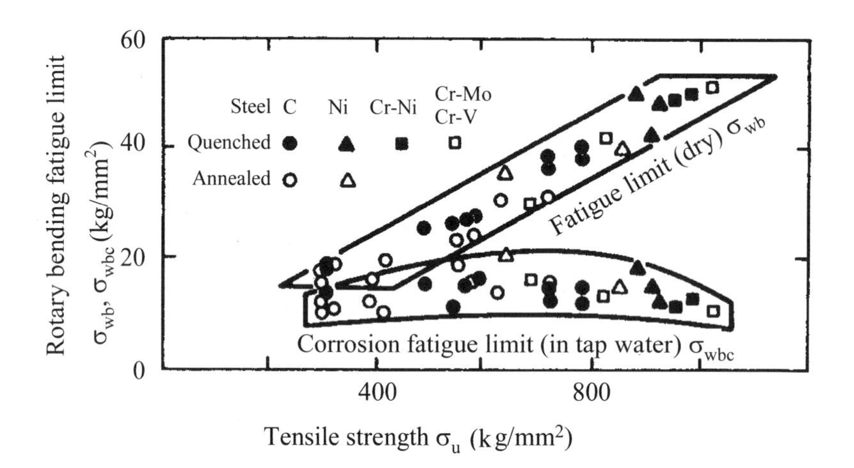

Figure 7.68 Relationship between fatigue limit and tensile strength for various steels in air and tap water. (From Kitagava [7.56] after McAdam.)

increases proportionally with the ultimate tensile strength σ_u in dry air, it is nearly independent of σ_u in tap water. Salt water has given a little lower range of fatigue strength than has tap water [7.56].

One more relationship is to be mentioned. In dry fatigue of smooth components, the initiation time makes the major part of the lifetime (e.g. 90%). For components with welds (not first class), with sharp notches or exposed to a corrosive environment, this relationship is turned upside down, so that the initiation period is relatively short (for instance 10% of the lifetime), while the crack growth forms the major part of the lifetime.

The crack development can be divided into four stages (Figure 7.69): 1. Initiation 2. Crack growth stage I, with crack growth direction about 45° relative to the tensile stress direction. 3. Crack growth stage II. 4. Fast fracture.

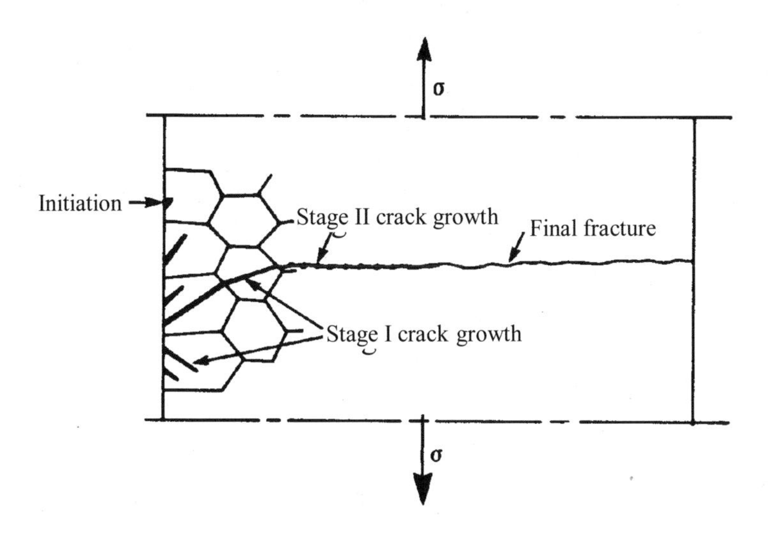

Figure 7.69 The stages of development of a fatigue fracture.

The observations dealt with above have been explained by various theories. The effect of corrosion is mainly ascribed to three different mechanisms:

i) Formation of corrosion pits and grooves leading to stress concentrations. This explanation has to some extent been supported by the observation that machined notches and corrosion attacks give similar effects on fatigue life. On the other hand, the significance of this mechanism has been questioned by some researchers. For some combinations of material, environment and loading it has been shown that pits at the crack mouth are rather a result of the crack formation than a reason for it.

ii) Mutual acceleration of corrosion and mechanical deterioration. This can be considered in connection with a common explanation of dry fatigue initiation mechanism for some materials. During the initiation period, local slip bands are formed by slip in crystallographic planes with maximum shear stress (Figure 7.70). This is a common mechanism of fatigue initiation, e.g. in aluminium at relatively low stresses. At higher stresses, fatigue is initiated at the grain boundaries. As a result of alternating stresses the slip along different parallel planes leads to extrusions (e) and intrusions (i), as shown in the figure. This implies stress concentrations at the intrusions followed by gradual crack formation in the plane of the slip band. When the crack has grown a length of one or a few grain diameters into the material (the order of 0.1 mm) the direction and plane of crack propagation is changed so it becomes normal to the tensile stress (Figure 7.69). The mutually accelerating effects of corrosion and mechanical deterioration in this case are explained in two ways: a) Corrosion-induced plastic deformation, i.e. corrosion induces or promotes the formation of slip bands, extrusions and intrusions by the dissolution of dislocation locks (these are localized at sites with high density of dislocations, which makes the material more active). b) Deformation-induced corrosion, which is caused by the extremely high activity of material under plastic deformation in action, and which implies formation and growth of microcracks and possibly macrocracks by intense localized anodic dissolution at the initiation site and successively at the crack front.

iii) Absorption and adsorption of species from the environment, including absorption of hydrogen and resulting embrittlement. One theory is that reduction of the surface energy at the crack surface (particularly at the crack front) caused by adsorbed elements promotes the formation and growth of cracks under fatigue loading. This theory has been supported by the observation that fatigue is accelerated in some non-corrosive liquids. However, in cases of CF in aqueous solutions, interest is usually concentrated on the absorption of hydrogen.

As we can see, the main mechanisms are, roughly speaking, the same as or closely related to those acting in SCC. However, the dynamic conditions in the material as well as in the local environment in a crack distinguishes CF from SCC.

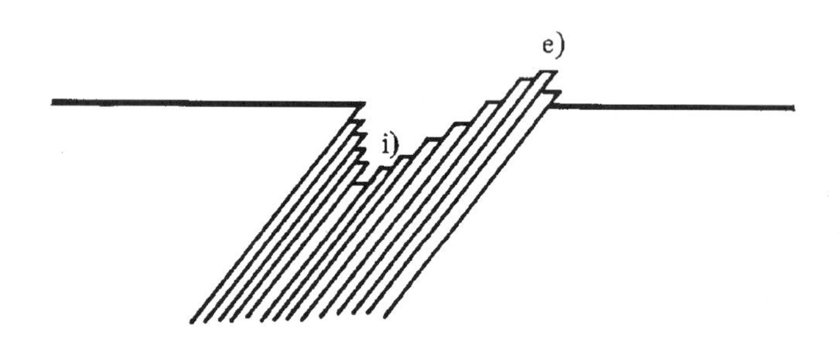

Figure 7.70 Slip bands, with extrusion (e) and intrusion (i) on the surface.

Mechanism i) is limited to the initiation period and is of minor importance if there are geometrical defects beforehand, e.g. weld defects. It is neither a prerequisite for crack initiation since this can occur by mechanism ii). The latter may act both in the crack initiation stage and under crack growth (particularly in the early stages of growth). The hydrogen mechanism iii) is mainly connected to crack growth in stage II.

The direct effect of accelerated anodic dissolution at the crack front is most important at low stress amplitudes, while reduced lifetime due to hydrogen embrittlement is associated with high stresses, particularly for higher-strength materials. There is, however, a certain relationship between the mechanisms, since accelerated anodic dissolution leads to an acidic local environment and thereby significant hydrogen development within the crack.

7.13.3 Factors Most Important for Crack Initiation and Early Growth

Loading frequency and *moderate variation of temperature* have little effect upon the number of cycles to fracture in dry fatigue. These factors are more important in CF. The effect of frequency is significant because the corrosion effect is time dependent, and the temperature is known to be an important factor in connection with corrosion. These two factors are mentioned at the same time because they both represent possible ways of accelerating CF tests. Reliable test results can, however, be obtained only if the fatigue acceleration factors are quantitatively known.

Since anodic dissolution is of major significance in the early stages of fatigue, both frequency and temperature considerably affects the fatigue damage in these stages [7.57]. The same applies for other corrosion affecting environmental factors, of which the *oxygen content* is particularly important. Uhlig and co-workers have shown that CF of steel is prevented by keeping the O_2 content below a certain level. *Flow conditions* may also have some effect. *pH* is an important environmental factor, but it affects the mechanism in different ways for different *materials*. For aluminium in 3.5% NaCl solution, CF is initiated faster at pH = 7–8 than at pH = 10. The reason is that the conditions for passivation of Al are more favourable at pH = 7–8, and that the attacks therefore are more concentrated in this case. On the other hand, the passivation conditions for steel are better in alkaline environment, and faster fatigue initiation and fracture has been observed at pH = 10 than at pH = 7 [7.58]. However, this is not a generally valid result, as seen in Figure 7.71. When pH becomes high enough, corrosion is in this case effectively prevented, so that CF is avoided. As indicated by this example, a stable and passivating oxide film on the material surface can prevent the corrosion effect on fatigue initiation. The effective passive regions defined by environmental parameters for a certain material may, however, be smaller in relation to CF than versus pure surface corrosion. In a similar way as for SCC, initiation of fatigue will in some systems develop faster at *inclusions and grain boundaries*. *Geometrical defects* and unfortunate *surface conditions* such as crevices or defects due to machining, welding or previous corrosion will accelerate initiation by – in addition to stress concentrations – faster development of an aggressive pit environment.

Cathodic protection is most beneficial in the early stages of fatigue (for later stages, see next section).

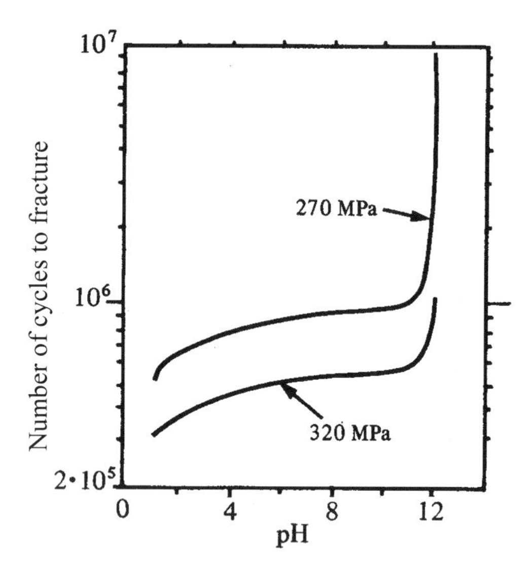

Figure 7.71 Effect of pH on the number of cycles to fatigue fracture of a 0.18% C steel in 3% NaCl solution at two levels of stress amplitudes. (After Duquette and Uhlig [7.57].)

7.13.4 Crack Growth Rate and Factors Affecting It

Provided that the crack length or the crack depth $a > 0.1$ mm, the crack growth rate is related primarily to fracture mechanics properties. In Section 7.12.3, the stress intensity factor $K = Y\sigma\sqrt{\pi a}$ was introduced. Correspondingly, for conditions with varying stress we make use of the stress intensity range ΔK:

$$\Delta K = Y\,\Delta\sigma\,\sqrt{\pi a}, \tag{7.13}$$

where $\Delta\sigma$ is the nominal stress range $= \sigma_{max} - \sigma_{min}$ for each cycle. An important parameter is also the stress ratio

$$R = \sigma_{max}/\sigma_{min}.$$

The crack growth rate (crack length increment/cycle) is usually expressed as a function of the stress intensity range

$$da/dN = f(K). \tag{7.14}$$

Different forms of this function are shown in Figure 7.72 for materials in corrosive and non-corrosive *environments*. The crack growth curves for an inert environment are useful as references. At low ΔK values, these curves approach the asymptote $\Delta K = \Delta K_{th}$ = the lower threshold value for fatigue crack growth. At the top, the curves approach $K_{max} = K_{IC}$ which gives unstable, fast fracture. The effect of the environment may be one of the following main types: a) "True corrosion fatigue" (the previously described mechanism (ii)), for material/environments not liable to SCC. b) "Stress corrosion fatigue" at $K_{max} > K_{ISCC}$, for materials sensitive to ordinary SCC under static loads, and without effect of environment otherwise. c) A combination of a) and b). For structural steels with yield strength > 400 MPa in seawater the last type is the most common one, particularly at higher stress ratio values. For mild steel and low R–values, and for aluminium alloys in water, type a) is common.

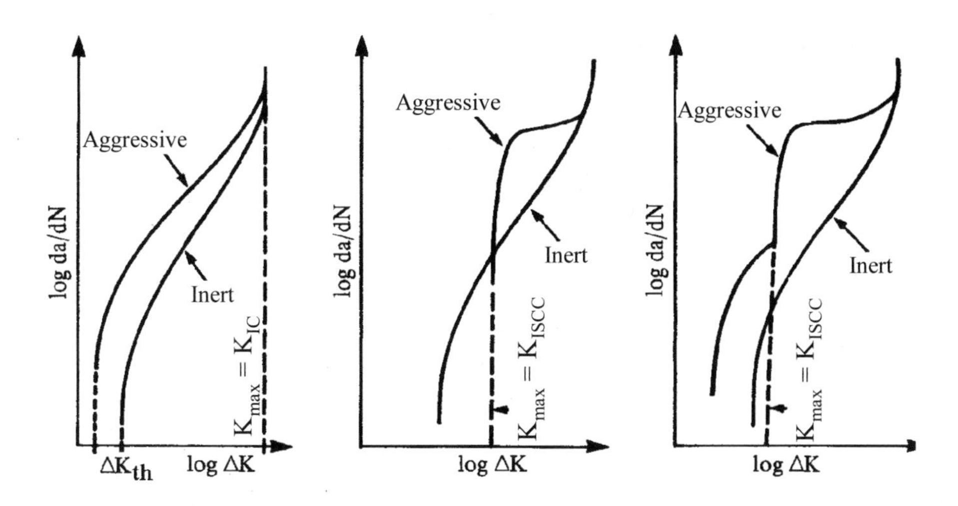

Figure 7.72 Different effects of environment on the crack growth curve.

The detrimental effects of a corrosive environment at higher stress intensity levels are caused by hydrogen absorption (curves b and c).

As seen, the log da/dN–log (ΔK) curve is a straight line in the middle ΔK range for case a) and for inert environments in the other cases. Here, the well-known fatigue crack growth equation, the Paris equation, is valid:

$$da/dN = C\,(\Delta K)^m. \tag{7.15}$$

In order to include the lower region, other formulas have been developed, e.g. the following one published by Lukas and Klesnil:

$$da/dN = C\,((\Delta K)^m - (\Delta K_{th})^m). \tag{7.16}$$

The *stress ratio* R is one of the most important factors. To take this into account, Equation (7.16) has been modified:

$$\frac{da}{dN} = C\left[\left\{\frac{\Delta K}{(1-R)^{\gamma}}\right\}^{m} - \left\{\frac{\Delta K_{th}}{(1-R)^{\gamma}}\right\}^{m}\right] \tag{7.17}$$

This means that an increasing stress ratio R moves the crack growth curve to the left in the log da/dN–log ΔK diagram.

Cathodic protection affects crack growth in steel in a beneficial or an adverse manner depending on the range of ΔK, as shown in Figure 7.73. The effects can be explained as follows.

Under corrosive conditions and particularly under cathodic protection, thinner or thicker deposits are formed on the crack surfaces. These can cause so-called crack closure, i.e. the crack is closed before it is completely deloaded or before the stress has reached its lowest nominal value. The real crack intensity range (ΔK) at the crack front becomes less and the real R value higher, while K_{max} is unchanged compared with cases without crack closure but with the same nominal stress cycle. As a consequence, crack closure causes smaller crack growth rates at given nominal ΔK and R values. This is assumed to be the reason for the crack growth rate at low ΔK values being lower in cases with cathodic protection in seawater than when the crack is exposed in air, as shown in Figure 7.73 [7.59]. The threshold value, ΔK_{th}, is also higher in the case with cathodic protection. Crack growth tests combined with subsequent measurement of deposit layer thickness on the crack surface near the crack front and calculations support the hypothesis that the increase of the threshold value ΔK_{th} is caused by the layer of deposits [7.60].

At higher ΔK values, cathodic protection may increase the crack growth rate, as shown in Figure 7.73. This detrimental effect is due to development of hydrogen, and it is most pronounced at strong degree of cathodic protection and high stress ratio. The reason for this is that the maximum stress intensity K_{max}, is highest at the highest stress ratio (at constant ΔK).

Regarding the importance of ΔK and $K_{max,}$ it should be noticed that the exponent γ in Equation (7.17) is between 0 and 1. If it was 0, da/dN would be completely determined by ΔK. If it was 1, da/dN would be given by K_{max}. With any value of γ between 0 and 1, both ΔK and K_{max} affect the crack growth rate.

A significant geometrical factor is the *thickness* of the material. According to some investigations, the crack growth rate increases considerably with increasing thickness, but the effect of this factor varies.

The *frequency* affects crack growth too. At high ΔK levels, the maximum values of da/dN have been found at f = 0.01–0.03 Hz. This finding has been related to results from stress corrosion tests proving that SCC occurs for a certain range of strain rate [7.58]. Generally it has been shown that da/dN increases when f decreases, e.g. in the range 10–0.1 Hz. The difference between 1 Hz and an average sea wave frequency of approximately 0.17 Hz is found to be small [7.59]. The form of the stress wave may have some influence because the environment is found to affect the deterioration process only in the phase of increasing K (increasing tensile stress).

If we consider *different materials,* an interesting relationship is that the crack growth rate in dry fatigue depends in a universal way on ΔK/E (E = the modulus of elasticity), although there are smaller or larger deviations for some materials. Roughly, this relationship is valid for many cases of corrosion fatigue too. The above-mentioned detrimental effect of cathodic protection at low potential and high stress intensity range (hydrogen embrittlement) has been observed both for ordinary structural steels and duplex stainless steel but not for austenitic stainless steel. This effect is assumed to increase with increasing steel strength.

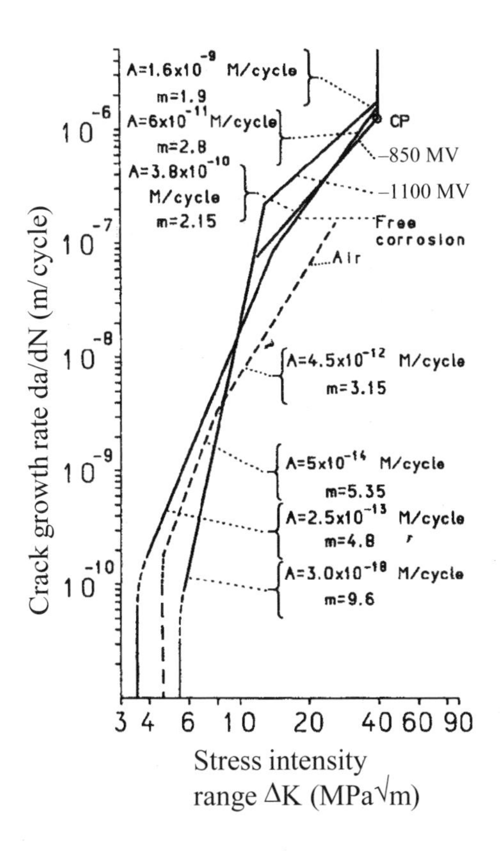

Figure 7.73 Crack growth curves for a St 52–3 steel in air, in seawater with free corrosion, and with cathodic protection at two different potentials. f = 0.17–1 Hz, t = 5–12° C, R = 0.5 [7.59].

7.13.5 Calculation of Number of Cycles to Failure of Welded Steel Structures

It is generally agreed that welded joints may contain crack–like defects of such a size that most of the fatigue lifetime is spent in crack growth. Therefore, we can roughly calculate the number of cycles to fracture by use of crack growth data. The necessary conditions are that one knows i) the initial defect size (crack length a_i), ii) the actual load spectrum, iii) crack growth data valid for this spectrum, and iv) the actual geometry function Y in Equation (7.13).

The calculation is based upon the crack growth function da/dN = f (ΔK). This function is integrated with respect to dN. The number of cycles N_f until the crack has grown to the critical length a_f giving unstable fracture, is

$$N_f = \int_0^{N_f} dN = \int_{a_i}^{a_f} \frac{da}{f(\Delta K)} = \int_{a_i}^{a_f} \frac{da}{f\left(Y\,\Delta\sigma\sqrt{\pi a}\right)} \tag{7.18}$$

If the stress amplitude is constant and Paris' equation (Equation (7.15)) is valid, i.e. $f(\Delta K) = C\,(\Delta K)^m$, Equation (7.18) can be integrated analytically. With other load spectra and deviation from Paris' equation, a numerical integration is necessary.

Examples of results from such calculations are shown in References [7.59]. Roughly speaking, the results show that free corrosion on structural steel St 52–3 in seawater reduces the fatigue lifetime with a factor of about 3 compared with that in air (others have found factors from 2 to 3). Cathodic protection in seawater gives longer lifetime than in air at long lifetimes (small initial defects), but is no better than free corrosion at short lifetimes (larger initial defects). Under unfavourable conditions, i.e. extra-low potentials and high stress ratios, cathodic protection may give even shorter fatigue life than under free corrosion conditions.

7.13.6 Prevention of Corrosion Fatigue

a) As indicated in the preceding sections, moderate cathodic polarization provides adequate protection against the environmental contribution to CF of ordinary structural steels, provided that the initial defects are reasonably small. High-strength steels are more sensitive to hydrogen and will normally be loaded with higher stresses, which also increases the risk for hydrogen–assisted failures.

b) Reduction of the tensile stress level, e.g. by stress annealing and good design, is efficient. If possible, it is important to localize welds where the stress concentrations for other reasons are small. Cast and wrought structural components may increase the possibilities in this context.

c) Reliable coatings on the areas where the fatigue stresses are highest can be utilized more than has been done (anodic coatings, e.g. Zn on steel, and organic coatings).

d) By reasonable materials selection combined with adequate design, CF damage can be avoided or reduced. It should be emphasized that the CF strength of high-strength steel is not higher than that of steel with lower tensile strength (Figure 7.68). Therefore milder and thicker steel may be beneficial.

e) Change of environment, e.g. de-aeration [7.57] or use of inhibitors, is sometimes applicable. De–aeration will not always hinder CF, but can at least in neutral waters prevent initiation of CF on steel surfaces without surface defects.

f) Anodic protection has been shown to increase the fatigue strength of carbon steel and stainless steel in oxidizing environments (it improves passivation)

[7.57]. However, it must be used carefully, since anodic polarization will increase the CF rate in many environments, such as chloride–containing and other solutions causing localized corrosion in any form.

References

7.1 Fontana MG, Greene ND. Corrosion Engineering. New York–Singapore: McGraw-Hill, 1967, 1978, 1986.

7.2 Roberge PR. Handbook of Corrosion Engineering. New York: McGraw-Hill, 1999.

7.3 Pludek VR. Design and Corrosion Control. London: The Macmillan Press, 1977.

7.4 Yamamoto K, Kagawa N. Ferritic stainless steels have improved resistance to SCC in chemical plant environments. Materials Performance, 20, June, 1981.

7.5 Valen S, Bardal E, Rogne T, Drugli JM. New galvanic series based upon long duration testing in flowing seawater. Stavanger: 11th Scandinavian Corrosion Congress 1989.

7.6 Bardal E, Drugli JM, Gartland PO. The behaviour of corrosion resistant steels in seawater. A review. Corrosion Science, 35, 1993: 257–267.

7.7 Mansfeld F, Hengstenberg DH, Kenkel JV. Galvanic corrosion of aluminium alloys I. Effect of dissimilar metal. Corrosion, 30, 1974: 343–353.

7.8 Wallen B, Anderson T. Galvanic corrosion of copper alloys in contact with a highly alloyed stainless steel in seawater. Stockholm: 10th Scandinavian Corrosion Congress, 1986.

7.9 Bardal E, Johnsen R, Gartland PO. Prediction of galvanic corrosion rates and distribution by means of calculation and experimental models. Corrosion, 40, 1984.

7.10 Gabe DR, Shirkhanzadeh M. Polarity reversal of the Al-Fe galvanic couple. British Corrosion Journal, 15, 1980; 4: 216–221.

7.11 NS 2672–2674. Aluminium/steel Screw Connections. Oslo: Norwegian Technology Standard Institution, 1984 (in Norwegian).

7.12 Ijsseling FP. Survey of Literature on Crevice Corrosion (1979–1998). London: The Institute of Materials, 2000.

7.13 Combrade P. Crevice corrosion of metallic materials. In: Marcus P, editor. Corrosion Mechanisms in Theory and Practice 2nd ed. New York–Basel: Marcel Dekker, 2002.

7.14 Oldfield JW. Test techniques for pitting and crevice corrosion resistance of stainless steels and nickel alloys in chloride containing environments. International Materials Review, 32, 1987; 3: 153–170.

7.15 Oldfield JW, Sutton MH. Crevice corrosion of stainless steels I, II. British Corrosion Journal, 13, 1978: 104.

7.16a Gartland PO. Modelling crevice corrosion of Fe-Cr-Ni-Mo-alloys in chloridesolutions. Houston: 12. International Corrosion Congress, 1993.

b Valen S, Gartland PO. Crevice corrosion repassivation temperatures of highly alloyed stainless steels. Corrosion, 51, 1995; 10: 750–756.

c Gartland PO. A simple model of crevice corrosion propagation for stainless steels in seawater. Corrosion'97, paper no.417. Houston: NACE, 1997.

7.17 Mollica A, Trevis A. The influence of the microbiological film on stainless steels in natural seawater. Proceedings 4. International Congress on Marine Corrosion and Fouling. Juan-les Pins, 1976.

7.18 Johnsen R, Bardal E. Cathodic properties of different stainless steels in natural seawater. Corrosion, 41, 1985; 5: 296–302.

7.19 Drugli JM, Bardal E. A short duration test method for prediction of crevice corrosion rates applied on stainless steel. Corrosion, 34, 1978: 419–424.

7.20 Drugli JM, Rogne T, Valen S, Johnsen R, Olsen S. Corrosion test methods for evaluating stainless steels for well flow applications. Corrosion/90, Las Vegas: NACE, Paper No 270, 1990.

7.21 Annual Book of ASTM Standards Part 10. Philadelphia: ASTM.

7.22 Ijsseling FP. Electrochemical methods in crevice corrosion testing. British Corrosion Journal, 15, 1980; 1: 51–69.

7.23 Strehblow H-H. Mechanisms of pitting corrosion. In: Marcus P, editor. Corrosion Mechanisms in Theory and Practice. 2nd ed. New York–Basel: Marcel Dekker, 2002.

7.24 Kaesche H. Die Korrosion der Metalle. Berlin–Heidelberg–New York: Springer-Verlag, 1966.

7.25 Pourbaix M. Electrochemical aspects of stress corrosion. In: Scully JC, editor. The Theory of Stress Corrosion Cracking in Alloys. Brussels: NATO, 1971.

7.26 Aziz PM. Application of the statistical theory of extreme values to the analysis of maximum pit depth for aluminium alloys. Corrosion, 12, 1980: 35–46.

7.27 Godard HP, Jepson WB, Bothwell MR, Kane RL. The Corrosion of Light Metals. New York–London: John Wiley & Sons, 1967.

7.28 Unpublished work at SINTEF Corrosion Centre, 1977.

7.29 Mattson E. Locaized corrosion. British Corrosion Journal, 13, 1978; 1: 5–12.

7.30 Uhlig HH. Corrosion and Corrosion Control. New York–London: John Wiley & Sons, 1971

7.31 Dundas HJ, Bond AP. Niobium and titanium requirements for stabilization of ferritic stainless steels. In: Steigerwald RF, editor. Intergranular Corrosion of Stainless Steel. STP 656. Philadephia: ASTM, 1978.

7.32 Friend WZ. Corrosion of Nickel and Nickel-Base Alloys. New York–London: John Wiley & Sons, 1980.

7.33 Unpublished work at SINTEF Corrosion Centre.

7.34 Hutchings IM. The Erosion of Materials by Liquid Flow. MTI Publ. No 25. Materials Technological Institute of the Chemical Processing Industry, 1986.

7.35 DNV. Recommended practice RP 0501. Erosive Wear in Piping Systems. Oslo: Det Norske Veritas, 1996.

7.36 Ives LK, Ruff AA. Transmission and scanning electron microscopy studies of deformation at erosion impact sites. Wear, 46, 1978: 149–162.

7.37 Elison BT, Wen CJ. Hydrodynamic Effects on Corrosion, AIChe, Symposium Series. The American Institute of Chemical Engineers, 1981.

7.38 Bardal E, Eggen TG, Stølan Langseth Aa. Combined erosion and corrosion of steels and hard-metal coatings in slurries of water and silica sand. Helsinki: Proceedings 12th Scandinavian Corrosion Congress/Eurocorr'92, 1992.

7.39 Bardal E, Rogne T, Bjordal M, Berget J. Rates and mechanisms of combined Erosion and corrosion of metallic–ceramic coatings/surfaces. Proceedings Conference on Organic and Inorganic Coatings for Corrosion Prevention, Nice 1996. London: Institute of Materials, 1997; 278–290.

7.40 Gartland PO, Bardal E, Andersen RE, Johnsen R. Effects of flow on the cathodic protection of a steel cylinder in sea water. Corrosion, 40, 1984, March.

7.41 Bernhardson SO, Mellstrøm R, Tynell M. Sandvik 2RK-65, a high alloy stainless steel for sea water cooling. Symp. Advanced Stainless Steels for Seawater Applications. Ass. Italiana de Metallurgia – Climax Mo. Co. 1980. (After Coit RL. Inco Power Conference, Lausanne, 1967.)

7.42 Gemmel G, Nording S. Monit and 904L – two high alloy stainless steels for seawater applications. Symp. Advanced Stainless Steels for Seawater Applications. Ass. Italiana de Metallurgia – Climax Mo. Co. 1980. (after Coit RL. Inco Power Conference, Lausanne. 1967.)

7.43 Efird KD. Effects of fluid dynamics on the corrosion of copper–base alloys in seawater. Corrosion, 33, 1977: 3–8.

7.44 Syrett BC. Erosion corrosion of copper–nickel alloys in seawater and other aqueous environments – A literature review. Corrosion, 32(6), June 1976.

7.45 Friedersdorf FJ., Holcomb R. Pin-on-disk corrosion wear test. Journal of Testing and Evaluation JTEVA, 26(4); 1998: 352–357.

7.46 Batchelor AW, Lam LN, Chandrasekaran M. Materials Degradation and Its Control by Surface Engineering. Singapore: Imperial College Press/World Scientific Publishing, 1999; 189–197, 43.

7.47 Waterhouse RB. The effect of environment in wear processes and the mechanisms of fretting wear. In: Suh NP, Saka N, editors. Fundamentals of Tribology. Massachusetts: The MIT Press, 1978; 567–584.

7.48 Sproles Jr ES, Gaul DJ, Duquette DJ. A new interpretation of the mechanisms of fretting and fretting damage. In: Suh NP, Saka N, editors. Fundamentals of Tribology. Massachusetts: The MIT Press, 1978; 585–596.

7.49 Lees DJ. Characteristics of stress corrosion fracture initiation and propagation. Metallurgist and Materials Technology, 14, 1982; 1: 29–38.

7.50 Scully JC. Stress corrosion cracking; Introductory remarks. In: Scully JC, editor. The Theory of Stress Corrosion Cracking in Alloys. Brussels: NATO, 1971.

7.51 Staehle RW. Stress corrosion cracking of the Fe–Cr–Ni alloy system. In: Scully JC, editor. The Theory of Stress Corrosion Cracking in Alloys. Brussels: NATO, 1971.

7.52 Sandoz G, Fujii CT, Brown BF. Solution chemistry within stress–corrosion cracks in alloy steels. Corrosion Science, 10, 1970; 1: 839–845.

7.53 Scully JC. Fractographic aspect of stress corrosion cracking. In: Scully JC, editor. The Theory of Stress Corrosion Cracking in Alloys. Brussels: NATO, 1971.
7.54 Ford FP. Stress corrosion cracking. In: Parkins RN, editor. Corrosion processes. London: Applied Science Publishers, 1982.
7.55 Edwards J. Internal corrosion offshore pipelines. Final report NTNF-project B.0601.T121. Part project 3, Trondheim: SINTEF.
7.56 Kitagava H. Corrosion fatigue, chemistry, mechanics and microstructure In: Devereux O, McEvily AJ, Staehle RW, editors. Corrosion Fatigue. Houston: NACE 2. 1972: 521–528.
7.57 Duquette DJ. Corrosion fatigue, chemistry, mechanics and microstructure. microstructure In: Devereux O, McEvily AJ, Staehle RW, editors. Corrosion Fatigue. Houston: NACE 2. 1972: 12–22.
7.58 Congleton J, Craig IH. Corrosion fatigue. In: Parkins RN, editor. Corrosion Processes. Applied Science Publishers, 1982.
7.59 Bardal E. Effects of marine environment and cathodic protection on fatigue of structural steels. In: Almar-Næss A, editor. Fatigue Handbook. Trondheim: Tapir, 1985.
7.60 Bardal E, Eggen TG, Grøvlen MK. A quantitative determination of crack closure effects on fatigue properties of cathodically protected steel in sea water. In: Scott P, Cottis RA, editors. Environment-assisted Fatigue. EGF Publication 7, London: Mech Eng Publ, 1990.

EXERCISES

1. Compare data in Table 7.3 and Figure 7.5. Chose aluminium as example of anodic material and carbon steel, copper, titanium and stainless steel as examples of cathodic materials.

   Express on this basis if the galvanic series is suitable for indication of galvanic corrosion rates. Which property determines the galvanic current density in Table 7.3? Explain the data in Table 7.3 by means of schematic polarization curves of the materials in question.

2. The steel plate in Exercise 2 in Chapter 6 is in metallic contact with a copper plate under conditions described in the exercise. Assume that the steel plate corrodes uniformly, that it is without deposits, that anodic dissolution of copper is negligible, that the overvoltage curves for oxygen reduction on steel and copper are identical, and that the conductivity of the solution is high. Draw the overvoltage curves for oxygen reduction and dissolution of steel and find the corrosion current density on steel under the following conditions:

   a) The surface areas of steel and copper are 100 cm^2 and 1 cm^2, respectively.
   b) The surface areas of the two materials are both equal to 100 cm^2.

c) The surface area of copper is 100 cm^2 and that of steel is 1 cm^2.
d) The metallic contact between steel and copper is broken.

Additional information:
Equilibrium potential of $Fe^{2+} + 2e^- = Fe$: $E_{0a} = -0.62$ V,
Exchange current density of $Fe^{2+} + 2e^- = Fe$: $i_{0a} = 0.01\ \mu A/cm^2$,
Tafel constant of $Fe^{2+} + 2e^- = Fe$: $b_a = 60$ mV/decade c.d.,
Tafel constant of oxygen reduction: $b_c = 120$ mV/decade c.d.,
Exchange current density of oxygen reduction: $i_{0c} = 0.1\ \mu A/cm^2$.

The overvoltage curves should be drawn with log current (log I) on the horizontal axis.

3. The outlet from a tank made of stainless steel includes a short unalloyed steel tube that is insulated from the tank and formed as shown in Figure 1. There is a water flux of 600 litres/min through the outlet. The corrosion rate dependence on flow rate is schematically shown in Figure 2. It is found by experiments that:

   $v_2 = 12$ m/s under the actual conditions,
   the corrosion current density at v_2 (critical current density) = 240 $\mu A/cm^2$,
   the passive current density = 0.1 $\mu A/cm^2$,
   the Tafel constant of dissolution of steel in the upper part of the active potential range is 200 mV/current decade,
   and the limiting current density of oxygen reduction in the actual flow velocity range is proportional to $v^{0.8}$.

   Additional data:

   Faraday's number = 96,500 C/mol e^-,
   Density of steel = 7.8 g/cm^3,
   Atomic weight of iron = 56.

   Assume that the cathodic reaction is diffusion controlled in the whole actual potential range.

   Determine the corrosion current density and the corrosion rate in mm/year in the narrow and in the wide part of the unalloyed steel tube under each of the following conditions a)–d). Illustrate each case with schematic anodic and cathodic potential–log current curves, and indicate on the curves the potentials in the respective parts of the tube.

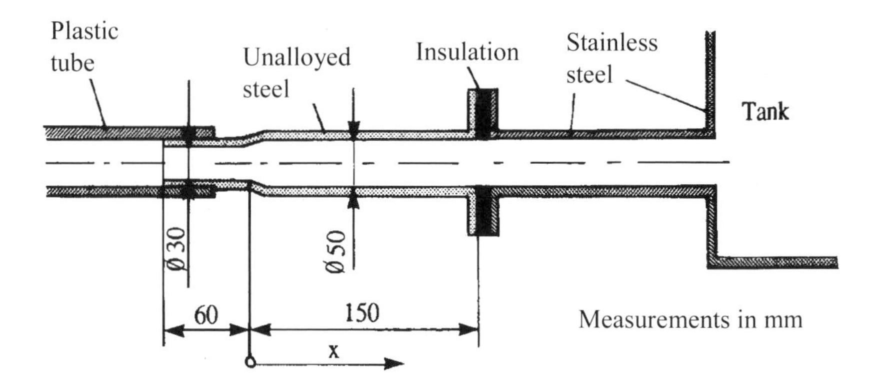

Figure 1 Outlet from tank.

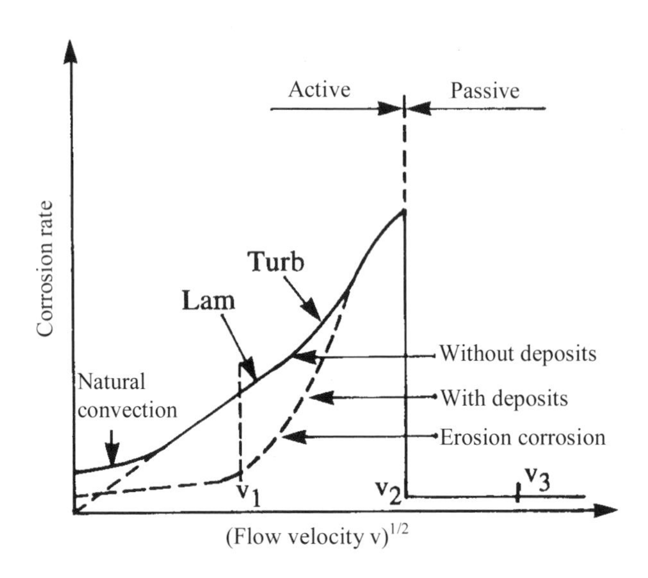

Figure 2 Corrosion rate as a function of (flow velocity v)$^{1/2}$.

a) The conductivity of the water is so low that any galvanic effect between the two parts of the unalloyed steel tube can be neglected.

b) The conductivity of the water is so high that the potential in the tube can be considered equal all over.

c) The conductivity is at a level causing a potential variation implying that the narrow part of the steel tube is in the passive potential range and the wide part is in the active potential range. For simplicity, assume that the potential is constant over the whole surface in the wide part.

d) The conditions are like those in item c) except that there is a potential drop of 50 mV from one end to the other of the wide part of the unalloyed steel tube. Assume that the potential in this part of the tube varies linearly with the distance x (see Figure 1). In this case, draw a curve showing the corrosion current density and the corrosion rate in mm/year as functions of x.

4. Propose a simple improvement of the design in Exercise 3 with respect to corrosion.

5. A copper pipe and a pipe of unalloyed steel are joined as shown in Figure 3. Water with a resistivity of the order of 300 ohm cm is flowing through the pipeline. Assume that the oxygen reduction reaction is diffusion controlled on both materials, and that the dissolution of steel is activation controlled. Compare with a couple of examples in Section 7.3 (Figures 7.13 and 7.11).
   Sketch schematically how the potential and the anodic as well as the cathodic current density vary in the region ± 20 cm from the joint.
   What role does the flow direction play?

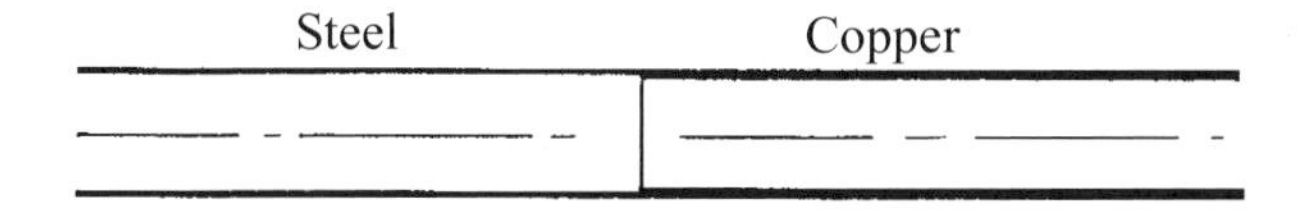

Figure 3. Joined steel and copper pipe.

6. A tank for storing an aqueous solution has a bottom made of stainless steel and a mantel made of unalloyed steel. The bottom and the mantel are welded together. The mantel and the weld are painted. Explain what happens if a small defect (a "holiday") arises in the paint coating on the unalloyed steel. How could this have been avoided? (There are various ways.)

7. In a pipe system carrying water with some salt dissolved, a section of a polymeric composite pipe is inserted between a pipe of stainless steel and a pipe

of aluminium. What is the role of the composite pipe section in the following cases?

a) No metallic connection exists between the stainless steel pipe and the aluminium pipe anywhere in the system.

b) Such a contact exists, but not near the composite pipe section.

8. In two test pipe systems for seawater, joints between a pipe of brass (CuZn37) and a pipe of copper–nickel (CuNi10) are designed in two different ways, A and B, as shown in Figure 4.

   By means of potential measuring probes as shown at positions a_1, a_2, b_1 and b_2, the following potential values (versus saturated calomel electrode, SCE) are measured early in the exposure period:

   a_1: –180 mV, a_2: –280 mV.
   b_1: –244 mV, b_2: –264 mV.

   By other measurements it is shown that the cathodic reaction at this time is limited by oxygen diffusion, with a diffusion-limiting current density i_L = 30 μm/cm^2 on both materials, and that the anodic Tafel constants are

   CuNi10: 70 mV/decade
   CuZn37: 90 mV/decade.

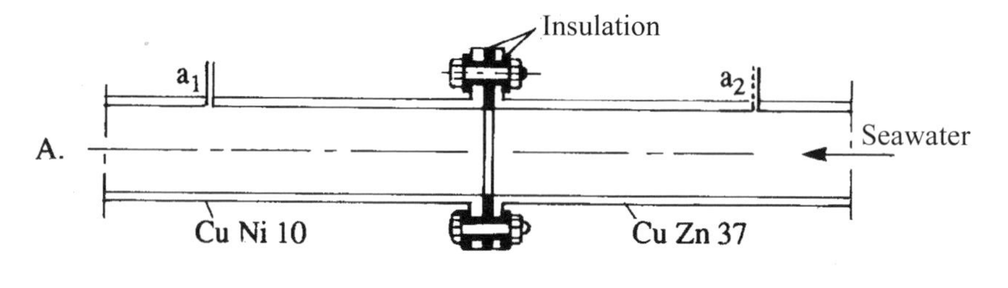

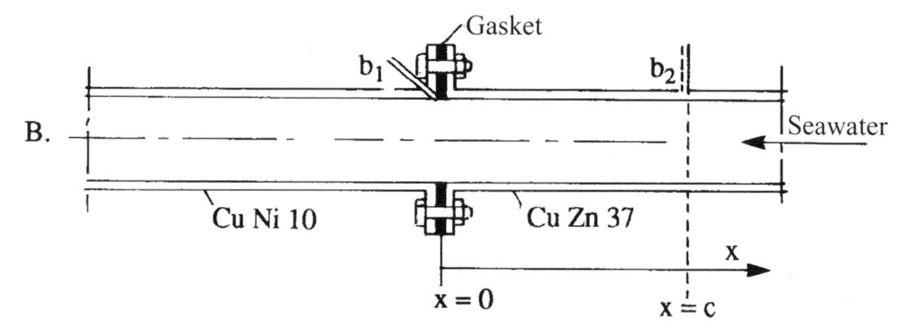

Figure 4 A joint between a pipe of brass and a pipe of copper–nickel designed in two different ways.

a) Draw on semi-logarithmic graph paper the actual part of the overvoltage curves for the two materials (use a large scale on the potential axis), and determine the corrosion current densities (I_{corr}) at the four points a_1, a_2, b_1 and b_2. At point b_1, the corrosion current density on both materials is to be determined.

b) Calculate on the basis of the situation at point b_1, the effective area ratio between copper–nickel and brass in joint B (i.e. the area ratio that corresponds to conditions under which the current densities everywhere on the respective materials were equal to the determined values at point b_1)

c) For simplicity, assume that the ohmic potential drop in the seawater (ΔE) from the joint to various positions in the brass pipe is proportional to the distance x (see the figure), and draw to scale a curve showing the corrosion current density on CuZn37 as a function of x in the region $0 \leq x \leq 3c$ (where $x = c$ is the position of the probe b_2). State how much of the corrosion current density is due to self-corrosion and to the galvanic couple, respectively.

d) Discuss the assumption in exercise c) and sketch schematically a real $x - i_{corr}$ curve in comparison with that determined in c).

e) Assume an internal diameter of $d_i = 5$ cm, and $c = 15$ cm, and determine on the background of the $x - i_{corr}$ curve drawn in item c) the total galvanic current I_g between the two materials.

f) Show by a schematic drawing the simplest way of checking the calculated value of I_g by direct measurement.

g) Calculate the average corrosion depth in the brass pipe at position $x = 0$ in joint B after one year of exposure, provided that the nominal electrochemical conditions do not change during the year and that the corrosion process is dezincification.

Density of brass = 8.6 g/cm^3,
Atomic weight of Zn = 65,
Faraday's number = 96500 C/mol e^-.

h) May the galvanic corrosion in joint B be avoided by making the gasket of an insulating material?

9. Through a pipe with the joint shown in Figure 5 cold seawater is flowing slowly. May corrosion occur in the stainless steel pipe if the material is a) conventional 18-8 steel (AISI 304); b) AISI 316; c) high-alloy austenitic stainless steel with 6% Mo?

10. A tank made of stainless AISI 304 steel (18–8 CrNi) contains stagnant seawater of 15 °C. In the course of few months, corrosion has penetrated a 4 mm thick plate in the bottom. State cause, corrosion form and how the conditions can be improved.

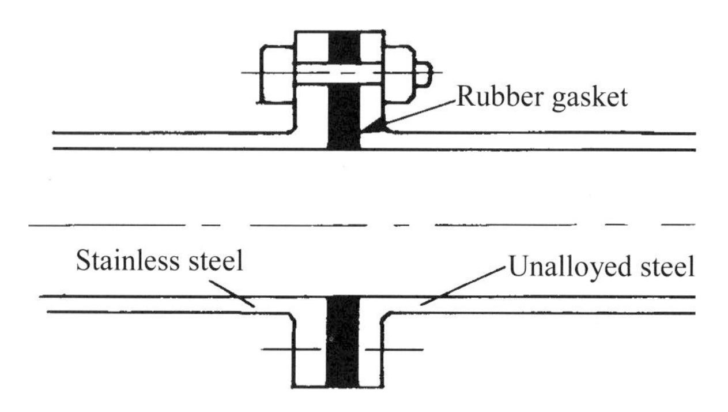

Figure 5 Joint between a pipe of unalloyed steel and a pipe of stainless steel carrying seawater.

11. The same steel (AISI 304) is used in a pipeline carrying water to which ferric chloride ($FeCl_3$) is added to such an extent that the free potential is about 600 mV versus the saturated calomel electrode (SCE). The temperature is 50°C. Explain what will happen. Propose alternative material selections.

12. In the steel dealt with in Exercises 10 and 11, cracks may arise even though there is no significant stress. What does this indicate about the composition and treatment of the material? How can we prove which corrosion form has acted and that the necessary conditions for this corrosion form have existed?

13. In a case of pitting corrosion on an aluminium sheet in seawater, the largest pit depth is 200 μm after 2 months. What will be the maximum depth expected after 1 year? After 10 years?

14. Water containing 0.01 wt% sand is flowing through a pipe at a flow velocity of 60 m/s. Describe how the annual erosion depth increment in the pipe wall can be calculated on the basis of short duration tests in a similar pipe where the sand concentration in the water is varied in the range 0.1–10 wt% and the flow velocity in the range 20–40 m/s. Assume in the first instance that the deterioration process is mainly erosion and only little corrosion.

    If there is a significant corrosion component, how can the magnitude of the corrosion contribution be found by experiments? How should a large corrosion contribution be taken into account when the results are roughly transferred from one set of conditions (velocity and sand concentration) to another?

15. A plate with width 300 mm and thickness 30 mm is made of steel with yield strength 900 MPa and fracture toughness 90 MPa. In a certain corrosive environment a critical stress intensity factor for stress corrosion cracking is found to be K_{ISCC} = 70 MPa.

By non-destructive testing, a crack-like defect has been detected, reaching 27 mm into the material from the edge, as shown in Figure 6.

a) Can linear elastic fracture mechanics be applied in this case, in dry air and in the said corrosive environment, respectively?

b) How large can the tensile force P be without leading to brittle fracture or stress corrosion fracture in the two cases?

Information:
The geometry function Y in the equation $K = Y \sigma \sqrt{\pi a}$ is set = 1.21.

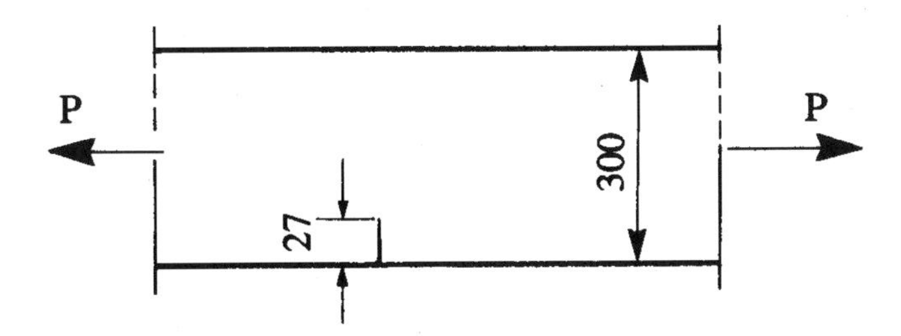

Figure 6. High–strength steel plate with an edge crack.

8 Corrosion in Different Environments

... on earth, where moths and rust destroy ...

8.1 Atmospheric Corrosion

8.1.1 Environmental Factors and Their Effects

Atmospheric corrosion is electrochemical corrosion in a system that consists of a metallic material, corrosion products and possibly other deposits, a surface layer of water (often more or less polluted), and the atmosphere. The general cathodic reaction is reduction of oxygen, which diffuses through the surface layer of water and deposits. As shown in Section 6.2.5, the thickness of the water film may have a large effect, but it is more familiar to relate atmospheric corrosion to other parameters. The main factors usually determining the accumulated corrosion effect are time of wetness, composition of surface electrolyte, and temperature. Figure 8.1 shows the result of corrosion under conditions implying frequent condensation of moisture in a relatively clean environment (humid, warm air in contact with cold metal).

The parameters that determine time of wetness and composition of surface electrolyte have been surveyed by Kucera and Mattson [8.1]. They present also a thorough description of the mechanism, with thermodynamic and kinetic aspects of corrosion on various materials. For instance, they consider potential–pH diagrams as a useful thermodynamic basis for understanding atmospheric corrosion.

The three main factors, i.e. time of wetness, composition of electrolyte, and temperature, depend for one thing on the type of atmosphere:

1. Rural (inland): dry environment with little or no pollution.
2. Marine: on and by the sea, with high humidity and chlorides.
3. Urban: polluted by exhaust, smoke and soot.
4. Industrial: highly polluted by industry smoke and precipitate.

Climatic conditions are also important. For instance, the corrosion conditions may be severe in the tropical zone due to high temperature and – particularly at the coast – high humidity. However, the picture is not clear. There are also tropical marine atmospheres which are moderately corrosive, and more corrosive atmospheres are

Figure 8.1 Corrosion damage of a 14-year-old ventilation outlet from a bathroom in a private house. (Photo: Sigmund Bardal.)

found in the typical industrial areas in the economically developed countries. The ranking also depends on the corroding material [8.2]. The effects of climatic conditions and other factors are described in various handbooks, recently by Roberge [8.3], who also presents the ISO classification system and standard.

If the metal surfaces are completely clean and smooth, and there is no pollution of the atmosphere, significant corrosion does not occur at a relative humidity below 100%. In reality, however, materials like steel, copper, zinc and nickel corrode when the relative humidity exceeds 50–70%. This is partly due to corrosion products. An example is shown in Figure 8.2. If the surface in addition is contaminated by chloride, soot or dust particles, which are hygroscopic, the corrosion rate is considerable at lower relative humidity than shown in the figure, e.g. when it exceeds 60%.

The significance of dust is mentioned above. Industrial and urban atmospheres contain more or less solid particles consisting of carbon, soot, sand, oxides, and salts, e.g. chloride and sulphate. Many of these substances attract moisture from the air; some of them also attract polluting and corrosive gases. The salts cause high conductivity, and carbon particles can lead to a large number of small galvanic elements because the particles act as efficient cathodes after deposition on the surface.

Rain affects the corrosion rate in different and opposite ways. Some effects stimulate corrosion, but rain may also wash away pollutants and thus reduce the corrosion rate. The total effect depends on, among other factors, how frequently and heavily it is raining, and on the degree of pollution of the atmosphere. Seldom and heavy rain showers are beneficial. Direct rain on a surface is most favourable at places with strongly polluted atmospheres [8.1].

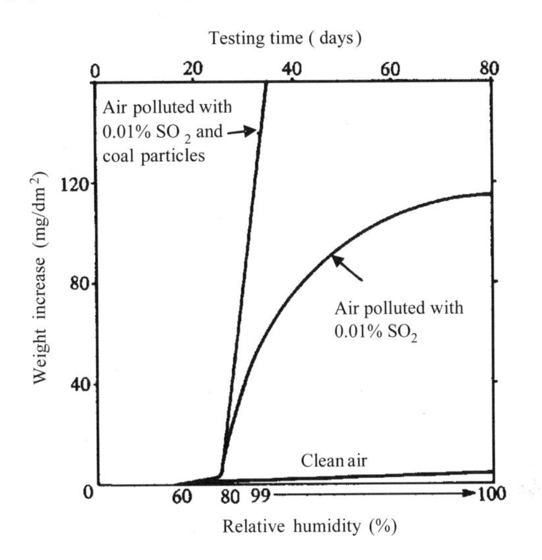

Figure 8.2 Effect of relative humidity and air pollution on atmospheric corrosion of iron. (From Shreir [8.4] after Vernon.)

The most significant pollutant in the form of a gas is SO_2, originating from combustion of oil, gas or coal containing some sulphur. Together with water, SO_2 forms sulphuric acid, H_2SO_4. Therefore, metallic materials that corrode strongly in acids are not suitable without a coating in industrial *atmospheres with high contents of* SO_2.

In *marine atmospheres* measured corrosion rates are related to the rates of chloride deposition.

In Table 8.1 it is shown how the type of atmosphere and the deposition rates of SO_2 and Cl^- at various places in Europe affect the corrosion rate of carbon steel. It is seen that *the corrosion rates within the actual climate zone correspond well to the deposition rates.*

The corrosivity of atmospheres in other parts of the world varies roughly within the same ranges as those shown for Europe.

Automobile corrosion worries a huge number of people and is of large economic significance. Internal cavities in doors and nearly closed sections are critical places. This is due to varying temperature and bad ventilation, leading to formation and conservation of condensate, i.e. long time of wetness. In an investigation carried out in Gothenburgh, Sweden [8.6], it was found that SO_2 plays a considerable role in the corrosion conditions in internal cavities, while chloride is more significant for the external corrosion of the car. Deposits are strongly polluted, and they attract and preserve moisture. Therefore, a dirty car corrodes continuously most of the time.

Table 8.1 Typical corrosion rates for carbon steel in different types of atmospheres. (Adapted from References [8.1, 8.5].)

Type of atmosphere	Deposition rates (mg/m^2 d)		Steady-state corrosion rates (μm/year)	Comments
	SO_2	Cl^-		
Rural	<20	<3	5–10	Measured at various places in Scandinavia and Eastern and Western Europe
Urban	20–100	<3–50	10–30	
Industrial	110–200		30–60	
Marine	<10	3 – >100	10–40	Measured after 4 years of exposure at various places in Scandinavia*
Arctic	<10	<3	4	Measured after 4 years of exposure, Gällivare, nothern Sweden

* Corrosion rates at higher atmospheric temperatures, such as in the tropical zone, may be several times higher.

8.1.2 Atmospheric Corrosion on Different Materials

As shown in the previous section, the corrosion rate of *steel* depends strongly on the type of atmosphere. Some alloying elements also have a large effect on the corrosion rate. For instance, small amounts of Cu, Cr and Ni cause the rust layer formed during service to be denser, so that the corrosion rate becomes relatively low after some time. Steels alloyed in this way are the so-called weathering steels, of which Cor-Ten steel is a well-known product. The rust layer is often decorative with nice colours, and such steels are used in front panels for buildings. Figure 8.3 shows corrosion measurements on different steels at four places in England, namely one rural site (a), one with a marine atmosphere (b), and two industrial sites (c and d, the latter one most polluted). It is seen from Figure 8.3 that use of weathering steel is most beneficial in an aggressive industrial atmosphere. It should also be noticed that the weathering steels are more corrosion resistant than ordinary structural steels only in the atmosphere.

Several *aluminium alloys* show very good corrosion resistance in various atmospheres. Some pitting occurs, but the pits remain small. Maximum depth seldom exceeds 0.5 mm during 6–20 years of exposure; it is usually in the order of 0.1 mm. Some alloys may, however, be attacked by intergranular corrosion or exfoliation corrosion (see Section 7.7). Extensive galvanic corrosion may occur on aluminium in contact with copper, mild steel (in marine atmosphere) and graphite, less in contact with stainless steel, while aluminium is compatible with zinc [8.2].

More about aluminium is presented in Sections 10.1.8 and 10.6.

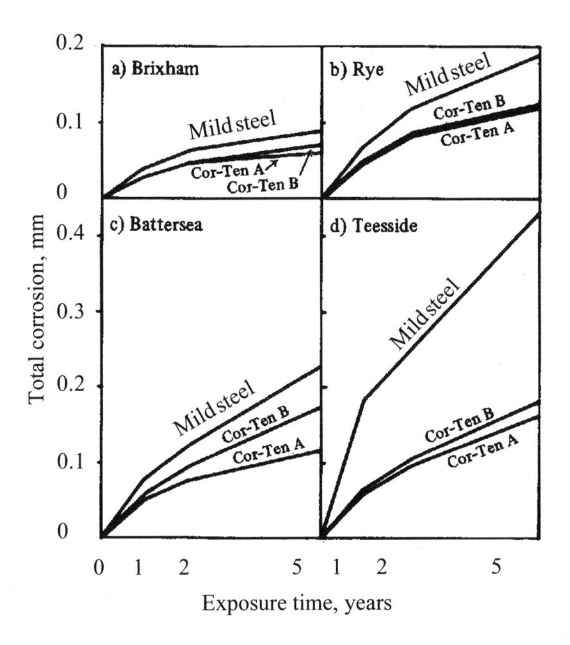

Figure 8.3 Corrosion–time curves for three different steels exposed four places in England [8.7].

Stainless steel behaves mainly in accordance with its name under atmospheric conditions.

The corrosion rate of *zinc* depends more upon the type of atmosphere, as shown in Table 8.2. For comparison, corrosion rates in different waters are also shown in the table. The low rates of corrosion on zinc are due to protecting surface films of basic zinc chloride and carbonate.

Copper and copper alloys are highly resistant to atmospheric corrosion because of surface films mainly composed of basic copper salts. The corrosion rate is below 2–3 μm/year [8.9]. *Tin* as well as *nickel* and *nickel alloys* also corrode at similar rates. *Lead* possesses excellent corrosion resistance in atmospheres due to surface-protecting films (insoluble sulphate, sulphide, carbonate and oxide).

The most important methods for preventing atmospheric corrosion are 1) to select a corrosion-resistant structural material, 2) to use a corrosion-resistant coating (metallic, inorganic, organic or composite), and 3) to close internal cavities in order to avoid access of humid air, or conversely, 4) to make sure that there is ample ventilation and drainage, and possibly 5) to supply oil, grease, or special substances such as tectyle to the cavities.

Table 8.2 Corrosion rate of zinc in different environments (μm/year). (Adapted from Reference [8.8].)

Rural atmospheres:		*River water:*	
Tempered climate	0.7–2.8	Soft	14–36
Tropical, humid climate	0.14–4.3	Hard	2.8–4.3
Urban atmosphere:		*Tap water:*	
Large cities	2.8–5.7	Soft 20°C	14–36
Small towns	0.7–2.8	Hard 20°C	14 (average)
Marine atmospheres:			
Coastal climate, Europe	0.7–2.8	*Seawater*	21–85
Onboard ships	6 (average)		
Urban and marine atmospheres:			
(West coast Sweden)	4.3 (average)		
Industrial atmosphere:			
Very aggressive	14–100	*Rain water:*	
Normal	4.3–14		
		pH = 4.3	14–17
Indoors	0.07–0.6	pH = 5.8	3.6–4.3
		pH = 6.1	2.8–3.6

8.2 Corrosion in Fresh Water and Other Waters

This group of environments comprises primarily fresh water in lakes, rivers and brooks, rain and ground water. In addition there are waters contaminated in some way, or with some matter added.

A common condition for corrosion in natural waters is the presence of oxygen. The rate-determining step is often diffusion of oxygen to the metal surface (see Chapter 6). This is particularly the case for corrosion of unalloyed or low-alloy *steels*. At least partly it applies also to *zinc* when pH < 8–9, where reduction of hydrogen may play an additional role. On the other hand, materials like *aluminium* and *stainless steels* (and Zn at higher pH values) are under anodic control (passivation). Contrary to atmospheric corrosion, corrosion in natural waters is hardly dependent on the grade of iron, unalloyed and low alloy steel, because *composition and treatment of the steel* in this case have little effect on the barrier properties of surface layers of corrosion products. In the pH ranges < 4 and > 10, however, the composition of the steel or iron is significant because the corrosion process involves hydrogen evolution and passivation, respectively. At low pH values, the corrosion rate increases with increasing contents of C, N, P and S, and with cold working. The reason is depolarization, partly of the hydrogen reduction and partly of the anodic dissolution reaction. The effects of some other elements, such as As, Mn, Cu, Cr and Ni, depend on the concentration [8.10]. The properties of steel and other metallic materials in acidic environments are of interest when applied to ground water with sulphate-reducing bacteria and water contaminated by H_2S.

Oxygen concentration, temperature and flow velocity have large effects on oxygen transport to the metal surface and consequently on the corrosion rate in natural waters. These factors have been dealt with quantitatively in Chapter 6. In addition, variation in corrosion rate is also caused by the composition of the water, since this may strongly affect the barrier properties of surface layers.

Natural waters contain different amounts of *Ca and Mg salts*. If the salt content is high, the water is called hard, in the opposite case it is soft. In Figure 8.4, the corrosion rate of iron in soft water is shown as a function of pH. The constant corrosion rate in the pH range 4–10 is caused by, and emphasizes, that the corrosion rate is controlled by oxygen diffusion, and that the deposits formed on the surface in this pH range (in soft water) have similar barrier properties. At pH < 4, iron corrodes faster because of hydrogen reduction, while the lower corrosion rate at pH > 10 may partly be due to diffusion-limiting deposition (see below) and partly to passivation.

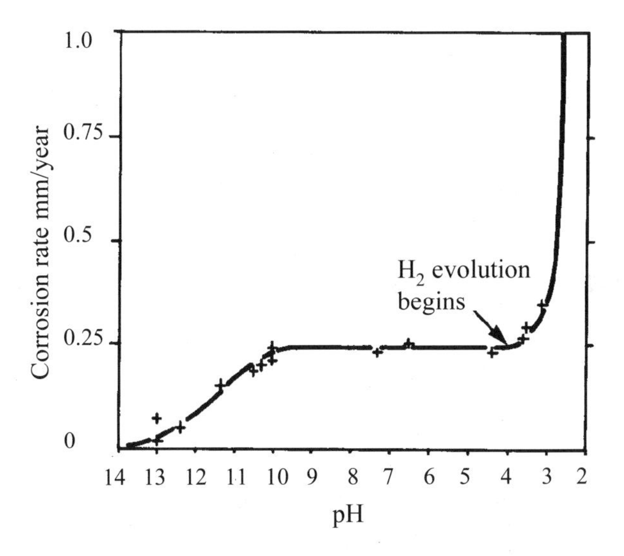

Figure 8.4 Effect of pH on corrosion of iron in aerated soft water at room temperature. (From Uhlig [8.10] after Whitman, Rusell and Altieri.)

Hard waters are known to be less corrosive than soft ones. The reason for this is that, when the salt content becomes high enough, deposits – particularly $CaCO_3$ – are formed, which hinder the oxygen diffusion to the metal surface. However, deposition of salts depends also on the pH (under otherwise equal conditions, pH must exceed a certain limit), on total alkalinity (which is determined by titration with methyl orange colour indicator), and on dissolved substances in the water. The liability to deposition of calcium carbonate is expressed by the Langelier saturation index SI [8.4]:

$$SI = pH - pH_s \tag{8.1}$$

where the term pH expresses the actual pH of the water, and pH_s is the pH value at saturation with respect to $CaCO_3$:

$$pH_s = pCa + pAlk + (pK_2 - pK_s) \text{ at constant temperature,} \tag{8.2}$$

where

pCa = –log (calcium concentration expressed as ppm $CaCO_3$),
$pAlk$ = –log (alkalinity to metyl orange expressed in ppm of equivalent $CaCO_3$),
pK_2 = ionization constant of HCO_3^-, $[H^+][CO_3^{2-}]/[HCO_3^-]$ and
pK_s = solubility product of $CaCO_3$.

When the measured pH is higher than pH_s, i.e. the saturation index SI is positive (the water is supersaturated with $CaCO_3$), $CaCO_3$ is deposited. Good correlation has been found between SI and corrosivity of the water. An SI value of 0.5 is considered satisfactory. Higher values may easily lead to excessive deposits. To obtain a positive SI in an originally soft water, $Ca(OH)_2$ or Na_2CO_3 can be added.

The saturation index increases usually with increasing temperature (the solubility of $CaCO_3$ decreases), but in the pH intervals given in Table 8.3, the effect of temperature is relatively small. Thus, in these intervals the liability to deposit formation is about the same in cold and hot water.

Table 8.3 Alkalinity and corresponding pH intervals giving nearly equal liability to $CaCO_3$ deposition at different temperatures. (From Uhlig [8.10] after Powell, Bacon and Lill.)

Alkalinity (ppm, as $CaCO_3$)	pH (measured at room temp.)
50	8.10–8.65
100	8.60–9.20
150	8.90–9.50
200	8.90–9.70

As a rule, the saturation index is a useful qualitative guide in connection with diffusion-controlled corrosion, but certain types of organic and inorganic particles as well as complex ions may prevent salt deposition on the surface and thereby make the SI invalid [8.10].

Soft waters attack pipeline materials and cause extensive leakage in many drinking water pipe systems. In this connection it has been recommended that the ratio between the concentration of bicarbonate HCO_3 and the concentration of SO_4^{2-} and Cl^-, respectively (in mg/l) should be higher than 1. The pH value should be between 7.4 and 8.3, the oxygen content about 6 mg/l, and the water velocity ≥ 0.5 m/s. The bicarbonate content should normally be increased to 50–60 mg/l, e.g. by addition of CO_2 and calcium compounds.

The effects of water hardness and pH on corrosion of Zn can be seen in Table 8.2.

In addition to calcium salts, water often contains varying amounts of other salts. The following brief comments on the various salts and their effects are mainly based upon Reference [8.10]. One group comprises those based on an alkali metal, NaCl being the most important example. Chloride content in different types of water and its effect on stainless steels is dealt with in Section 10.1.5. Figure 8.5 shows how the corrosion rate of iron depends on the NaCl concentration in the water. There are two reasons for this dependence:

1. The solubility of oxygen in water decreases with increasing chloride content, which tends to reduce the corrosion rate over the whole chloride concentration scale.

2. At low chloride content, item 1 is, according to Uhlig, overshadowed by the following aspect: the conductivity of the water is so low that the anodes and cathodes are located very close to each other, with the result that iron from the anodes and OH^- from the cathodes are combined to iron hydroxide in close contact with the metal surface. Therefore it forms a more efficient barrier against oxygen diffusion than in the case of high conductivities, where the hydroxide is formed less close to the metal. There is some uncertainty whether this is a complete explanation, but anyway there is a more pronounced positive effect the lower the conductivity is. Other alkali metal salts, such as Na_2SO_4, NaBr and KCl, show about the same effects as does NaCl.

Salts of earth alkali metals (e.g. $CaCl_2$) are a little less corrosive than the alkali metal salts.

Acid salts, i.e. salts that hydrolyze and give acid solutions, lead to corrosion under combined hydrogen evolution and oxygen reduction to the extent that may be expected at the actual pH. The oxygen reduction reaction may be more efficient than usual. Examples of such salts are $AlCl_3$ and $FeCl_3$.

Ammonium salts (e.g. NH_4Cl) are also acidic, and they give higher corrosion rates than other salts at the same pH. This applies particularly to ammonium nitrate, which in the presence of NH_3 may give extreme corrosion rates.

Alkaline salts (e.g. Na_3PO_4, Na_2SiO_3, and Na_2CO_3), which hydrolyze and give alkaline solutions, act as corrosion inhibitors by passivation of iron and steels under access by oxygen. Oxide is formed, but in some cases one get a more or less passivating salt layer in addition (phosphate, silicate).

Oxidizing salts may either lead to rapid corrosion by providing an extra cathodic reaction (as do $FeCl_3$, and $CuCl_2$), or to inhibition and passivation (Na_2CrO_4,

$NaNO_2$). The latter two are dealt with under inhibitors in Section 10.2.

When copper piping is used in tap water systems, one will, particularly in hot water pipes, and also in cold water pipes in periods with little or no tapping, get an increased concentration of copper ions, sometimes to a great extent. After tapping of such water into containers, pots or pans of less noble metals (e.g. aluminium), copper is deposited, and heavy pitting corrosion may occur.

The chemical composition of water may deviate from its natural level due to addition of chlorine. When chlorine is added to tap water, it is normally used at less than 1 mg Cl_2/l, and the corrosivity is assumed not to be affected. In swimming pools, higher concentrations are used, and at some places above the water, local increase of chlorine concentration has taken place, which has caused attack on stainless steels. Concentrations of 10–50 ppm have led to increased pitting corrosion on aluminium after a short time [8.2].

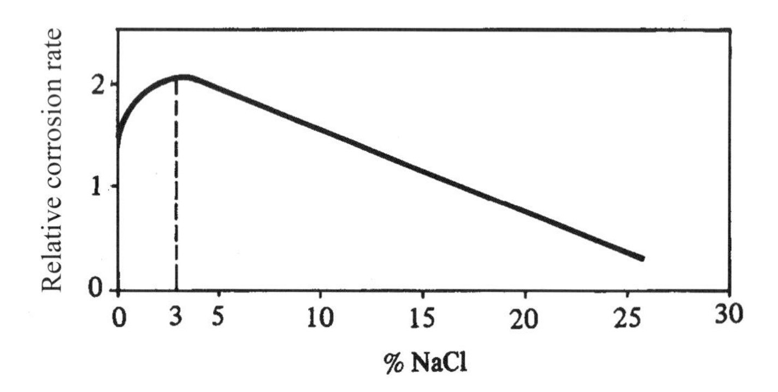

Figure 8.5 Effect of NaCl concentration on corrosion rate of steels in aerated solutions at room temperature [8.10]. (Reproduced from Uhlig HH. Corrosion and Corrosion Control. 2nd Ed. New York: John Wiley & Sons, 1971.)

Organic matter in the water may affect corrosion in different ways. Substances that exist suspended or dissolved may be deposited and form more or less voluminous layers in regions with low flow velocity. This may give a reduced average corrosion rate, but increased and sometimes heavy localized corrosion, i.e. deposit corrosion. In hot tubes, such layers may also result in local overheating and crack formation.

Other types of surface layers are formed due to organic growth. The organisms may, for example, be algae or various types of bacteria. In addition to the usual deposit effect, these organisms may change the chemical composition of the corrosive environment on critical places. Algae can consume CO_2 and produce O_2, while other organisms consume oxygen. Oxygen depletion may be followed by action of anaerobe, sulphate-reducing bacteria. These bacteria depolarize the hydrogen reduction reaction and promote localized corrosion (see Section 6.4). They may also produce H_2S, which has special effects (Section 6.6).

The so-called iron bacteria lead to voluminous deposits in the form of tubercles composed of iron compounds, calcium compounds and organic matter. This is a frequent problem in hydro-power plants. Familiar examples of iron bacteria are *Gallionella* and *Sphaerotilus*, which affect iron, steel and stainless steels at pH = 7–10 and temperature 20–40°C. In pipes with smaller diameters, such deposits may reduce the effective cross–section strongly, as shown with an example in Figure 8.6, and sometimes block the pipe completely. Deposit corrosion takes place underneath the tubercles. Several cases of corrosion attacks in which bacteria have played a significant role have been experienced, e.g. in cooling water systems and heat exchangers [8.11].

Corrosion in public water piping systems is responsible for large economic loss. In pipes of cast iron, steel and other metallic materials, corrosion may be prevented by use of coatings (Section 10.6) or by water treatment (addition of calcium compounds, alkalization or carbonation). Water distribution systems are further dealt with in Section 8.4, Corrosion in Soils.

Water treatment for minimizing corrosion and possibly for other purposes is dealt with in various handbooks, e.g. References [8.3, 8.11].

Figure 8.6 Rust tubercles in a 2-inch cooling water pipe at a hydro-power plant after 15 years of service. (Photo: Erling Abusland, SINTEF Corrosion Centre.)

8.3 Corrosion in Seawater

The importance of seawater as a corrosive environment has increased during the last few decades because of offshore exploration of oil and gas. There is also an increasing application of seawater for production of fresh water in parts of the world where this is in short supply. Corrosion is a major problem in desalination plants.

Seawater has a chloride content that gives maximum corrosion rate (Figure 8.5). However, the corrosion rate is usually not particularly large because the water contains Ca^- and Mg^- salts (see Table 8.4). The bulk pH in seawater is commonly 8–8.3, but due to cathodic production of OH^-, the pH value at the surface increases sufficiently for deposition of $CaCO_3$ and possibly to a smaller extent some $Mg(OH)_2$ together with iron hydroxide. The deposits form a surface layer that reduces oxygen diffusion so much that the average corrosion rate on steel in seawater (0.1–0.15 mm/year) is less than in soft fresh water (see Figure 8.4). Table 8.5 shows the corrosion rate of various materials in stagnant and slowly flowing seawater. The corrosion rate at different flow velocities was presented in Section 7.9.

Because the corrosion rate depends on the supply of oxygen, it will be highest in the splash zone where a thin water film exists a major proportion of the time, and where parts of the corrosion product layer frequently are washed away (see Figure 8.7 and compare with the significance of deposits and water film thickness dealt with in Sections 6.2.2 and 6.2.5). On offshore structures the splash zone is the most difficult zone to protect satisfactorily, because it is not possible to make use of cathodic protection. Thick organic coatings (e.g. Neoprene) or Monel sheathing have been used in many cases. For corrosion prevention in general, reference is made to Chapter 10.

The oxygen content in air-saturated water at 10°C is approximately 6.5 cm^3 STP/l. The content is often given simply in millilitres per litre. Over the world it varies from about 8 ml/l for surface water in the Arctic to about 4.5 ml/l in the tropics. In the North Sea there is relatively efficient mixing of water from different depths, thus the water is nearly air-saturated even at large depths. Typical concentrations of dissolved oxygen at the actual temperatures 4–16°C are therefore within the range 5.5–7 ml/l, the highest ones at the lowest temperatures.

Table 8.4 Typical contents of inorganic species in natural seawater (a) [8.12], and composition of artificial seawater according to ASTM D 1141 (b) [8.13].

(a) Species	Conc. g/kg	(b) Substances	Conc g/l
Na^+	10.77	NaCl	24.53
Mg^{2+}	1.30	$MgCl_2$	5.20
Ca^{2+}	0.409	Na_2S_4	4.09
K^+	0.338	$CaCl_2$	1.16
Sr^{2+}	0.010	KCl	0.695
Cl^-	19.37	$NaHCO_3$	0.201
SO_4^{2-}	2.71	KBr	0.101
Br^-	0.065	H_2BO_3	0.027
H_3BO_3	0.026	$SrCl_2$	0.025
		NaF	0.003

Table 8.5 Corrosion rate in μm/year for metals and alloys in stagnant or slowly flowing seawater. (Adapted from References [8.8, 8.11, 8.14,].) As regards composition of materials, see Section 10.1.

Material	Corrosion rate (μm/year)
Hastelloy C	≈ 0
Titanium	≈ 0
AISI 316 stainless steel	Insignificant, except for deep pits
AISI 304 stainless steel	Insignificant, except for deep pits
Nickel–chromium alloys	Insignificant, except for deep pits
Aluminium	1–5, except for pits
Nickel–copper alloys	Usually <25, except for pits
Nickel	Usually <25, except for deep pits
70–30 Copper–nickel 0.5 Fe	3–12
90–10	3–12
Copper	13–90
Admiralty bronze	13–50
Aluminium brass	13–30
Bronze	25–50
Nickel–aluminium bronze	25–50
Nickel–aluminium–manganese bronze	25–50
Manganese bronze	25–75
Zinc	20–85
Austenitic nickel–alloyed cast iron	50–70
Carbon steel	100–160

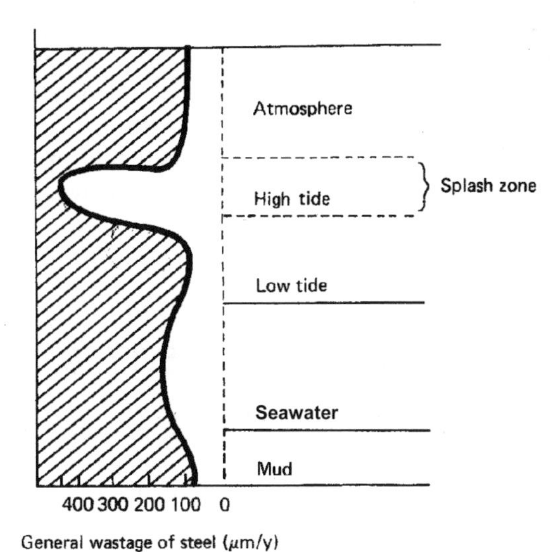

Figure 8.7 Corrosion rate on steel in seawater as a function of depth. (Reproduced from Marine and Offshore Corrosion by K.A. Chandler. Reprinted by permission of Elsevier Limited.)

Metals in seawater are exposed to marine growth (fouling). The fouling tendency is very different for different metals and increases in the following order:

1. Copper and 90–10 copper–nickel.
2. Brass and bronze.
3. 70–30 copper–nickel, aluminium bronze and zinc.
4. Nickel–copper.
5. Carbon steel, low-alloy steels, stainless steels, nickel alloys and titanium.

Groups 4 and 5 show much lower resistance to fouling than do the copper alloys. In order to prevent fouling in seawater pipe systems, chlorination is frequently carried out, primarily by adding sodium hypochlorite. In stainless steel systems, amounts corresponding to 1–2 ppm Cl_2 are often added continuously. By this, slime formation is prevented, but at the same time the potential is raised, and the risk for initiation of crevice corrosion is increased (see Section 6.7). Recent research [8.16] shows that the resistance to crevice corrosion initiation on high-alloy stainless steel increases with the time in chlorinated seawater. If moderate chlorination is applied in the beginning, with increasing addition of sodium hypochlorite after some time, the risk for initiation of corrosion will be lower than if full chlorination was applied from the start. For 6 Mo stainless steel (see Section 10.1) a maximum Cl_2 concentration of 2 ppm has been considered acceptable. An alternative is discontinuous chlorination in such a way that biological growth is prevented at the same time as the potential is kept below the critical level. In this case, the maximum Cl_2 level may be raised to 10 ppm in 6 Mo steel piping [8.16].

Corrosion in seawater may also occur on metallic materials that originally are protected by thinner or thicker organic coatings. The reason for this is that water, oxygen and chlorides diffuse through the coating. Undermining corrosion starts and the coating will slowly be broken down. One example of this is corrosion on sea cables. The outer layers of a sea cable consist of a metallic reinforcement – usually steel wires with hot-dip zinc coating – and an external layer of jute and asphalt or polypropylene and asphalt. After 25 years such a cable may look as shown in Figure 8.8.

Corrosion in seawater is also dealt with in Chapters 6, 7 and 10.

8.4 Corrosion in Soils [8.10, 8.17, 8.18]

Soils can be classified on the basis of particle size. Gravel contains the coarsest particles (> 2 mm) and clay the finest ones (< 0.002 mm), with sand and silt in between. Soils containing the finest particles, with ample distribution of small particle sizes are very dense and prevent supply of oxygen (but not of water), while gravel allows oxygen to be transported easily. Most metallic materials that are used in soils corrode under cathodic control, i.e. under control of oxygen transport. Thus, the *density of the soil* is important. The relationship is, however, somewhat

Figure 8.8 Deposit corrosion underneath remnants of jute on a 24-year-old sea cable. (Photo: Steinar Refsnæs, EFI, Trondheim.)

complicated: the corrosion rate is often lower in porous than in more compact soil, because preventive deposits may be formed more easily in the former case. Where the density varies along the metal surface, differential aeration cells may be established with concentrated corrosion at areas with a dense soil.

There are also other important environmental factors: other conditions affecting water and oxygen supply, resistivity, dissolved salts, pH, microbiological activity and stray electric currents.

Generally, aggressive soils have low *resistivity*. In corrosive sediments it may be up to 2000–4000 ohm cm, in little or non corrosive 4000–6000 ohm cm or more. Low resistivity may, for one thing, contribute to the establishment of large macro–galvanic cells (with extension up to kilometres). An example of the significance of resistivity in marine sediments is shown in Figure 8.9, which is based upon investigation of 10–70-year-old steel piles.

Various biological factors affect corrosion in soils. Organic acids originating from humus are relatively corrosive versus steel, zinc, lead and copper. In certain soil types, particularly acidic clay, very high corrosion rates have been experienced under anaerobic conditions. This is commonly caused by sulphate-reducing bacteria, which is indicated by ferrous sulphide found as part of the corrosion products. The mechanism of this type of corrosion has been dealt with in Section 6.4.

Stray currents may occur, as shown in Figure 8.10. Here, a steel pipeline interacts with a cathodic protection system (regarding cathodic protection, see Section 10.4).

On the critical surface area of the pipeline, the current runs from metal to environment by anodic dissolution of the metal, giving high corrosion rates under unfavourable conditions. Stray currents may also originate from railways, electrolysis plants, electric cables and large electric plants. Generally, direct current causes the greatest problem. Alternating stray current has usually little effect on steel, but it may attack lead to a great extent.

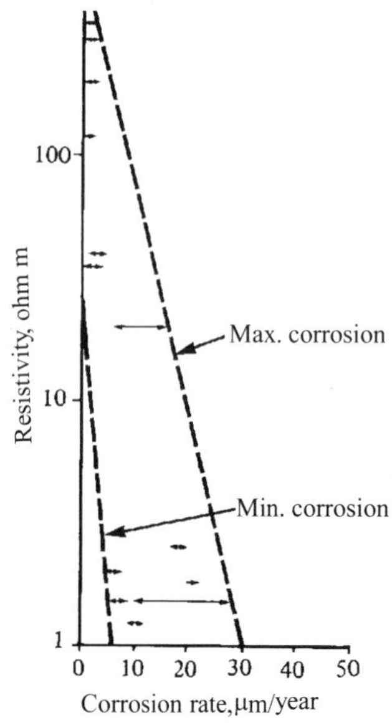

Figure 8.9 Corrosion rate on steel piles as a function of sediment resistivity. (From Fisher and Bue [8.19]. Reprinted with permission of ASTM International, 100 Bar Harbor Drive, West Conshohochen, PA 19428.)

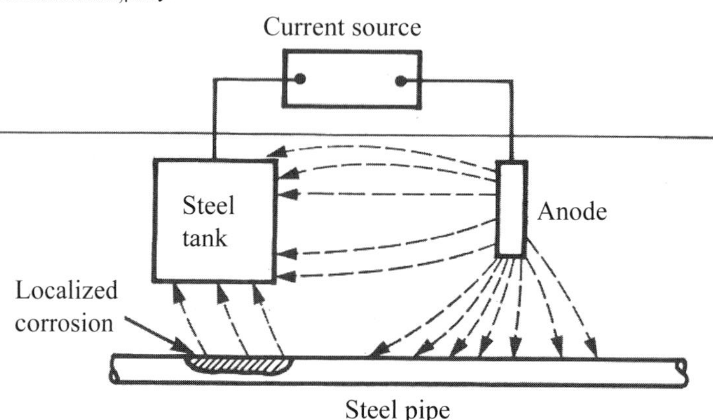

Figure 8.10 Stray current originating from a cathodic protection system may cause localized corrosion on steel pipelines. (Adapted from reference [8.14].)

Composition of carbon steel and low-alloy steel is of little significance for corrosion in soils, as is the case for corrosion in seawater and other waters.

The National Bureau of Standards in the USA has investigated corrosion of various materials in a large number of environments [8.17]. Some of the results are shown in Table 8.6. The results have been selected in order to show the variation in corrosivity versus iron and steel. As shown, the ranking of the environments is different in relation to lead and zinc.

It can be added to Table 8.6 that the average corrosion rate over the whole surface of steel was about 0.1 mm/year in the worst case (tidal zone sediment), and about 0.02 mm/year in average for all steel samples. Corresponding data for copper, zinc and lead were < 0.02 mm/year. One can draw various conclusions, e.g. that steel corrodes relatively uniformly in tidal zone sediment, and that copper in general corrodes much more uniformly than the other materials.

On commercially pure aluminium, depths from 0.1 to 1.6 mm have been measured after exposure for five years in four different soil types in England. For a fifth type of soil no attack was found. Similar variations have been recorded for 18–8 CrNi steel. AISI 316 (with Mo) (see Section 10.1) showed good resistance, but no general guarantee can be given.

The maximum depth of pitting attack is sometimes expressed by an equation of the following form (Section 7.6.5):

$$d = kt^n .$$

For steel, it has been reported that n varies from about 0.1 in well–aerated soil to 0.9 in poorly aerated soils [8.9]. The latter value means that the corrosion rate is only slightly reduced with time.

Table 8.6 Corrosion of steel, copper, lead and zinc in soils. Maximum corrosion depth (mm) after given periods of exposure. (Adapted from Uhlig [8.10].) The width and length of the samples are from a few inches to one foot.

Soils	3 diff. irons/steel 12 years exp.	Copper 8 years exp.	Lead 12 years exp.	Zinc 11 years exp.
Average of several soils	1.5–1.8 (44 soils)	< 0.15 (29 soils)	> 0.8 (21 soils)	> 1.3 (12 soils)
Tidal marsh, Elizabeth, N.J.	2.0–2.5	< 0.15	0.33	0.9
Montezuma clay, Adobe, San Diego, Cal.	< 3.3	< 0.15	0.25 (9.6 years)	
Merrimac gravel–sandy loam, Norwood, Mass.	0.5–0.7	< 0.15 (13.2 years)	0.48	

Under anaerobic conditions with sulphate-reducing bacteria the corrosion rate may attain much higher values than those mentioned above. In England, corrosion rates up to 6 mm/year have been measured under such conditions [8.18].

A problem of particularly large economic significance is corrosion on buried pipelines, and there are special reasons to worry about the external corrosion. Examples from some communities show that a major part of the water in the public fresh water pipeline system gets lost due to leakage, mainly because of corrosion.

The leakages cause damage to other property, the capacity demand of the waste water system increases, costly repair is needed, and such work may conflict with other activities and traffic. To prevent this, one of the external coating systems that has been recommended for water pipelines in soil is 25 μm zinc + asphalt or bitumen [8.18].

Generally, the most important measures for corrosion prevention in soils are use of organic coatings (coal tar, plastics or rubber (6 mm), wound tape), inorganic coatings (cement mortar or enamel, but these are brittle and may easily be damaged mechanically), fibre-reinforced cement mortar, metallic coatings (hot-dip or thermal spray zinc), filling by limestone or other calcium-containing matter, and cathodic protection. The combination of reinforced coal tar coating and cathodic protection has been used extensively to protect pipelines and tanks.

8.5 Corrosion in Concrete

The development and building of large concrete structures for offshore oil and gas production in the 1970s called for increased attention to corrosion of steel reinforcement, and of seawater-exposed steel parts in metallic contact with the reinforcement. Research projects were established at various laboratories, e.g. as dealt with in References [8.20, 8.21].

Concrete constitutes an alkaline environment for the embedded steel reinforcement, which leads to passivation of the steel. As dealt with elsewhere, for instance in Reference [8.22], passivated steel is liable to localized corrosion in environments with sufficient concentration of an aggressive substance like chloride. A crucial corrosion mechanism is therefore the transport of chloride and oxygen through the outer layer of concrete, which covers the reinforcement. At an early stage of the actual research it was assumed and shown by laboratory experiments that intensive corrosion might occur on i) reinforcement steel bars at cracks in the concrete cover, and ii) on seawater-exposed steel parts that were metallically connected to the reinforcement, because these cases most often implied a large area ratio between the reinforcement (cathode) and the exposed steel (anode). Sites without cracks but with extensive penetration of chloride through the concrete cover were also assumed to be strongly liable to corrosion due to local breakdown of the passive oxide layer on the steel bars. There is still some uncertainty as regards risk for galvanic corrosion on exposed steel parts. This problem may exist, particularly due to efficient aeration of the concrete in the splash zone. Regarding "drainage" of current from the cathodic protection system, see Section 10.4.2.

In harbour structures, bridge structures and buildings, many cases of extensive damage due to corrosion of reinforcement have occurred during recent years. For structures exposed to the atmosphere there is, in addition to possible localized breakdown due to chloride, a second important mechanism: carbonation of the concrete, which reduces the pH of the solution in pores and leads to general breakdown of passivity. Besides localized or general surface corrosion of the reinforcing bars, stress corrosion cracking or corrosion fatigue may take place, which emphasizes the severity of the attacks. A major cause of corrosion in these structures is that they have not been subject to the same strict rules as offshore structures with respect with concrete quality and cover. One of the problems is connected with concrete decks on road bridges, where salt containing a high proportion of chlorides may originate from de-icing or from the sea. A large demand for repair of structures has developed, and the annual cost of corrosion in this sector is very high, e.g. in the USA it was estimated at USD 150–200 billion in the early 1990s [8.3]. In addition to mechanical repair (replacement of old concrete and application of surface coatings), electrochemical methods are used for rehabilitation of concrete structures nowadays. Electrochemical methods comprise chloride removal, re-alkalization and cathodic protection. The principle of chloride removal is shown in Figure 8.11. Practical experience with these electrochemical methods is described in References [8.24, 8.25] and other publications presented in the proceedings from the same congress (Eurocorr'97). Another advancement is that improved concrete with low permeability has been developed in recent years.

In addition to the mentioned protection methods, i.e. use of high-quality concrete, sufficient cover, external organic coating or sealing, and cathodic protection, possible methods for prevention of reinforcement corrosion in new concrete structures include also the application of a protection measure directly on to the reinforcement: i) hot-dip zinc coating, with varying experience as to the effectiveness, ii) powder epoxy coating, e.g. for bridge decks, iii) reinforcing bars

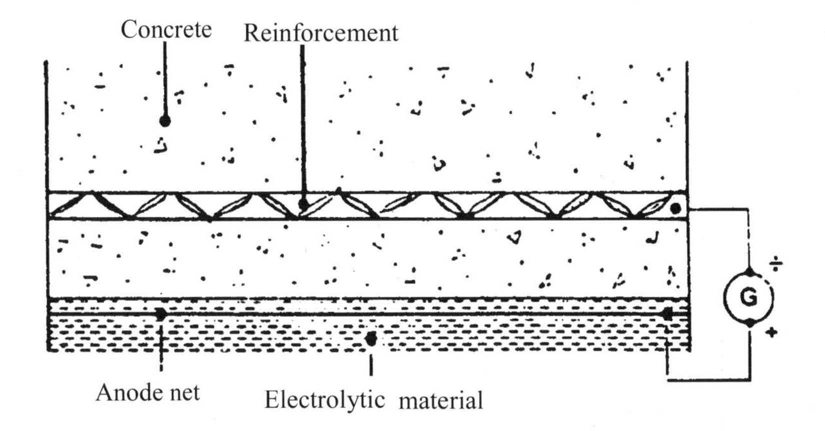

Figure 8.11 Schematic sketch of electrochemical method for chloride removal [8.23].

of stainless steel (to a limited extent), and iv) corrosion inhibitors, primarily in high-quality concrete with a thick cover. For further study of corrosion and protection of concrete reinforcemen,t reference is made to a comprehensive state-of-the-art report [8.26].

8.6 Corrosion in the Petroleum Industry

In oil production plants, many cases of extensive corrosion have occurred in production tubing, valves, and in flow lines from the wellhead to the processing equipment. The reason for this is that oil and gas from the well contain varying amounts of water, which can be precipitated as a separate phase in contact with the material surface, and that this water contains gases such as CO_2 and possibly H_2S, as well as salts. In most cases of severe corrosion, CO_2 plays a major role.

The mechanism of the main cathodic reaction in CO_2 corrosion has briefly been dealt with in Section 6.5. The carbonic acid reacts with steel, and a layer of reaction products, to a large extent $FeCO_3$, is formed on the steel surface. The deposit is cathodic relative to steel, and when small defects occur in the deposit layer, pitting corrosion is developed. The conditions may be particularly corrosive in the production tubing, which carries the oil/gas up from the well. In production tubing of carbon steel, corrosion rates in excess of 10 mm/year may occur under unfavourable conditions [8.27]. Various factors contribute to the high corrosivity:

- *water content* large enough to give a separate water phase in contact with the tube wall. For oil wells with low pressure it has been found that the oil stream prevents contact between water and steel when the water content is below 25–35% of the total produced fluid. On the other hand, in high–pressure wells such as in the North Sea, 1% water may be sufficient to make the well corrosive [8.27].
- *Salt content* in the water phase similar to that in seawater, sometimes higher and sometimes lower.
- *High total pressure.* The pressure may exceed 400 bar in some wells.
- *Considerable concentration (%) of CO_2* in the gas, which together with high total pressure gives high partial pressure of CO_2. Usually, the oil and gas are considered corrosive when the partial pressure of CO_2 is higher than 2 bar; the corrosivity may vary much depending on various other factors when the partial CO_2 pressure is between 0.5 (0.2) and 2 bar, and it may be insignificant when the CO_2 pressure is below 0.5 (0.2) bar.
- *High temperature*, which usually contributes to high reaction rates, particularly in combination with:
- *High flow rate*, which above a certain level causes removal of $FeCO_3$ from the tube wall and thereby activates the steel surface. Experience from Ekofisk field in the North Sea indicates that erosion corrosion occurs when the fluid velocity on top of the well exceeds 7–10 m/s. Figure 8.12 shows a similar result from a gas/condensate well [8.28].

In Figure 8.13, extensive corrosion attacks with pits penetrating the tube wall are shown. This production tubing has been in service at a field in the North Sea for a period slightly more than 3 years. At the wellhead, the temperature and pressure were on average about 90°C and 90 bar, both varying considerably. Calculated fluid velocity was 9.4 m/s. The amount of produced water was 2.7% of produced oil, and the partial pressure of CO_2 varied from 5 bar in the lower end of the tubing to 1.3 bar at the wellhead during production.

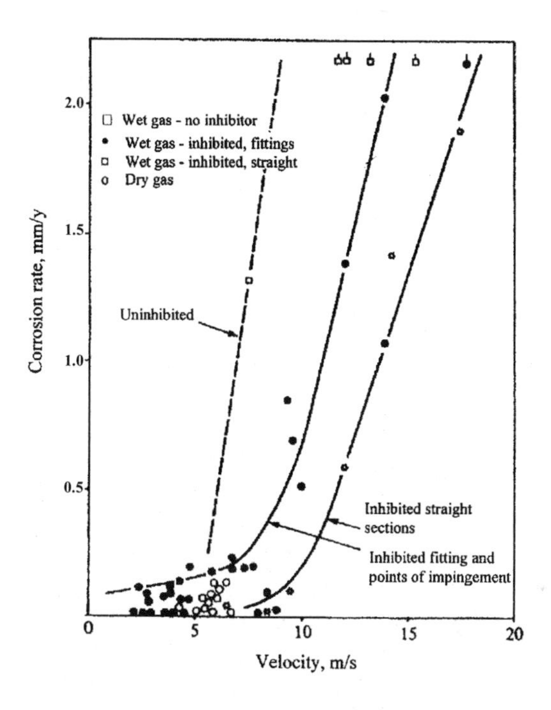

Figure 8.12 Corrosion rate as a function of gas velocity in a gas/condensate well. From Duncan [8.28]. Total pressure at the wellhead 350 bar; partial pressure of CO_2 20 bar, of H_2S 0.2 bar. Reprinted with permission of NACE.

In pipe systems on the platform, the most severe corrosion attacks have been found between the wellhead and the first–stage separator, where water is precipitated, and where pressure, temperature as well as flow velocity are highest [8.29]. To a great extent, the attacks are localized in and at welds, in pipe joints, bends, and at places with reduced pipe diameter. (In many cases the attacks at welds could have been avoided by proper selection of welding consumables.) When the water content exceeds a critical level, the attacks become more severe. This is probably the reason for increased corrosion when the producing wells get older (the content of water increases with time due to water injection).

Corrosion attacks in a flow line from a field in the North Sea are described in

Section 7.9 and shown in Figure 7.44.

Water is separated from oil and gas in separators ("produced water system"). In cases with high water content, it has corroded much more than expected. Penetration of the material after a few years of service has been reported [8.29]. This has happened because the fluid goes from high to low pressure; the water is therefore supersaturated with CO_2 while it is in the produced water system (the "seltzer effect"). Consequently, when the water content is high the design should be based upon the conditions in the high-pressure system. When the water content is low, the time for the water to be present in the low-pressure system is much longer; equilibrium is reached and the mentioned problem is avoided. The corrosion rate may increase because of small amounts of oxygen in the water. Inhibitors added at the wellhead are not efficient in the "produced water system" [8.29].

The actual measures to avoid unacceptable rates of CO_2 corrosion in general are to use inhibitors and/or stainless steels, and possibly coatings. The use of inhibitors may be expensive, and stainless steels become more and more common.

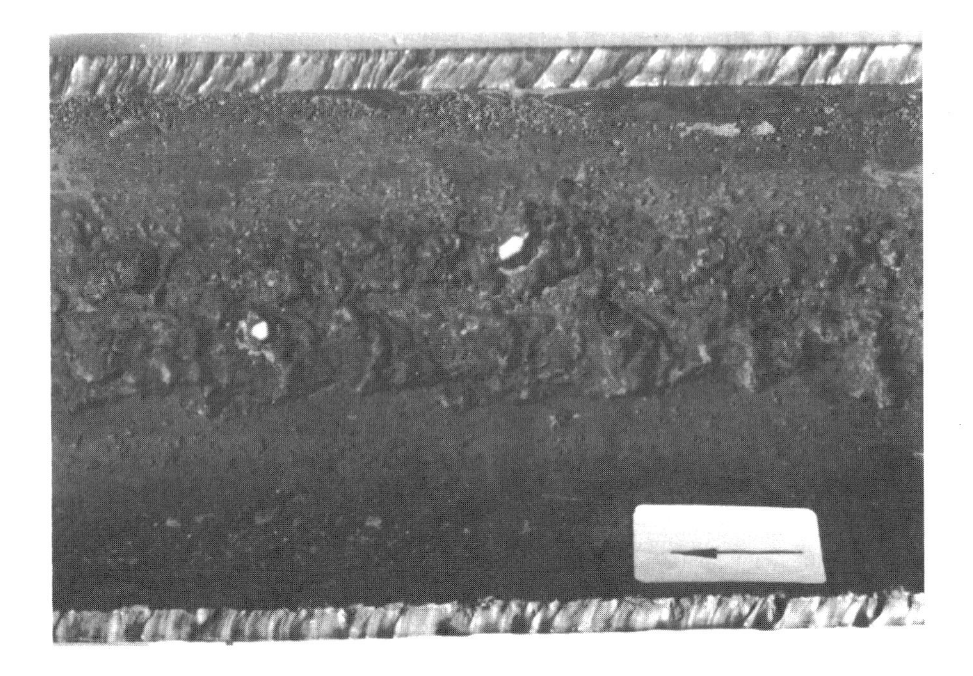

Figure 8.13 Corrosion attacks with penetrating pits in a production tubing after about three years of service. (Photo: John M. Drugli, SINTEF Corrosion Centre, Trondheim.)

Development of weldable stainless steels that are cheaper than the conventional ones has been subject of a good deal of attention in recent years. The application of stainless steels is dealt with in Section 10.1.5, where examples of materials selection for offshore process systems are given.

If the gas/oil contains a significant amount of H_2S, it is commonly called "sour" (Section 6.6), and so is the corrosion caused by H_2S (contrary to pure CO_2 corrosion, which is called "sweet" corrosion). The largest risk due to H_2S is the occurrence of stress corrosion cracking (sulphide stress cracking). Cracks have been observed in steel after exposure to water containing as little as 0.1 ppm H_2S and with H_2S partial pressure in the gas phase down to 0.001 bar, but the critical concentration of H_2S varies (see also Section 6.6). For carbon steel it has been required that the Rockwell hardness R_c does not exceed 22 for H_2S-containing wells, while R_c up to 26 has been accepted for low-alloy CrMo steel [8.30]. The risk of H_2S stress corrosion cracking in stainless steels depends on the actual steel grade (Section 10.1.5).

Other corrosion problems in oil and gas production are corrosion in the water injection system, particularly at welds, galvanic elements and irregularities that disturb the flow, and external corrosion on hot process piping, with or without heat insulation [8.27]. Corrosion in seawater is dealt with in Section 8.3 and at various places in other chapters of the book.

Corrosion may also take place in systems for transportation or storage of oil, e.g. when the fluid for some reason has absorbed water that is precipitated afterwards. This is a known problem in various types of storage tanks. To prevent corrosion, inhibitors and (particularly in larger tanks), coatings and/or cathodic protection are used.

In oil refineries, various corrosion problems exist, usually due to inorganic species, such as H_2S, CO_2, H_2SO_4 and NaCl [8.14].

References

8.1 Kucera V, Mattson E. Atmospheric corrosion. In: Mansfeld F, editor. Corrosion Mechanisms. New York: Marcel Dekker, 1987.

8.2 Godard HP, Jepson WB, Bothwell MR, Kane RL. The corrosion of light metals. New York: John Wiley & Sons, 1967.

8.3 Roberge PR. Handbook of corrosion engineering. New York: Mc. Graw-Hill, 1999.

8.4 Shreir LL. Corrosion. Vol. 1. London–Boston: Newnes–Butterworths. 2nd Ed. 1978.

8.5 Haagenrud S, Kucera V, Atteraaas L. Atmospheric corrosion of unalloyed steel and zinc – 4 years exposure at test sites in Scandinavia. Copenhagen: 9th Scandinavian Corrosion Congress 1983.

8.6 Henriksen JF. Norwegian Inst. of Air Research. Private communication, 1985.

8.7 Kilcullen MB, McKenzie M. Weathering steels. In: Corrosion in Civil Engineering. London: Institute of Engineering 1979.

8.8 Thomas R. Varmförzinkning som korrosionsskydd (Hot-dip galvanizing for corrosion protection), Nordisk Förzinknings Förening, Stockholm, 1969. In Swedish.
8.9 Carter VE. Atmospheric corrosion on non-ferous metals. In: Parkins RN, editor. Corrosion Processes. London: Applied Science Publishers 1982.
8.10 Uhlig HH. Corrosion and Corrosion Control. 2nd, Ed. New York: John Wiley & Sons 1971.
8.11 Metals Handbook. 9th Ed. Vol. 13th Corrosion. Metals Park, Ohio: ASM International, 1987.
8.12 Rogers TH. Marine Corrosion. London: George Newnes, 1969.
8.13 ASTM D1141. Standard Practice for Substitute Ocean Water. Philadelphia: American Society for Testing and Materials, ASTM, 1998.
8.14 Fontana MG, Greene ND. Corrosion Engineering. New York: McGraw-Hill, 1967, 1978, 1986.
8.15 Chandler KA. Marine and Offshore Corrosion. London: Butterworths & Co, 1985.
8.16 Bardal E, Drugli JM, Gartland PO. The behaviour of corrosion–resistant steels in seawater. A review. Corrosion Science, 35 (1–4). 1993: 257–267.
8.17 Romanoff M. Underground Corrosion. Circ. 579. National Bureau of Stdandard, (US). 1957.
8.18 Iverson WP. An overview of the anaerobic corrosion of underground metallic structures. Evidence for a new mechanism, In: Escalante E. editor. Underground Corrosion. ASTM STP 741. American Society for Testing and Materials, 1981.
8.19 Fisher KP, Bue B. Corrosion and corrosivity of steel in Norwegian marine sediments. In: Escalante E, editor. Underground Corrosion. ASTM STP 741. American Society for Testing and Materials, 1981.
8.20 Gjørv OE, Vennesland Ø, El–Busaidy AHS. Corrosion/76. Paper No 17. Houston, Texas: National Association of Corrosion Engineers, 1976.
8.21 Arup H. Galvanic action of steel in concrete. Glostrup, Denmark: Korrosionscentralen, Report, 1977.
8.22 Nürnberger U. Chloride corrosion of steel in concrete. Fundamental relationships–practical experience 1–2. Betonwek und Fertigteil-Technik, 601–704, 1984.
8.23 Vennesland Ø. Private communication. NTNU. Trondheim, 1994.
8.24 Elsener B, Zimmermann L, Bürchler D, Böhni H. Repair of reinforced concrete structures by electrochemical techniques – field experience. Proceedings Eurocorr' 97. Trondheim, 1997.
8.25 Haldeman Ch, Schreyer A. 10 years of cathodic protection in concrete in Switzerland. Proc. Eurocorr'97. Trondheim, 1997.
8.26 COST 509 Corrosion and protection of metals in contact with concrete. Draft final report. Workshop September 1–3, Heriot-Watt University, Edinburgh, 1996, 127.
8.27 Houghton CJ, Westermark RV. Downhole corrosion mitigation in Ekofisk (North Sea) field. In: CO_2 Corrosion in Oil and Gas Production, Selected Papers, Abstracts, and References. Houston: NACE Task Group T-1-3, 1984.

8.28 Duncan RN. Materials performance in Khuff gas service. Materials Performance, July 1980: 45–53.
8.29 Rogne T, Drugli JM. Unpublished work at SINTEF Corrosion Centre, Trondheim, 1993.
8.30 MR 01–75. Sulfide Stress Cracking Resistant Metallic Material for Oil Field Equipment. Houston: NACE.

EXERCISES

1. Plot the data in Table 8.1 (the two data columns to the left) in a diagram showing steady-state corrosion rate as a function of the SO_2 deposition rate, under which you assume that the lowest and the highest corrosion rate in a deposition interval correspond, respectively, to the lowest and highest deposition rate in the same interval.
   Draw a curve through the data points, and briefly discuss and characterize the relationship.

2. Explain the curves in Figures 8.4 and 8.5.

3. Arrange the materials in Table 8.5 in six groups in such a way that the corrosion characteristics are similar within each group. Describe in a few words what characterizes each group. In which group would you place stainless steel with composition:

   a) 22Cr5Ni3Mo,
   b) 25Cr7Ni4Mo,
   c) 20Cr18Ni6Mo,
   assuming a temperature < 30–35°C?

   (See Section 10.1.5 for information about the steels.)

4. Figure 8.13 shows a section of production tubing with corrosion attacks, and corresponding environmental data are given in the text. Use average environmental data at the wellhead and the nomogram in Figure 6.14, and determine the expected corrosion rate. Compare this with corrosion rates in the real case. Discuss and try to explain deviation or possible similarity between real corrosion rates and the rate determined from environmental data, taking into account factors such as velocity, deposits, water content, corrosion rate variation along the surface etc.

9 Corrosion Testing, Monitoring and Inspection

If you won't follow my advice, then be ruled by a wiser counsellorr: time.
PERICLES

In this chapter, first a brief introduction regarding general objectives, methods and procedures of corrosion testing is given. This is followed by a presentation of the most important principles of electrochemical testing, and finally some actual methods of monitoring and inspection are reviewed. For further study, reference is made to References [9.1–9.4] and other sources indicated at the respective places in the text.

9.1 Corrosion Testing in General

9.1.1 Objectives

Typical *main objectives* of corrosion testing are:

1. Evaluation and selection of material or protection method for a specific environment and a certain application. Testing for this objective is initiated by the user, often by engagement of consultants, testing laboratories or research institutions. People determining test conditions are often faced with the following dilemma:

    - The testing time has to be sufficiently *short* for the material evaluation and selection to be made before a deadline set by the practical application.
    - The test results have to be *reliable* as a basis for prediction of the material performance in the *long run* (see under "Laboratory testing", next section).

2. General information about the behaviour of certain materials and groups of materials in the main types of environment. Testing is often done by the material producer or vendor, and the results aid the selection of materials to be tested for a certain application (objective 1).

3. Routine control of materials including coatings, e.g. for acceptance or reclamation of deliveries. Test methods and conditions may be given in specifications of the actual delivery. Various standardized tests are used, such as the salt spray test and others referred to in Sections 7.5, 7.6, 7.7 and 7.12.

4. Investigation of corrosion mechanisms and fundamental effects of various conditions, development of test methods, e.g. reliable model experiments for the prediction of lifetime (see next section), and other studies in research laboratories. This includes also contributions to the development of materials and coatings with special corrosion properties.

9.1.2 Test Methods

To cover the objectives stated in the previous section, numerous test methods are available. The methods can be arranged in the following groups:

1. Laboratory testing, comprising accelerated testing, model testing, and special tailor-made investigations in research. Accelerated test methods give measurable attacks after short time, which can be obtained by using a more aggressive environment, higher potential, higher temperature or higher mechanical stress than under the actual service conditions. These methods are suitable only for qualitative comparison and ranking of various materials with respect to specific properties, and for routine control and approval of delivered materials. Accelerated testing cannot answer questions about lifetime without knowing the degree of corrosion acceleration.

   Model testing aims at realistic conditions. Possible acceleration must be limited to avoid change of the corrosion mechanism. The testing takes longer than the accelerated test dealt with above, and will often require a sensitive measuring technique. For the extrapolation of results to a realistic service period or lifetime, a thorough understanding of the corrosion mechanism and the change of corrosion conditions and rate with time is necessary. New measuring techniques contribute to improvements in this direction.

2. Service (plant) and field testing, including exposure of coupons, test specimens or components in process environments of industrial plants as well as in special natural fields (in seawater, atmospheres, soils etc.)

3. Testing in a pilot plant, which is a true model (up to half scale) of the actual industrial plant. Pilot plant testing is used as a basis for appropriate design and selection of materials in cases where the condition cannot be satisfactorily modelled in laboratory tests. Pilot plants for modelling are suitable for relatively expensive installations, in which the corrosion conditions are complex and aggressive and the consequences of failure are great.

9.1.3 Testing Procedure

A typical testing procedure includes most of the following items.

1. Selection and pre-treatment of materials and test specimens.
   Attention must be paid to orientation of the specimen in relation to the rolling direction and texture, the surface of the component and welds. This is particularly important for testing of the susceptibility to corrosion forms such as stress corrosion cracking, corrosion fatigue, intergranular and pitting corrosion. It should be taken into account that the structure of the surface material of a rolled component may be different from that of the bulk material. The analysis and production history of the component must be considered, e.g. whether it has been formed by continuous or die casting, by hot or cold forming etc. Sometimes the specimen must be given an additional heat treatment. Parallel test specimens in an investigation should be taken from the same charge of material. Thus, for a larger test program, a sufficient amount of material from the actual charge must be available from the beginning. The geometry of the specimens is in many cases most important. Undesired crevices must be avoided. In order to reduce misleading effects of cut edges, which may be subject to a corrosion rate which is different from that of rolled surfaces, the specimens should have large area ratio between faces and edges. The specimens must be sufficiently large to allow reliable weight loss measurements. In some cases the shape of the specimen must correspond to the shape of a certain component. For later identification it is necessary to carefully mark the specimens, in such a way that the marks are visible after the testing and post-treatment.

2. Surface preparation.
   For prediction of the performance of a material in a particular service, the ideal surface of the test specimen would have to duplicate the surface of the component in service, both with respect to cleanliness, roughness and material structure. However, usually it is necessary to deviate from this ideal demand in order to get reproducible specimens and be able to measure the degree of corrosion with sufficient accuracy. Therefore, a metallically pure surface, and a roughness and state of surface material defined by a certain grade of grinding or mechanical polishing, or a certain procedure for pickling or electrolytic polishing, are commonly used. Increased reproducibility is obtained by removing at least 10–20 μm of the original surface material. Metallographic studies may also necessitate a smoother surface than in practice. The surface preparation must be the same for all specimens in a test series except when the preparation is a planned and defined variable. It is important to use clean emery papers and polishing belts or discs, and particularly to ensure that these are not successively used on different materials, as they may contaminate the surface of the specimen.

3. Measurement of surface area and possibly thickness.

The accuracy of thickness measurement must correspond to the expected reduction of thickness during the test.

4. Degreasing, e.g. by acetone.

5. Weighing.
The accuracy of the weight must correspond to the expected weight loss during the test.

6. Masking.
Surfaces that are not to be exposed, must be masked, by means of materials such as wax, lacquer, paint, tape, glue or enamel.

7. Exposure.
In some laboratory testing and particularly in field testing, standardized racks are used with prescribed position and orientation of the specimens. In other cases, adjustments are made with the actual practical purpose in mind.
In laboratory testing, a number of environmental parameters must usually be controlled:

 - Oxygen concentration, often controlled by bubbling air (in special cases oxygen) through the electrolytic solution until saturation, or by bubbling nitrogen for removal of oxygen.
 - pH of the electrolyte, particularly in long-term testing.
 - Concentrations in general. Mechanical agitation to obtain a homogeneous solution.
 - Temperature.
 - Relative humidity, in various test chambers containing air and vapour.
 - Relative velocity between corrosion medium and specimen.

8. Inspection of the specimens after exposure, including a preliminary visual inspection (with the naked eye or a microscope), followed by surface cleaning (mechanically, chemically or electrochemically), weighing and/or thickness measurement, and finally metallographic investigation using an optical or electron microscope.

9. Determination of corrosion rate, on the basis of weight loss or thickness reduction measurements, or electrochemical measurements during the exposure.

Numerous test methods have been standardized. Examples of extensive collections are the NACE (National Association of Corrosion Engineers) standards and the ASTM (American Society for Testing and Materials) standards. Some of these are mentioned in Chapter 7 in connection with crevice, pitting, intergranular and stress corrosion. Further reference is made to Shreir et al. [9.2] and Heitz et al. [9.3] who present surveys of various standards (British, American, German, ISO), as well as to the NACE series [9.5], the ASTM annual book [9.6], and *Metals Handbook* Vol. 13 [9.7] in which various methods are described completely.

9.2 Electrochemical Testing

A simple form of electrochemical testing is to measure the corrosion potential, applying a reference electrode and a voltmeter with high internal resistance. In other respects, modern electrochemical testing most often makes use of a potentiostat. The normal wiring diagram for potentiostatic experiments is shown in Section 4.10 (Figure 4.9), where it is also described how polarization curves are recorded, and how overvoltage curves for the different reactions can be determined under favourable conditions. As dealt with at the respective places in Chapters 5, 6 and 7, polarization curves can be recorded with the aim to investigate passivation properties of materials, to clarify environmental effects, and to determine corrosion rates in galvanic and crevice corrosion, critical potential for pitting corrosion, as well as potential ranges where the material is liable to stress corrosion cracking. Potentiostatic tests (i.e. tests where the potential is kept constant at certain values) can be used to determine intergranular corrosion tendency. Potentiostatic tests combined with stepwise increase of the temperature are applied to determine critical crevice corrosion temperature (CCT) and critical pitting corrosion temperature (CPT) (see Sections 7.5.4 and 7.6.5). Another method is simply to measure current between two test samples that are different in some way, for example in chemical composition. Electrochemical methods are also used for testing painted surfaces [9.8]. Both direct and alternating current (DC and AC) methods are used in electrochemical corrosion testing. Recently, advanced methods, such as the scanning vibrating electrode technique (SVET), have been developed [9.9]. SVET is used to determine the variation of current density over the surface being studied.

We shall look a little closer at the two most frequently applied electrochemical methods for determination of corrosion rate. They were originally developed for general corrosion, but with some precautions they may also be used for the determination of average corrosion rate in pitting corrosion, selective corrosion, and – depending on the geometry – galvanic corrosion. One of the methods is the determination of polarization curves and *extrapolation of linear parts of the polarization curves,* as described in Section 4.9. The intersection point between the extrapolated Tafel regions, in other words the overvoltage curves, gives the corrosion current. Unfortunately, the overvoltage curves are not always pure Tafel curves. But if either the anodic or the cathodic overvoltage curve has a sufficiently long linear part around the corrosion potential, the necessary basis for the determination of corrosion rate is present: the linear part of the actual polarization curve can be extrapolated *to the corrosion potential.* Thus, the corrosion rate is determined regardless of the other polarization curve. In Figure 9.1 four different cases are shown, in which the corrosion rate is determined by extrapolating linear parts of both or one of the polarization curves.

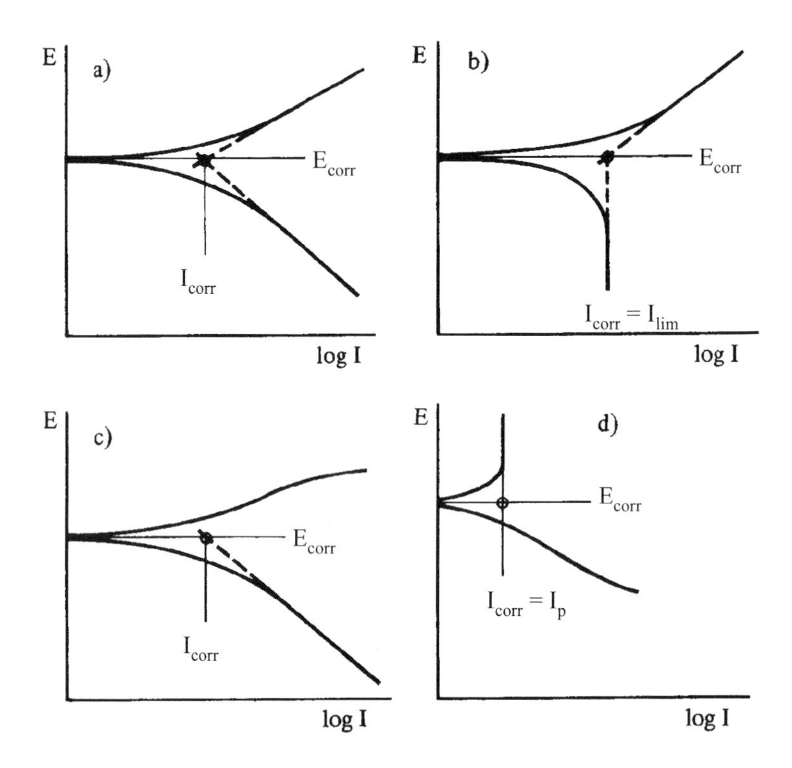

Figure 9.1 Determination of corrosion current density by extrapolation of linear parts of the polarization curves. a) Both the cathodic and the anodic reaction are under activation control (the overvoltage curves are Tafel lines). b) The cathodic reaction is diffusion controlled and the anodic reaction activation controlled. c) The cathodic reaction is activation controlled, the anodic curve is irregular. d) The cathodic curve is irregular, the metal is passive, i.e. the corrosion current equals the passive current.

The other main electrochemical method for determination of corrosion rates is the (linear) polarization resistance method (the LPR method). In a limited potential range around the corrosion potential (up to ±20 mV) a linear relationship exists between the potential increment ΔE and the increment in external current ΔI_e, as shown in Figure 9.2. It can be shown mathematically that the slope of the curve in this potential range is given by Stern–Geary's equation

$$\frac{dE}{dI_e} = \frac{b_a \times b_c}{2.3(b_a + b_c)I_{corr}}, \tag{9.1}$$

where b_a and b_c are the anodic and the cathodic Tafel constants. By polarization to reasonable potential distances such as ±10 mV from the corrosion potential E_{corr}, the slope $\Delta E/\Delta I_e$ is determined at $E = E_{corr}$, and the corrosion current I_{corr} can be calculated from the equation when the Tafel constants are known. The method can also be used when the metal approaches a true passive state ($b_a \rightarrow \infty$), and when the cathodic reaction approaches pure diffusion control ($b_c \rightarrow \infty$). For these two cases, the slope of the curve has the following limiting values:

$$\frac{dE}{dI_e} = \frac{b_c}{2.3 I_{corr}} \quad \text{(passive state)} \tag{9.2}$$

and

$$\frac{dE}{dI_e} = \frac{b_a}{2.3 I_{corr}} \quad \text{(diffusion-controlled cathode rection).} \tag{9.3}$$

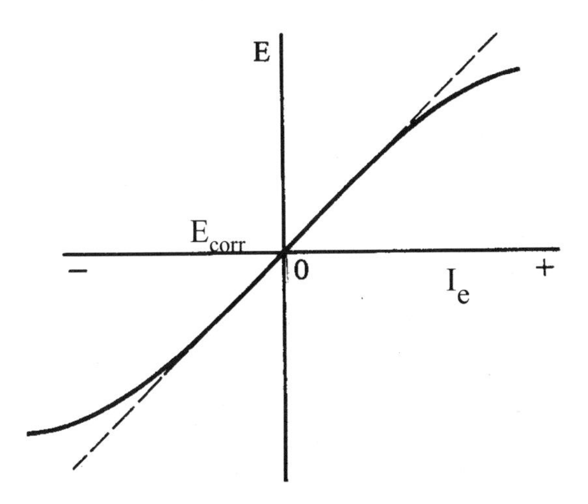

Figure 9.2 Linear potential–external current curve around the corrosion potential.

The main benefit of the LPR method is that the corrosion rate is determined without significant disturbance of the natural corrosion conditions (the electrode is polarized only slightly from the natural corrosion potential). Because a condition is that the Tafel constants are known, it is often preferable to combine the LPR method with the first described method (extrapolation of linear parts of the polarization curves). To avoid disturbance of the natural conditions during the test, the complete polarization curves should be recorded at the end of the exposure period.

Another important electrochemical corrosion test method is the *measurement of current* between an area with dominating cathodic reaction and an area with dominating anodic current. This is done by separating the two areas, using two specimens, and then connecting these specimens to each other via a current-measuring instrument. To avoid a potential drop in the instrument, a zero-resistance instrument can be used. For this purpose a potentiostat can be applied, coupled as shown in Figure 9.3. The potential difference (the voltage) between W and R on the potentiostat is set to zero, and the current in the circuit (i.e. the current flowing between the two specimens W and C) can be determined from the voltage drop over the resistance R_o or with the built-in ammeter on the potentiostat. In order to understand the function of the circuit shown in Figure 9.3, it should be kept in mind that the current flowing through the terminal R on the potentiostat is negligible because of very high internal resistance in the potentiostat between W and R. Special zero-resistance ammeters are also available commercially.

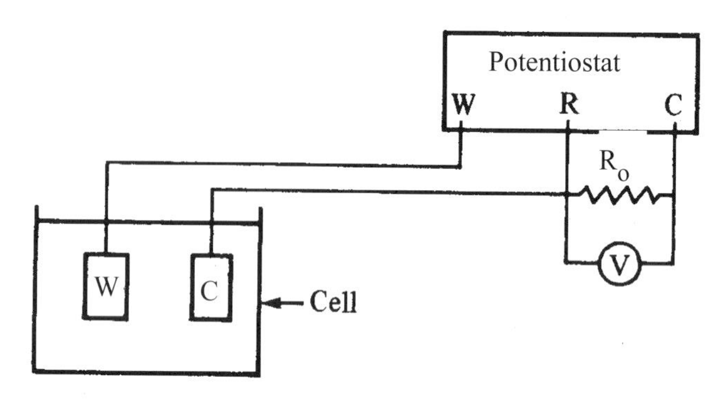

Figure 9.3 Potentiostat used as a zero-resistance ammeter.

Examples of measurements of current between dominating anodic and dominating cathodic areas are shown in connection with galvanic corrosion (Section 7.3.1, Table 7.3). The method can also be used on thermogalvanic corrosion, crevice corrosion and corrosion at damaged parts of paint coatings. In the last case the "galvanic current" (i.e. the current flowing between intact and damaged parts of the coating) is often very low. In research work, currents as low as 2×10^{-12} A have been measured on painted specimens. "Galvanic" current measurements, determination of polarization curves, and current measurements under potentiostatic polarization have all contributed to a better understanding of the behaviour of paint systems [9.8].

9.3 Corrosion Monitoring and Inspection

Corrosion rates and distribution cannot always be predicted. But in many cases it is important to be aware of the state at any time in order to avoid unforeseen failures and to take or adjust necessary protective measures. Systems that require monitoring are often more or less inaccessible for visual inspection, and therefore various equipment and methods for instrumented monitoring have been developed. A benefit of these methods is that quantitative results are usually obtained, which indicate corrosion rate and/or degree of protection, development of the state with time etc. Some monitoring methods are dealt with in the next sections.

9.3.1 Monitoring of Cathodic Protection

Cathodic protection (CP) is described in Section 10.4. The simplest and most important method for monitoring structures protected by CP is *potential measurement.* This is used to check that the potential is below a certain limit at which the corrosion rate is assumed to become significant (see Section 10.4). As described earlier (Chapter 3) the potential is measured with a voltmeter with high internal resistance. One terminal on the voltmeter is connected to the reference electrode, which is held close to the actual part of surface of the structure to be monitored. The other terminal on the meter is connected to the structure.

On offshore platforms, the measuring equipment has usually been operated by a diver. Two types of equipment are shown in Figures 9.4 and 9.5. In the case illustrated in Figure 9.4, the contact between the voltmeter and the structure is made on deck. With the equipment in Figure 9.5, the same contact is made by the sharp point shown to the right, which is pressed against the surface of the structure at the measuring spot (metallic contact to the structure is needed). The reference electrode is usually an Ag/AgCl electrode. In recent years, an instrument for digital reading built into the pistol-shaped probe has commonly been used. Some of these probes have in addition a cable connected to an instrument on deck for control [9.10].

Probes with the same principle as described above, but with a design adapted to the method of operation, are also applied either a) fixed to the structure or b) moved and positioned by a remote-operated vehicle. In fixed probes, both a Zn electrode and an Ag/AgCl electrode are frequently used to make sure as far as possible that reliable results are obtained during long periods. In some modern equipment the measuring signals are transferred directly from the measuring probe to instruments above the sea by means of ultrasonic equipment, thus making cables superfluous.

In order to get more information about the state of cathodically protected surfaces in the sea, *current density measurements* have also sometimes been carried out in addition to potential measurements. Current density determination is usually done by measuring the electric field and converting this to current density. The field is determined by measuring the potential difference ΔP between two reference electrodes separated by a certain distance L. If the current density in the direction through both electrodes is constant over the distance L, the current density is given by

$$i = \Delta P/(\rho L), \tag{9.4}$$

where ρ is the resistivity of the water.

The electrodes may in principle be fixed. However, with the reference electrodes located near the cathode (the surface of the structure) where the current density is relatively low, the potential drop ΔP over a reasonable distance L will be rather small. To get sufficient accuracy with fixed reference electrodes it is necessary that they are very stable relative to each other. In a probe developed by Morgan Berkeley & Co., this problem was solved by accurate balancing of the two electrodes. A different way to solve the problem is to use a rotating pair of electrodes, as done in the so-called T-sensor, a principle and equipment developed at SINTEF Corrosion Centre [9.11] and adapted for field operations by CorrOcean. When the probe rotates, so that the two reference electrodes R_1 and R_2 are exchanging position with each other twice per revolution, the electrodes are steadily calibrated relative to each other. The measured potential difference between the electrodes as a function of time is expressed by a sine curve. With the amplitude of the sine curve denoted $\pm\Delta P$, the current density is derived from Equation (9.4).

With both types of equipment referred to, the potential of the structure can be measured as the voltage between the structure and one of the reference electrodes, disregarding the ohmic potential drop in the water, which is sufficiently accurate

Figure 9.4 Diver with reference electrode and a cable leading up to a voltmeter on deck [9.10].

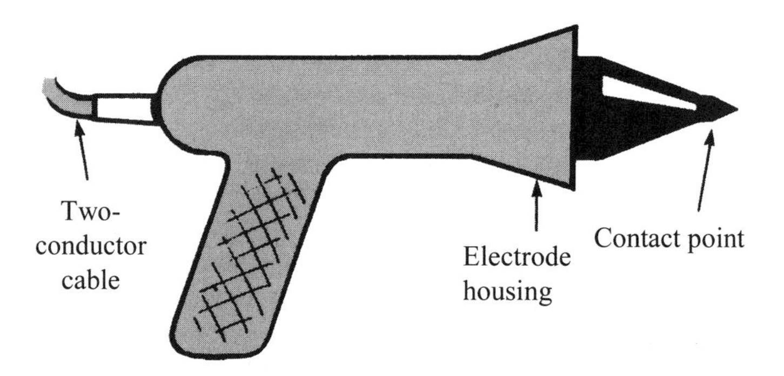

Figure 9.5 Diver probe with two conductors leading from the contact point and from the reference electrode, respectively, to a voltmeter on deck [9.10].

close to the structure. In order to *measure the total current* from an anode, the anode may be insulated from the structure and then connected to it through a zero-resistance ammeter. This is done only on very special occasions.

Methods like those dealt with in this section can be used on different structures, such as steel platforms, concrete platforms and pipelines. The best method to use depends on the type of structure. For pipelines, special problems exist and special adaptations of the methods are used [9.10].

9.3.2 Inspection and Monitoring of Process Plants

During the last few decades, instrumented internal monitoring of process plants has become more common in the petroleum and chemical industries. In some cases the plants are intended to function continuously for several years. If the process has to be stopped for periodical inspection or because of damage, the cost of production loss may be large. According to Dawson [9.4] this type of cost may, in the chemical industry, form 20–25% of the total maintenance cost, the latter being of the same order of magnitude as the capital cost. It has been shown and experienced that much can be saved by the use of a good monitoring system under such conditions. It is preferable to apply two or three different methods that are complementary to each other.

One of the most important considerations is the choice of appropriate positions for the measuring probes, which makes it possible to follow the performance at the most critical sites. Such places are:

1. At abrupt changes in direction of flow or in pipe diameter, and at obstructions and irregularities of any kind that create local changes of velocity, turbulence and possibly erosion.
2. At crevices and areas with stagnant water, which may cause crevice or deposit corrosion.
3. At junctions of dissimilar metals, which may create severe galvanic corrosion.
4. Positions with high local stresses, fluctuating temperature or varying pressure.

Note that weld joints may be particularly critical elements, partly because galvanic corrosion or in some cases pitting/crevice corrosion can occur, and partly because of local flow disturbance due to the weld.

The selection of monitoring positions should be based upon thorough knowledge about the process and the behaviour of the materials under the actual conditions, and upon experience from similar plants. In process systems on offshore installations, it must, for instance be taken into account where the probability of water precipitation is highest.

Inspection and monitoring methods can be divided into two groups: A) those indicating the state of the plant (Numbers 1–7), and B) those indicating the function of the plant, and possible corrosion in action (Numbers 8–11). Numbers 1, 3, 4, 5 and 6 are considered to be inspection methods, while the others are monitoring methods.

1. *Visual inspection,* to be done at all places accessible at production stops.
2. *Leak detection,* e.g. on pressure vessels, which can be carried out by a number of methods. Thermography can be used if the leaking fluid has a temperature different from the ambient. Leakage can also be detected by acoustic emission techniques. In order to indicate when the corrosion allowance of pressurized equipment has been consumed by uniform corrosion or erosion, a narrow hole is drilled from the outside of the wall (at the installation of the equipment) to a depth where only the corrosion allowance is left. Leak detection cannot of course be used for explosive fluids.
3. *Radiography* is useful for detecting defects such as corrosion pits. The method has been widely developed for inspection of welds and has a number of advantages, but a limitation is that the component must be accessible from both sides. Pipes can be checked without access to the inside.
4. *Eddy currents* can be used to detect pits, cracks, intergranular attacks and selective corrosion.
5. *Ultrasonic techniques* have the advantage compared with radiography that access from one side is sufficient. Cracks and other forms of attack can be discovered. The surface must be prepared carefully to obtain good contact with the probe.
6. *PIG (pipeline inspection gauge) devices,* i.e. recording equipment following the flowing liquid in pipelines. Different devices based on ultrasound, eddy currents, spring-loaded callipers, magnetic flux and induction have been developed for recording variation in wall thickness and the existence of cracks.
7. *The field signature method (FSM)* is a relatively new method developed by CorrOcean ASA from the principle commonly used in electrical resistance (ER) probes. An electric current is sent through the part of the installation being tested (often a pipeline), and the potential differences between a considerable number of points on the external surface are measured. The potential differences depend on reduction of the material thickness and how this reduction is distributed along the pipe wall [9.12].
8. *Weight loss coupons,* made of the same material as the actual part of the plant, placed in the process environment and removed after a certain time. Standardized coupons are described, e.g. in ASTM G4–68, and cleaning after exposure is dealt with in ASTM G3 [9.6].
9. *Probes for measuring electrical resistance (ER probes).* These have been extensively applied in the oil industry during the last few decades. Based upon his experience from the Ekofisk field in the North Sea, Houghton characterizes the method as the most reliable one for continuous monitoring of corrosion rate and effect of inhibitor additions in oil production installations [9.4]. When using the method, one must, however, be careful as regards selection of the type of probe, and in this connection take the flow conditions into account. The principle of the ER probes is that an electric current is sent through a probe of the same material as the one in the actual part of the process system, and the voltage drop from one point to the another in the probe is measured. As a result of corrosion the thickness of the probe is reduced and the resistance and consequently the voltage drop is increased. The probes have to be mechanically

robust, but at the same time sufficiently sensitive, their lifetime must be adequate and they must indicate corrosion rates that are representative for the system at the actual position. The probe may be positioned flush with the pipe wall, as for the type shown in Figure 9.6, or it may stick into the pipe.

10. *Linear polarization resistance (LPR) measurement* is based upon the principle described in Section 9.2. Probes with three electrodes (with wiring diagram in principle as shown in Figure 4.9) as well as with two electrodes have been developed. In the two-electrode probes both electrodes are made of the same material as the actual part of the process system. Three-electrode probes have in addition a reference electrode that is usually made of stainless steel. A probe with three electrodes is shown in Figure 9.7.

    An advantage of the LPR probes is that they measure the instantaneous corrosion rate. Changes in the corrosion rate are therefore rapidly recorded, and can easily be related to changes in the process or service conditions. In addition to measuring general corrosion rate they can indicate localized corrosion expressed by a "pitting index". LPR probes for direct current require good conductivity of the corrosion medium. Alternating current probes are used for measuring electrode impedance in research laboratories. These may also be used in industrial plants where the electrolytic conductivity is low.

11. *Acoustic emission*, a technique that involves analysis of ultrasonic waves generated by cracking (e.g. stress corrosion and corrosion fatigue) and deformation of material.

Figure 9.6 ER probe developed by CorrOcean AS.

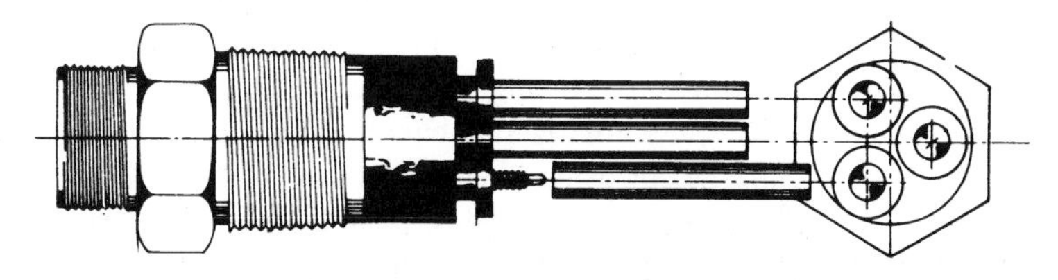

Figure 9.7 Linear polarization resistance probe (Petrolite Corporation).

Other methods include i) ordinary potential measurement with a reference electrode and a voltmeter with high internal resistance, ii) hydrogen probes that measure the rates of hydrogen development and absorption of hydrogen in steel, and iii) analysis of the process fluid, for example with respect to ferrous ions, which indicate that dissolution of steel has taken place.

In modern corrosion engineering, the concept of *Corrosion Management* is often used, which includes *Risk-based Inspection Planning (RBI)* and *Process Monitoring* in addition to corrosion monitoring and inspection execution. All these information sources are coupled together in order to design an inspection and monitoring system as good as possible seen from a technical /economic point of view.

9.3.3 Monitoring and Testing in Other Environments

The methods described in the previous sections can be used in waters and other liquid corrosion media under various conditions. Some of them can be applied in the atmosphere as well, some in soils and even in concrete. There are, however, also methods specially designed for such environments.

For atmospheric field and service testing, weight loss coupons have traditionally been used. Modern testing includes in addition electrochemical methods, e.g. with the cell shown in Figure 9.8 [9.13], the quartz crystal microbalance utilizing the frequency response to mass changes [9.14], and fibre optics [9.15].

In the electrochemical cell shown in Figure 9.8, two sets of metal plates are arranged alternately close to each other with an insulation layer between neighbouring plates. Two different methods are used: a) (as in the figure) the same metal (steel, zinc or copper) is used in both sets; an external voltage is impressed and the current flowing between the two sets is measured. b) Different metals are used in the two sets (copper/steel or copper/zinc); the galvanic current is measured, and there is no impressed voltage.

By the use of method a) with a noble metal like copper, the measurements are mainly expressing the wet time, while probes including a metal such as steel or zinc (also in method b) give results with a certain relation to corrosion rates in the actual environment. The ratio between corrosion rates determined with weight loss coupons and electrochemical probe, respectively, stays constant under varying conditions at a given site, but varies from one site to another [9.13].

Corrosivity in soil is determined by weight loss and thickness measurements, but electrochemical probes have also been developed for this purpose, for example as described in Reference [9.16]: a) a galvanic probe, where the current between a magnesium point and a steel tube is measured. The current is determined by the efficiency of the cathodic reaction on steel and the resistivity of the soil. b) A galvanostatic polarization probe consisting of a reference electrode, a counter electrode and a working electrode. Polarization curves are recorded on the basis of a stepwise change of current. The corrosion rate is measured by the LPR method or by extrapolating a possible linear part of the cathodic polarization curve to the corrosion potential. Measured corrosion rates are higher than average long-term corrosion rates, and the measurements are therefore divided by a factor for the estimation of long-term corrosivity.

Electrochemical monitoring methods have also been developed for application on steel reinforcement in concrete. These methods include potential measurement on the concrete surface, linear polarization (LPR) and determination of polarization curves [9.17]. Electrical resistance probes (ER) and probes embedded in the concrete for measuring galvanic current have also been used.

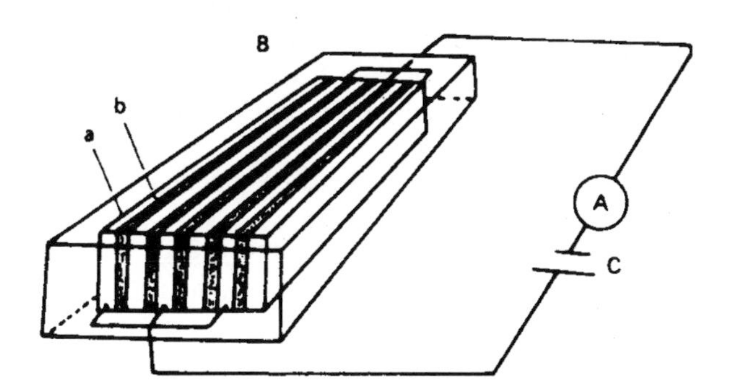

Figure 9.8 Principle of electrochemical measurement of wet time in the study of atmospheric corrosion. (A) is a zero-resistance ammeter, (B) is the electrochemical cell, ((a) electrodes, (b) insulation layers), and (C) is a direct current source with constant voltage. (After Kucera and Gullman [9.13].)

References

9.1 Fontana MG, Greene ND. Corrosion Engineering. New York: McGraw-Hill, 1967, 1978, 1986.

9.2 Shreir LL, Jarman RA, Burnstein GT, editors. Corrosion, Vol. 2. Oxford: Butterworth Heinemann, 3rd. Ed., 1994.

9.3 Heitz E, Henkhaus R, Rahmel A. Corrosion Science. An Experimental Approach. Chichester, England: Ellis Horwood, 1992.

9.4 Wanklyn J, editor. Corrosion Monitoring. Oyez Scientific and Technical Services, 1982.

9.5 NACE TM Series of Corrosion Test Methods. Houston, Texas: NACE.

9.6 Annual book of ASTM standards. Wear and Erosion; Metal Corrosion. Philadelphia: American Society for Testing and Materials (ASTM), 1999.

9.7 Metals Handbook, 9th Ed. Vol. 13 Corrosion. Metals Park, Ohio: ASM International, 1987.

9.8 Steinsmo U, Bardal E. Use of electrochemical methods for evaluation of paint films on steel, aluminium and zinc. Corrosion, 48, 1992; 11: 910–917.

9.9 Isaacs H. The use of the scanning vibrating electrode technique for detecting defects in ion vapor-deposited aluminium on steel. Corrosion, 43, 1987; 10: 594–598.

9.10 Backhouse GH. Equipment for offshore measurements. Conference on Cathodic Protection, Theory and Practice – the Present Status, Coventry, April 1982.
9.11 Eggen TG, Strømmen RD, Bardal E. In-situ current density measurements on cathodically protected structures in seawater. Proceedings 8th Scandinavian Corrosion Congress, Helsinki, 1978.
9.12 Daaland A. Investigation and modelling of corrosion attacks for developing the FSM technology, dr.ing. thesis, NTH, University of Trondheim, 1994.
9.13 Kucera V, Gullman. In: Mansfeld F, Bertocci U, editors. Electrochemical Corrosion Testing, STP 727. Philadelphia: ASTM, 1981.
9.14 Forslund M, Leygraf C. Journal of the Electrochemical Society, 143, 3, March 1996; 839–844.
9.15 Smyrl WH, Batler MA. Corrosion Sensors. Pennington, New Jersey: The Electrochemical Society Interface, 1993: 35–39.
9.16 Fischer KP, Bue B. Corrosion and corrosivity of steel in Norwegian marine sediments. In: Escalante E., editor. Underground Corrosion, STP 741. Philadelphia: ASTM, 1981.
9.17 Schell HC, Manning DG. Evaluating the performance of cathodic protection systems on reinforced concrete bridge substructures. Materials Performance, July 1985.

EXERCISE

In a laboratory, a thermally sprayed coating with the composition 17Cr3.5B4Si1C4FebalNi is exposed to flowing seawater containing some silica sand. The samples are attacked by erosion as well as erosion corrosion (see Section 7.9.3).

After cleaning and removal of corrosion products the samples are weighed (both before and after exposure). During a test lasting six hours, linear polarization measurements are carried out after two and four hours. At the end of the test, the samples are polarized stepwise in the cathodic and anodic directions, and the current at the various potential steps is recorded.

For one of the samples the following data are obtained:

Surface area: 6.5 cm^2.
Weight before exposure: 13.4435 g.
Weight after exposure: 13.4278 g.
Polarization resistance = dE/dI_e at E_{corr} after 2 hours: 477 ohm.
Polarization resistance = dE/dI_e at E_{corr} after 4 hours: 494 ohm.

Polarization data at the end of the test:

E (mV): Cat	–444	–464	–500	–550	–600	–650	–700	–750	–800
I (μA):	0	–52	–161	–519	–1199	–2138	–2882	–3476	–3976

E (mV): An.	–444	–424	–400	–350	–300	–250	–200	–150	–100
I (μA):	0	29	64	136	236	398	662	962	1414

Draw the polarization curves. Determine the corrosion current by extrapolation of linear parts of the curves recorded at the end of the test, and by using the LPR measurements obtained at two and four hours, respectively. Assume that the Tafel gradients do not change during the exposure period. Compare and discuss the results. Explain the differences between the corrosion currents determined by the two methods.

Use the data obtained after two and four hours of the exposure period, assume that the corrosion current decreases linearly with the time, and determine the corrosion contribution in percent of the total weight loss. Determine finally the average rate of thickness reduction in mm/year.

On average for the coating:
Atomic weight 57.3
Valence, z 2.19
Density, ρ 8.5 g/cm^3

10 Corrosion Prevention

Five different main principles can be used to prevent corrosion:

1. Appropriate materials selection
2. Change of environment
3. Suitable design
4. Electrochemical, i.e. cathodic and anodic protection
5. Application of coatings

The choice between these possibilities is usually based upon economic considerations, but in many cases aspects such as appearance, environment and safety must also be taken care of. Two or more of the five principles are commonly used at the same time. It is important to decide upon corrosion prevention at the design stage. In addition to the sections in the present chapter, reference is made to recommended protection methods under the treatment of various corrosion forms in Chapter 7.

10.1 Materials Selection

'Gold is for the mistress – silver for the maid –
Copper for the craftsman cunning at his trade.'
'Good!' said the Baron, sitting in his hall.
'But Iron – Cold Iron – is master of them all.'
RUDYARD KIPLING

10.1.1 General Considerations

When selecting materials, each component must be considered with respect to design, manufacture and its effect on the total geometry. However, it is also important that the materials in adjacent components are compatible. With regard to corrosion, compatibility often means that detrimental galvanic elements must be avoided (Section 7.3). Not only the main structural materials, but also insulation and other secondary materials must be taken into account to prevent galvanic corrosion. In many cases it is possible to avoid other forms of corrosion by using a favourable combination of materials, e.g. to include a material that implies cathodic protection (Sections 7.3.3 and 10.4) against uniform, selective, pitting, crevice or erosion corrosion on critical regions, even against stress corrosion cracking or corrosion fatigue. Not only do the grades of structural, load-bearing materials have to be specified, but also surface treatment and coatings.

The corrosion properties and other functional properties of materials depend on several external factors such as geometry, manufacture, surface conditions, environmental factors and mechanical load conditions. For each functional property, these factors have to be evaluated. The final materials selection is often a result of compromises between various properties and their dependence on external factors.

A proper selection of materials depends on sufficient knowledge on how the actual practical conditions affect each material candidate. To ensure that important aspects are not forgotten, check lists should be used, which should include risk of different forms of corrosion for each candidate, relevant conditions affecting each form of corrosion, possibilities of changing the corrosion properties of a certain material, possibilities for application of various protection methods, accessibility for maintenance, environmental conditions, loads and special requirements during various parts of the lifetime etc. Aspects related to other functional properties than the corrosion properties should of course be treated in similar ways.

The best tools for weighing the various aspects are quantitative expressions of properties and performance data valid under various conditions, such as corrosion rate and distribution, lifetime in corrosion fatigue, mechanical or electrochemical threshold values (K_{ISCC}, K_{th}, E_p etc.), compared with corresponding quantified requirements or service conditions, i.e. specified lifetime, actual stress intensity factors and functions, and corrosion potential.

The spectrum of different kinds of material data increases, but so do the number of possible material candidates. During planning of large and expensive structures and plants, the amount of relevant data can be increased by standardized or more or less tailor-made testing. However, in many cases, the materials have to be selected partly on the basis of qualitative information or stipulated characters representing corrosion properties.

The simplest way is to choose the material that has been used before for the same or some similar purpose. If the material selection plays a less important role, or if the risk and consequence of changing to a new material are uncertain, it may be preferable to stick to the same as before. However, as a general principle it is not acceptable. Nevertheless, previous selection for similar purposes is a part of the empirical bases that must be taken into account. Other factors are the policy of the company, market conditions, commercial relationships, standards, specifications and public requirements. Besides these aspects, and sometimes in conflict with some of them, the engineers have the responsibility to find the best technical–economic solution, which often must be based upon thorough knowledge about materials properties and a systematic analysis of the service conditions.

As a general guideline for materials selection primarily dictated by corrosion aspects, the reason for the corrosion resistance of the respective material candidates may be used. If the resistance is based upon:

1. Passivity, the alloy is suitable for oxidizing environments (only in the absence of species that promote localized corrosion).

2. Immunity, the alloy is suitable for reducing environments.

This means that reducing environments are compatible with relatively noble metals or alloys (copper, lead, nickel and alloys based upon these metals). When metallic materials are to be used in oxidizing environment, on the other hand, their corrosion resistance must be based upon passivity (e.g. titanium and alloys that contain sufficient amounts of chromium). These relationships are easy to understand when the Pourbaix diagrams for metals such as Cu and Ti (Figure 10.1) are considered, and it is kept in mind that reducing environments lower the corrosion potential and oxidizing environments lift it. Irrespective of the mentioned rule, a metal is usually most corrosion resistant when it contains the smallest possible amounts of impurities. Some "natural" combinations of environment and material are listed in Table 10.1.

When selecting materials that will be in contact with chemicals of various kinds, the References [10.1–10.8] are useful. Trade names and compositions of many corrosion resistant alloys are also given. Iso-corrosion diagrams, e.g. like that shown in Figure 10.2 for various combinations of chemical and material represent one of the applied methods for displaying corrosion data [10.1, 10.3].

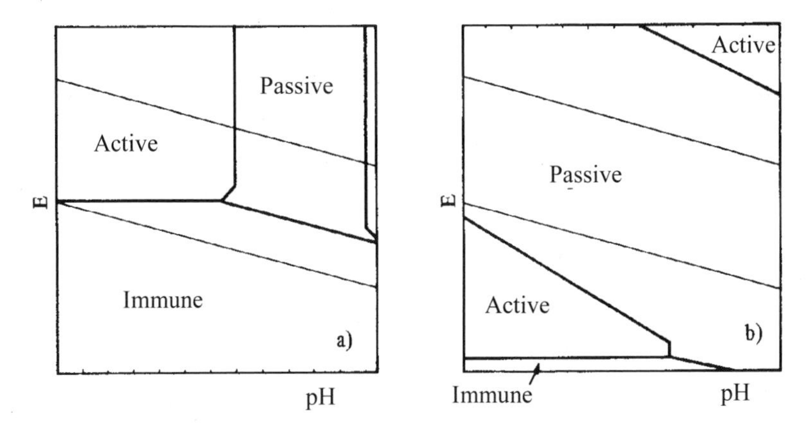

Figure 10.1 Pourbaix diagrams of copper (a) and titanium (b) in water.

Table 10.1 Some natural combinations of environment and material [10.1].

Environment	Material
Nitric acid	Stainless steels
Caustic solutions	Nickel and nickel alloys
Hydrofluoric acid	Monel
Hot hydrochloric acid	Hastelloys (Chlorimets)
Dilute sulphuric acid	Lead
Non-staining atmospheric exposure	Aluminium
Distilled water	Tin
Hot, strongly oxidizing solutions	Titanium
Ultimate resistance	Tantalum
Concentrated sulphuric acid	Steel

In several cases, non-metallic materials such as polymers, rubbers, ceramics, wood or concrete must also be taken into consideration (see Section 10.1.11). With respect to materials selection for screws, bolts, nuts, rivets or other small parts for use under possible galvanic corrosion conditions, it should be checked that they are a little more noble (have a little more positive corrosion potential in the actual environment) than the components they are binding together, or are in metallic contact with in any way. In addition, one should be particularly careful to avoid catastrophic deterioration forms such as hydrogen embrittlement, stress corrosion cracking and corrosion fatigue in such parts.

In Sections 10.1.3–10.1.12 a survey of the most important materials is given. They are characterized in relation to natural environments, such as the atmosphere, fresh water and seawater. As an example of a challenging industrial sector, also with respect to corrosion problems, particular attention is paid to offshore oil and gas production.

10.1.2 Some Special Aspects of Materials Selection for the Offshore Industry

With the huge and expensive installations in deep seawater, and with production equipment and pipelines carrying very aggressive mixtures of hydrocarbons and salt water, strict demands are naturally made on the corrosion technology for this industry. During the last few decades, oil companies have started using more corrosion-resistant materials in production and seawater systems. There are several reasons for this:

i) Large inhibitor expenses can be eliminated or reduced strongly.
ii) Demands for inspection, maintenance and replacement is reduced. This is of particular importance in sub-sea production systems.
iii) More rational localization of separators for removal of corrosive species from oil and gas can be obtained.
iv) More pollution and higher flow rates may be accepted in seawater systems.
v) Weight may be reduced by reducing thickness (omitting corrosion allowance) or by using lighter materials.

The items ii), iii) and v) have drawn much attention in connection with development of fields in deep water. One of the actual concepts comprises a central production platform that receives hydrocarbons through pipelines from several production units on the seabed at the respective wells. Under these conditions, corrosion-resistant and strong materials are useful for the sub-sea systems and strong, light and possibly corrosion-resistant materials for the installations on the platform. Several materials properties are more or less important, depending on the purpose: in addition to corrosion resistance, strength, density, and the ratio between strength and density, these properties may include erosion resistance, weldability, heat resistance, inflammability and thermal properties such as heat conductivity and expansion coefficient. Examples of materials selection for fields in the North Sea are described in the last part of Section 10.1.5.

Also when non-corrosion-resistant steels are selected in combination with cathodic protection and/or a coating system, various corrosion-related conditions may be of great significance. Particularly important is that materials possessing high strength and hardness are more sensitive to hydrogen development (due to cathodic protection or corrosion) than are those with low strength.

For extensive description of the various materials with respect to corrosion and related properties, see *Metals Handbook* Vol. 13 [10.6].

10.1.3 Unalloyed and Low-alloy Steels and Cast Irons

As described in Section 8.1.2, atmospheric corrosion rates can be reduced considerably by use of special low-alloy steels, i.e. weathering steels. In other natural environments such as soils and waters the differences between various unalloyed or low-alloy steels are small as regards the surface corrosion forms not affected by mechanical forces and conditions. When selecting materials within these groups, some kind of corrosion protection must be considered as an integrated part (coatings, cathodic protection, corrosion allowance, inhibitors). Regarding corrosion forms that interact with mechanical effects, i.e. erosion, abrasion, cavitation, fretting, static tensile stresses or fatigue, we must distinguish sharply between the different steels, since strength and hardness, and in some cases ductility, toughness and elasticity, have great significance (see Sections 7.9–7.13 regarding the corrosion forms in question). In oil production, increasing the strength of steel in order to reduce weight is of considerable interest, but the risk for hydrogen embrittlement (as a mechanism in stress corrosion cracking or due to cathodic protection) limits the permissible strength of material in production tubing, casings, production equipment, and welded structures affected by fatigue. The selection of material is closely related to design, not least for welded structures. Higher material strength may be tolerated when the welds are localized in less critical regions and when the weld quality is improved.

Contrary to corrosion in most natural environments, i.e. under O_2 reduction, CO_2 corrosion is considerably affected by relatively small amounts of alloying elements. As shown in Table 10.2, the effect of alloying steel for a gas/condensate well with Ni and Cr is strong. Alloyed steels have therefore been used sometimes in production tubing instead of carbon–manganese steel, which was commonly used previously.

Table. 10.2 Effect of alloying with nickel and chromium on the corrosion of steel in gas/condensate [10.1].

% Ni	Corrosion rate, mm/y	% Cr	Corrosion rate, mm/y
0	0.9	0	0.9
3	0.1	2.25	1.25
5	0.075	5	0.525
9	0.05	9	0.04
		12	0

Unalloyed steel and cast iron without external protection resist concentrated sulphuric acid (>70%) fairly well at room temperature under stagnant conditions (Figure 10.2). Carbon steel is therefore often used in tanks, pipes, tank vehicles, transport containers and production equipment exposed to this environment. The corrosion rate is higher at high flow rate. Hot acid (made by dissolving SO_3 in 100% acid) may attack grey cast iron along graphite flakes and cause cracking.

Usually, carbon steels are also resistant in alkaline environments such as NaOH and KOH solutions without contaminations, but may be subject to stress corrosion cracking (caustic embrittlement) at certain concentrations and temperatures (Section 7.12).

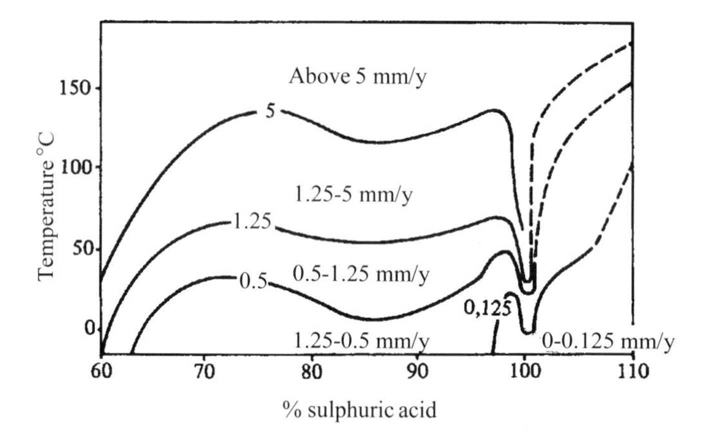

Figure 10.2 Corrosion of steel in sulphuric acid as a function of concentration and temperature. (Adapted from Reference [10.1].)

10.1.4 High-alloy Cast Irons

Of these materials we shall briefly consider two groups that possess very good corrosion resistance in many environments: a) silicon iron and b) cast iron with Ni and/or Cr. Tables 10.3 and 10.4 show trade names and compositions of various grades within these groups.

Table 10.3 High-alloy silicon irons. Adapted from Reference [10.2], with information in brackets adapted from Reference [10.1].

Material	% Alloying elements
Durasid	Min. 14.5 Si, with or without Mo
Durichlor	14.5 Di, (3) Mo
Duriclor 51	14.5 Si, Mo, Cr
Duriron	14.5 Si, 0.5 (0.95) C
Thermisilid	18 Si

When the Si content in grey cast iron is increased beyond 14%, the corrosion resistance in many environments becomes very good because a passive film of SiO_2 is formed on the iron surface. The material does not resist fluoric acid, but, according to Reference [10.1], in other respects it has the best universal resistance of all commercial (non-precious) metallic materials. The hardness of Si-irons is also high (e.g. HB = 520 for Duriron and Durichlor 51), and they resist erosion and cavitation corrosion relatively well. On the other hand, the tensile strength is low (the ultimate strength may for instance be 140 MPa), and their weldability is poor (components with simple shapes may be welded when thorough precautions are taken).

The high-alloy silicon irons are used in draining pipelines, pumps, valves, other process equipment, and anodes for cathodic protection with impressed current.

The most common high-alloy cast irons with Ni and Cr are the socalled Ni-Resist alloys (Table 10.4). These alloys have an austenitic structure, and therefore they are much tougher than the Si-irons and other cast irons. They are also produced as ductile irons; in this form they may have an ultimate tensile strength of 500 MPa and fracture elongation of 40%. Ni-Hard is an alloy with extra high hardness (HB = 550–725) and good resistance against erosion (and abrasion) corrosion, and it is therefore used in flow systems carrying mixtures of corrosive liquids and solid particles. Ni- and Cr-alloyed cast irons show superior heat resistance to other cast irons.

Table 10.4 High-alloy cast irons with Ni and/or Cr. Adapted from Reference [10.2] with information in brackets from Reference [10.1].

Material	% Alloying elements
Ni-Resist 1	15 Ni, 6 Cu, 2 Cr, 2.8 C
Ni-Resist 2	20 Ni, 2.3 Cr, max. 2.0 C
Ni-Resist 3	30 Ni, 3 Cr, max. 2.6 C
Ni-Hard	4 Ni, 2 Cr, white cast iron
Guronit G52	25–30 Cr
Wegucit	30 Cr
U.S. type 446 L(AISI)	23–27 Cr, max. 0.35 C, max. 0.25 N

10.1.5 Stainless Steels

A survey of different stainless steels is given in Table 10.5, comprising composition and structure, yield strength, reduction of area, and numbers of various standards. Some relatively new steels are listed in Table 10.6.

As dealt with in previous chapters, the corrosion resistance of stainless steel is due to passivation by a surface film of *chromium* oxide. The chromium content is higher than about 11%, and the low-temperature corrosion resistance as well as the resistance to oxidation and mill scale formation at high temperature increase with increasing content of chromium. Pure chromium steels are either ferritic (low C-content, non-hardenable by heat treatment) or martensitic (traditionally higher C-content for most grades, hardenable by heat treatment). With sufficient content of *Ni* the structure becomes austenitic, which gives increased formability, weldability,

Table 10.5 Composition, mechanical properties and standard numbers of stainless steels. (Extracts from References [10.7, 10.8].) Added type of steel (F = ferritic, M = martensitic, F–A = ferritic–austenitic (duplex), and A = austenitic), AISI-, Werkstoff- and UNS numbers, SS = Swedish Standard.

Steel type		Max%C	%Cr	%Ni	%Mo	Others	Min. yield strength 0.2% offset MPa	Min. reduction of area A_5 %	SS	AISI	W.st	UNS
13Cr	(F)	0.08	12.0–14.0	max. 0.5	–	–	250	20	2301	405	1.4002	S 40500
	(M)	0.25	12.0–14.0	max. 1.0	–	–	490	16	2303	–	1.4021	S 42000
17Cr	(F)	0.10	16.0–18.0	max. 0.5	–	–	250	18	2320	430	1.4016	–
17Cr1.5Ni	(M)	0.25	16.0–18.0	1.25–2.5	–	–	–	–	2321	431	1.4057	S 43100
18Cr2Mo	(F)	0.25	17.0–19.0	max. 0.5	2.0–2.5	Ti	340	25	2326	444	1.4521	S 44400
16-5-1Mo	(M-A)	0.05	15.0–17.0	4.0–6.0	0.8–1.5	–	620	15	2387	–	1.4418	–
23-4	(F-A)	0.03	23	4		N	400	25	–	–	1.4362	S 32304
22-5-3Mo	(F-A)	0.03	21.0–23.0	4.5–6.5	2.5–3.5	N[1)]	418	25	2377	–	1.4462	S 31803
25Cr7Ni4Mo	(F-A)	0.03	25	7	4	N	550	25	–	–	–	S 32750
18-9		0.07	17.0–19.0	8.0–11.0	–	–	210	45	2332	304	1.4301	S 30400
		0.05	17.0–19.0	8.0–11.0	–	–	210	45	2333	304	1.4301	S 30400
	(A)	0.08	17.0–19.0	9.0–12.0	–	Ti	210	40	2337	321	1.4541	S 32100
		0.030	17.0–19.0	9.0–12.0	–	–	190	45	2352	304L	1.4306	S 30403
		0.030	17.0–19.0	7.0–10.0	–	N[2)]	290	40	2371	(304LN)	1.4311	S 30453
18-9-0.5Mo	(A)	0.12	17.0–19.0	8.0–10.0	0.60	–	210	35	2346	303	1.4305	S 30300
17-12-2.5Mo		0.05	16.0–18.5	10.5–14.0	2.0–2.5	–	220	45	2347	316	1.4401	S 31600
		0.030	16.0–18.5	11.0–14.0	2.0–2.5	–	210	45	2348	316L	1.4404	S 31603
		0.08	16.0–18.5	10.5–14.0	2.0–2.5	Ti	220	45	2350	(316Ti)	1.4571	S 31635
	(A)	0.05	16.0–18.5	10.5–14.0	2.5–3.0	–	220	45	2343	316	1.4436	S 31600
		0.030	16.0–18.5	11.5–14.5	2.5–3.0	–	210	45	2353	316L	1.4435	S 31603
		0.030	16.0–18.5	9.0–12.5	2.5–3.0	N[2)]	310	40	2375	(316LN)	1.4429	S 31653
25-20		0.08	24.0–26.0	19.0–22.0	–	–	–	–	2361	310S	1.4841	S 31008
18-14-3.5Mo	(A)	0.030	17.5–19.5	14.0–17.0	3.0–4.0	–	220	40	2367	317L	(1.4438)	S 31703

Table 10.6 Some high-alloy modern stainless steels.

Steel type	%C	Typical			Others	$\sigma_{0.2}$ min.	A_5% min.	UNS no.
		%Cr	%Ni	%Mo				
25-4-4 (F)	≤ 0.025	25	4	4	Ti	550	20	S 44635
23-4 (FA)	≤ 0.03	23	4	–	Si,Mn,N	400	25	S 32304
W.st.1.4462								
22-5-3 (FA)	≤ 0.03	22	5.5	3.0	Si,Mn,N	430	25	S 31803
25–7–4 (FA)	≤ 0.03	25	7	4	Si,Mn,N	550	25	S 32705
20-18-6 (A)	≤0.02	20	18	6.1	Cu,N	300	35	S 31254
24-22-7 (A)	≤ 0.02	24	22	7.3	Cu,N	430		S 32654
27-31-3.5 (A)	≤ 0.02	27	31	3.5	Si,Mn,Cu	220	35	

toughness and heat resistance. Contrary to the steels with ferritic, martensitic or ferritic–austenitic structure, the austenitic steels are non-magnetic.

Mo gives improved corrosion resistance to certain acids, although no universal resistance exists versus acids. *Nitrogen* (N) increases the strength of austenitic steels and affects the structure in the same direction as Ni, i.e. it is an austenite former. An N-content of 0.2% gives markedly better corrosion resistance of austenitic steels to seawater than does 0.02–0.08%. *Cu* improves the resistance versus certain acids. *Ti* and *Nb* stabilize the structure by preventing precipitation of chromium carbide and subsequent corrosion at the grain boundaries, i.e. intergranular corrosion. The stabilization can also be obtained by reducing the content of *carbon* to below 0.03% (Section 7.7).

The *martensitic* steels are less corrosion resistant than other stainless steels. Since they can be hardened and tempered, they are used where it is suitable to combine high strength and hardness with moderate corrosion resistance and/or relatively good resistance to cavitation, erosion or other wear mechanisms. Examples of applications are ball bearings, turbine blades, valve guides, tools, cutlery, and surgical instruments. *Ferritic* stainless steels consist of α-iron from room temperature up to the melting point, which explains the non–hardenability. Grades with 18–27% Cr resist oxygen and sulphur at high temperature. The ferritic steels have good resistance to atmospheric corrosion, and 16–18% Cr-steel are used in unpainted automobile parts. These steels resist nitric acid and have been used in tanks for transport and storage of HNO_3. Because of better weldability, the *austenitic* 18–9 steels have to a great extent taken over this field of application. In addition, the latter alloys are ductile and usually more corrosion-resistant than the pure Cr-steels. The 18-9 steels are widely used in the process industry, in food manufacture, kitchen and architecture.

It is very important that the oxide film is dense. On clean and smooth surfaces the oxide is formed rapidly. Conversely, if the surface is rough or contaminated, with deposits, mill scale, welding slag etc. the passivation is hindered, and pitting or deposit corrosion may occur.

If the gas protection during welding has been incomplete, it is important to clean and repassivate the surface. This can be done by pickling and subsequent washing in water.

The most common pickling bath contains 0.5–5 vol% HF (it is strongly etching) and 8–20 vol% HNO_3 (poisonous gases), t = 25–65 °C. The higher alloyed the steel is, the higher the HF concentration and temperature used. The cleaning may also be carried out by fine grinding, succeeded by repassivation by washing in dilute HNO_3, e.g. 20 vol% for 20–30 minutes, and finally washing in water. Brushing and blasting can be used instead of grinding, but it does not always clean as well as grinding.

In extreme cases of inadequate gas protection, as shown by an example in Figure 10.3, cleaning the surface is not sufficient. As a best case, the problems may be solved by plane grinding of the weld. Without such treatment, crevice corrosion is very probable (compare Section 7.5).

Depending on the conditions, stainless steels may be attacked by uniform corrosion (in reducing acids, in extremely oxidizing strong acids and in strong alkalies), pitting, crevice corrosion, intergranular corrosion or stress corrosion cracking (Chapter 7). As regards the resistance to various acids and other chemicals, reference is made to iso-corrosion diagrams and tables in References [10.1–10.3].

In natural waters, the chloride content is crucial for the performance of stainless steels. The following guidelines are given in Reference [10.7]: if the water is very pure (de-ionized, distilled), the only risk for 18-9 steel is the probability of intergranular corrosion in sensitized grades above 100°C. In ordinary *tap water* (cold and hot) and *fresh water,* 18-9 steels without molybdenum can normally be used for chloride contents up to 200 mg/l. For the range 200–500 mg/l it is recommended to use a 17-12 steel with molybdenum, and at even higher chloride concentrations special steels may be needed..

In *waste water* systems (low temperature) the chloride content is often up to 300 mg/l, and 18-9 steel without Mo can normally be used with good results. In some cases, for instance if it is difficult to replace pipe sections, Mo-alloyed 17-12 steel is recommended.

In *swimming pools,* the water may have high contents of chlorine and chloride, which may cause discolouring of 18-9 steel, and therefore Mo-alloyed steels are often used and particularly recommended for purification and heating systems.

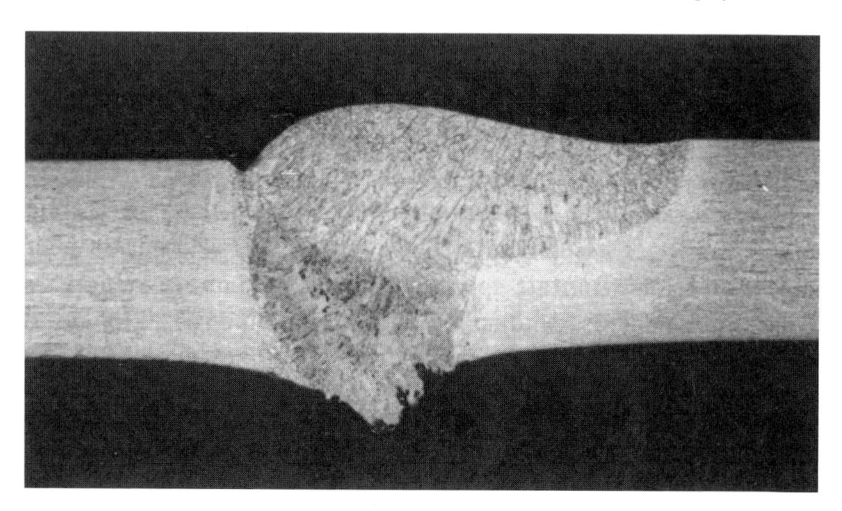

Figure 10.3 Steel of the type UNS S 31803 welded with insufficient gas protection. 5×. (Photo: T. Rogne, Sintef Corrosion Centre.)

Brackish water, i.e. a mixture of fresh water and seawater, contains varying amounts of chloride. For cold water and conditions where crevices and deposits are avoided, steel of the type 17-12-2.5Mo can be used. One should, however, be aware of the possibility that a bacterial slime layer (similar to that developing in seawater) may be formed, particularly at low water velocities (see below). Under such conditions, or if the water is contaminated, e.g. by H_2S (which is often the case in harbours) or the temperature is high, the conditions are more aggressive and higher-alloyed steels may be necessary.

In *natural seawater,* the risk of corrosion is high for most stainless steels. This is due to the effects of the bacterial slime layer that is usually formed in this environment (see Section 6.2.1). Conventional steels with alloying up to 17Cr12Ni2.5Mo are generally not to be recommended for seawater. This applies also to ferritic–austenitic steel with composition 22Cr5Ni3Mo.

As mentioned, special weak points and defects may occur in or at welds. In the guidelines for selection of steel that are given in the preceding paragraphs, high-quality welds, sufficient backing gas, and removal of a possible non-passivating surface layer by pickling or grinding, are presupposed.

The *weldability* varies much from one type of stainless steel to another and must be checked when steels for specific applications are selected. In older types of ferritic steels, embrittlement and intergranular corrosion could occur in the heat-affected zone (HAZ). These problems can be avoided with newer ferritic steels with low contents of C and N. Some of these new steels may be subject to grain growth in the HAZ. Ferritic–austenitic steels show usually somewhat better weld properties than do the ferritic steels. The weldability of austenitic steels is generally good, but there is some tendency to high-temperature cracks in those with highest content of alloying elements. Regarding intergranular corrosion, see Section 7.7.

In Table 10.7 a comparison is made between different steels as regards the resistance to *localized corrosion* in a chloride environment. The resistance to chloride stress corrosion cracking is low for steels with 8–9% Ni, but it is improved strongly when the Ni content is increased further.

Table 10.7 Comparison between different stainless steels as regards resistance to localized corrosion forms [10.7].

Steel type	Resistance to pitting and crevice corrosion *		Resistance to stress corrosion *
	Initiation	Growth	
17Cr	1	1	10
18Cr–2Mo	6	5	10
25–5–1.5Mo	7	6	9
18–9	2	4	1
17–12–2.5Mo	6	7	3
20–25–4.5Mo–1.5Cu	10	10	8

* Characteristics: 1 = poorest, 10 = best.

The steel that is ranked highest in relation to pitting and crevice corrosion in Table 10.7 is also represented in Figure 10.4 (20-25-4.5, i.e., 20Cr25Ni4.5Mo). The figure shows that the *newer austenitic steels* 20-18-6 and 24-22-7 as well as the *ferritic–austenitic steel* 25-7-4 are even better. The columns in the figure indicate critical pitting temperature and critical crevice corrosion temperature for rolled products as tested in a 6% $FeCl_3$ solution.

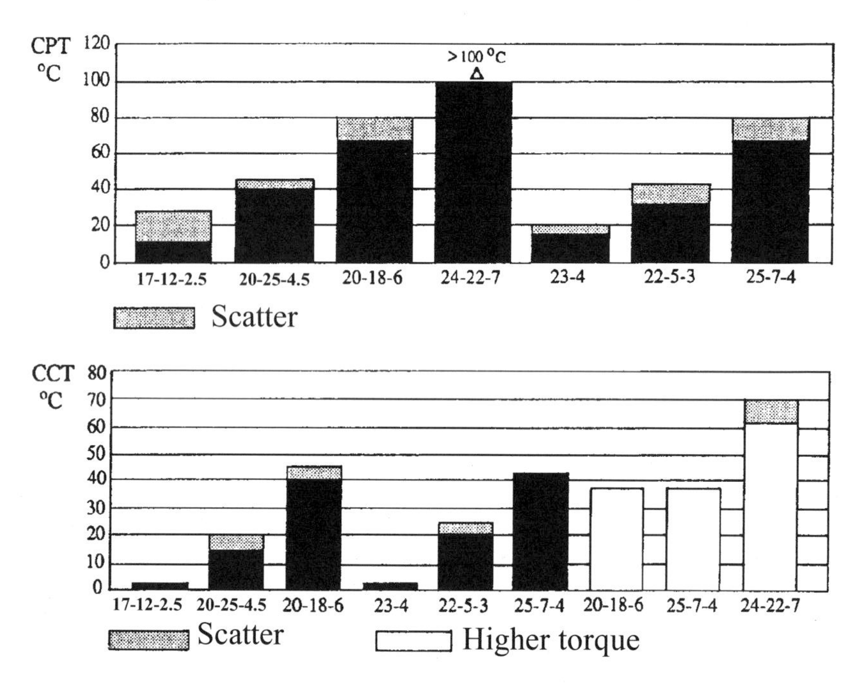

Figure 10.4 Critical pitting temperature (CPT) and critical crevice corrosion temperature (CCT) for various stainless steels in 6% $FeCl_3$ solution [10.9]. The figures below the columns show the contents of Cr, Ni and Mo, respectively (compare Tables 10.5 and 10.6).

The latter steels have been developed during the last few decades. (For compositions, properties and standard numbers, see Table 10.6.) A benefit of the ferritic–austenitic steels is their higher strength compared with the austenitic. Ferritic–austenitic 25-7-4 steel and austenitic 20-18-6 have both shown very good crevice corrosion properties in seawater, but also on these steels attacks may develop when the temperature is above a limit that depends on various conditions. Of these two steel types, the ferritic–austenitic steel may have somewhat lower corrosion resistance in welds than the austenitic 6 Mo steel. With first–class welds, pipes made of the latter material are considered safe to use up to 30–35°C in seawater. However, on flanges that are cast or produced by powder metallurgy, attacks have been found at a temperature as low as 10–15°C. For use in chlorinated seawater with a residual chlorine content of 1.5 ppm, the NORSOK standard [10.10] recommends a maximum temperature of 15°C for components with crevices and 30°C for components without any crevice. The standard may be changed in the light of more recent experience.

Pitting, crevice and stress corrosion in stainless steels are avoided in the absence of oxygen and other oxidizers.

Chloride and oxygen contents are crucial factors for stress corrosion cracking (SCC) of stainless steel (see Figure 7.64). SCC in chloride-containing environments is most relevant for the conventional austenitic stainless steels (see below). Austenitic steels with extraordinarily high Ni-content as well as ferritic steels are usually resistant to SCC [10.7].

The use of stainless steel in *oil production* has increased, and is of particular significance for production in deep waters. This applies both to production and process equipment carrying oil and gas and to seawater systems. For seawater systems the experience from nuclear power plants has been useful. In these plants, some high-alloy steels have performed well in equipments such as seawater pumps and heat exchangers. But the performance depends on service conditions, design and combination of materials. On the oil/gas side, good experience has been obtained with 13% Cr steel in production tubing, in components at the wellhead and in well valves. The same can be said about the use of other stainless steels in process systems offshore, but there have been some exceptions. For instance, SCC has occurred on duplex steel (22–5–3) under external heat insulation (compare Section 7.12) [10.11].

Materials selection for a large sub-sea production system is illustrated by the following example from a field in the North Sea [10.11]: for the production tubing and the well equipment (which are not to be welded) a 13% Cr steel was selected, and in the Xmas tree (valve tree) a martensitic stainless steel of the type ASTM A 182 F6NM (13% Cr, max 1% Si). The manifold module, which is a welded pipe system, was made of a ferritic–austenitic stainless steel with the composition 22Cr5Ni3Mo, which also was selected for pipelines for supply of hydraulic oil and methanol (used for preventing hydrate formation) to the sub-sea stations. The well frame and gas pipeline were made of a non-stainless steel.

An example of materials selection for the main process system for oil/gas/water at a typical installation in the North Sea is also described in Reference [10.11]. The primary and crucial parameter(s) for the materials selection is the degree of CO_2 corrosion, and a possible content of H_2S. If the corrosion rate on carbon steel according to de Waard and Milliams' nomogram with modifications is higher than 0.3 mm/y, a stainless steel is selected, namely duplex 22Cr5Ni3Mo if the temperature is > 60°C and a AISI 316 steel at lower temperatures. This distinction is made because of the conditions for SCC. If the content of hydrogen sulphide exceeds a certain level depending on the chloride content and the degree of cold work, an austenitic steel with 6% Mo can be selected instead of the duplex steel.

General recommendations for materials selection in the offshore sector are given in the NORSOK standard [10.10].

Stainless steels are widely used in the process industry. A comprehensive description of applications and corrosion of materials in this sector is given in Reference [10.12]. For further study of corrosion of stainless steels, see also Reference [10.13].

10.1.6 Nickel Alloys

This group of materials includes some of the most corrosion-resistant engineering alloys, thus it is an important group for applications in aggressive environments and for purposes where long lifetime is required. Partly, the high corrosion resistance is due to good properties of Ni itself, but more important is that the metal can be alloyed with other elements such as Cu, Cr, Mo, Fe and W over considerable concentration ranges without the formation of unstable, unsuitable or detrimental phases. Each of these alloying elements has unique properties, and by these it becomes possible to obtain Ni alloys that together can match a wide spectrum of aggressive environments. In addition, the austenitic structure gives the materials good technological properties and applicability.

Generally speaking, nickel alloys are resistant to alkaline environments (to a particularly high degree), neutral and slightly acidic solutions, and foodstuffs. They tolerate relatively high temperatures and show high resistance to SCC. Most of them are not resistant to strongly oxidizing media such as nitric acid and ammonia solutions, or to sulphur-containing gases at high temperature.

Table 10.8 shows trade names, standard numbers, compositions and application examples for a selection of nickel alloys. A more extensive survey of trade names and compositions is given by Roberge [10.3].

Monel shows very good resistance to fluoric acid. Inconel 625 and the two Hastelloy C alloys included in the table are three of the generally most resistant alloys commercially available.

Nickel, nickel–copper and nickel–chromium alloys tolerate seawater at high flow velocity, but are more or less liable to deposit corrosion and pitting in stagnant or slowly flowing seawater. However, the pit growth rate in NiCu (e.g. Monel) decreases markedly with time. NiCrMo and NiCrMoCu alloys are supposed to be passive in seawater, without significant localized corrosion. The latter ones are found at a high level in the galvanic series for seawater (see Figure 7.5).

Ni alloys are practical materials for process equipment in the oil industry, but the best ones are more expensive than the high-alloy stainless steels. Ni and Ni alloys constitute an important material group within other process industries [10.12].

Extensive information about nickel and its various alloys is given, e.g. in References [10.14, 10.15].

10.1.7 Copper and Its Alloys

Copper possesses good corrosion resistance primarily because it is relatively noble (corrosion resistant by immunity), and it is most suitable for reducing environments. The potential range where the material is active in aqueous solutions (Figure 10.1) is so high that reduction of hydrogen ions is not a possible cathodic reaction. Therefore, copper is immune for example in oxygen-free sulphuric acid. Corrosion can only occur if there is some other oxidizer present that causes a cathodic reaction.

The passive region shown in the Pourbaix diagram, Figure 10.1a, reaches from pH ≈ 7 to pH ≈ 13. This contributes to good surface corrosion resistance of copper and

Table 10.8 Nominal composition and examples of applications of some nickel-based materials. (Selected information from References [10.3, 10.5].)

Alloy	UNS	Approximate content of main elements (%)	Examples of applications
Monel 500 (K500)	N05500	66Ni30Cu3Al1Fe	High-strength components with good corr. resist, e.g. pump and propeller shafts, impellers, valve components, for ships and oil/gas production.
Incoloy 825	N08825	42Ni22Cr30Fe3Mo2Cu1Ti	Chemical processing, oil and gas well piping, acid production, nuclear fuel reprocessing, pickling equipment.
Hastelloy C22	N06022	59Ni 22Cr 13Mo3Fe3W	Suitable for welded or cold-worked components, welding consumables. Outstanding resistance to oxidizing and reducing media and local. corr.
Hastelloy C276	N10276	59Ni16Cr16Mo5Fe4W	Components in chemical. and petrochemical. organic chloride processes, pulp and paper industry, food processing, flue gas desulphurization.
Inconel 625	N06625	63Ni22Cr9Mo2Fe4(Nb+Ta)	High-strength components, in seawater, oil/gas production where H_2S and S exist above 150°C, flue gas/flue gas desulphurization, flare stacks offshore, hydrocarbon proc.

copper alloys in neutral and not to alkaline environments, except ammonia solutions. For instance, these materials withstand atmospheric corrosion very well (compare Section 8.1.2). But the different copper alloys show varying liability to corrosion forms such as selective corrosion, SCC and erosion corrosion, as shown for wrought alloys in Table 10.9 (see also Chapter 7). If the corrosion resistance is crucial for a certain application, further information should be collected. Copper is often used for roofing and for hot and cold water piping.

Copper alloys find widespread application in heat exchanger tubes, tube sheets, valves, piping and fitting in seawater and fresh water systems. Brasses, aluminium brasses or copper–nickel alloys are used, depending on the conditions. The alloys are sensitive versus sulphur-containing species, such as H_2S, and stagnant conditions as well as high flow velocities (compare Sections 7.8 and 7.9). Of the wrought alloys, copper–nickel grades are the best ones in seawater, but they are usually not recommended for flow velocities below 0.5–1.0 m/s (biological growth and possible sulphide production) and above 2.5 m/s (erosion corrosion), and not in galvanic coupling to significantly more noble materials in natural seawater, such as stainless steels. Of interest in connection with heat exchangers is that copper alloys are available together with other materials (steel, aluminium, stainless steels) in the form of so-called duplex tubing (one material inside the other). The mechanical properties vary considerably from one material to another and depend on the degree of cold work and dimension.

Numerous copper alloys exist in cast condition. Some of these are dealt with in Table 10.10, which comprises typical properties, corrosion resistance and applications. The table indicates also that the alloys possess good mechanical properties. However, strength, hardness and ductility of cast copper alloys vary a good deal from alloy to alloy and depend on the product form. Typical ranges comprise yield strength from 60 to 250 MPa, tensile strength from 150 to 700 MPa, Brinell hardness from 45 to 160 and elongation from 3 to 20%. Wrought alloys have usually somewhat higher strength and considerable better ductility. An interesting cast alloy is nickel–aluminium bronze (e.g. CuAl10Fe5Ni5), whose good corrosion properties in fresh water and seawater are retained up to high water velocities, e.g. 20 m/s. It is used in pumps, propellers and valves. Like other copper alloys, it is attacked by selective corrosion in heavily polluted seawater, where H_2S is developed.

Table 10.9 Corrosion resistance of some wrought copper alloys, given by a characteristic on a scale from 1 to 5. 5 is the best characteristic for copper alloys in general, i.e. not compared with other material groups. Adapted from reference [10.16] with UNS numbers according to reference [10.3].

Alloy[1]	UNS	Stress corr. cracking	Selective corrosion	Erosion corrosion	Comment
Cu-ETP	C11000	5	5	1	2
Cu-FRHC	C11020	5	5	1	2
Cu-OF	C10200	5	5	1	
Cu-FRSTP	C12900	5	5	1	2
Cu-DHP	C12200	5	5	1	
CuAg-OFS	C10400–10700	5	5	1	
CuZn10		4	4	2	
CuCn15	Group	4	4	2	
CuZn20	from C20500	4	4	2	
CuZn30	to C28580	3	3	3	
CuZn37		2	2	3	
CuZn43Pb1Al		1	1	3	
CuZn39Pb3		1	1	3	
CuZn39Pb2	Group	1	1	3	
CuZn40Pb	from C31200	1	1	3	
CuZn38Pb1	to C38590	1	1	3	
CUZn36Pb1		2	2	3	
CuNi10Fe1Mn	C70600	5	5	4	
CuNi30Mn1Fe	C71500	5	5	5	
	In group				
CuNi18Zn20	from C73200	4	4	4	
CuNi12Zn24	to C79900	4	4	4	

[1] ETP = electrolytic, tough pitch. FRHC = fire-refined high conductivity. OF = oxygen-free. OFS = oxygen-free with Ag. FRSTP = fire-refined tough pitch with Ag. DHP = Phosphorus deoxidized, high residual phosphorus.

[2] Risk of hydrogen embrittlement.

Table 10.10 Characteristic materials properties and suitable applications of some cast copper alloys. Adapted from Reference [10.16] with UNS numbers according to Reference [10.3].

Alloy	UNS Group	Characteristic materials properties and fields of applications
Tinbronze CuSn10	From C90200 to C94500	Seawater–, corrosion– and cavitation–resistant, tough alloy with good wear resistance. Well suited for pumps and turbine components, fittings for high mechanical loads, and thin–walled castings.
Gunmetal CuSn10Zn2		Seawater–resistant, hard alloy with very good wear resistance. Suitable for machine parts for high mechanical loads, hard bushes and sliding surfaces, worm gears for low velocities, and for propeller sockets and screw shaft bushes.
CuSn6ZnNi		Alloy with high strength and elongation, good castability and seawater resistance. Suitable for fittings, pump housings and dense castings.
Brass CuZn39Pb2Al	From C85200 to C85800	Material with good castability, toughness and Corrosion resistance in the atmosphere. Suitable for complicated components of various kinds in electro-mechanical and machine constructions.
High-strength brass CuZn35AlFeMn		Seawater- and cavitation-resistant, hard and tough alloy with relatively high strength. Well suited for ship propellers, components in rolling mills, couplings, and worm gears exposed to high load but low velocity.
Nickel–aluminium bronze CuAl10Fe5Ni5	From C95200 to 95810	Seawater-, corrosion- and cavitation-resistant alloy with relatively very high strength. Well suited for components in ships and machines exposed to high mechanical loads and corrosion, e.g. ship propellers, worm and worm gears, gears, transport screws, high–pressure steam fittings, and parts exposed to heavy mechanical wear conditions.
Aluminum bronze CuAl10Fe3		Seawater-, corrosion-resistant tough alloy. Very well suited for components in machines and equipment in chemical and foodstuffs industry, for fittings etc.

10.1.8 Aluminium and Its Alloys

This is a material group whose application has increased strongly due to a number of beneficial properties: low weight, excellent formability (in rolling, drawing and extrusion), high corrosion resistance in many environments, colourless and non-poisonous corrosion products, high electrical and thermal conductivity. Some of the alloys have relatively high strength and therefore high strength/weight ratio, the latter in fact at the same level as high-strength structural steels. On the other hand, the modulus of elasticity of aluminium materials is low. This may be more or less compensated by their good formability, which makes it possible to obtain suitable profiles with high moments of inertia. Regarding fatigue strength, there is only a slight increase with increasing tensile strength, and for welded structure it is almost independent of the tensile strength. The ratio between the fatigue strength and the density is about the same as for ordinary structural steel (e.g. St. 52). The ratio between the fatigue strength and the modulus of elasticity is also about the same for these material groups. The strength of aluminium materials decreases fast when the temperature is increased above 350°C. Conversely, the cryogenic properties are very good due to the face-centred cubic structure.

Common product forms, tensile strength, and resistance to atmospheric corrosion of some wrought aluminium materials are shown in Table 10.11. The strongest grades of the various alloys imply lower ductility, i.e. elongation mainly within the range 2–12%, compared to 15–35% for the softest grades. A number of fields of application of the same materials are listed in Table 10.12.

Functional properties of aluminium in general are described in a book by Hatch [10.17], which also includes a comprehensive chapter on corrosion properties.

Table 10.11 Standard designations, product forms, strength, and resistance to atmospheric corrosion of some wrought aluminium materials. (Adapted from References [10.5, 10.16].)

Material designation			Product form					Res.
AA	ISO	BS 1470–75	Plate, sheet, strip	Tube	Rod, wire	Section	Tensile strength MPa [1]	Res. to atm. corr. [2]
1050	Al99.5	1B	X	X	X		55–135	V
2014	AlCu4SiMg	H15	X			X	370 min	F
3003	AlMn1	N3	X				90–175	V
5052	AlMg2.5	N4	X		X		160–225	V
5083	AlMg4.5Mn	N8	X			X	275–345	V
6063	AlMgSi	H9		X	X	X	125 min	V
6351	AlSi1Mg	H30	X	X	X	X	200 min	G
7005	AlZn4Mg1		X		X	X	200–350	

[1] Depending on degree of strain-hardening or heat-treatment.
[2] V = very good, G = good, F = fair.

Table 10.12 Fields of application of materials presented in Table 10.11. (Adapted from Reference [10.16].)

Fields of application	Al99.0	Al99.5	Al99.7	AlCu4SiMg	AlCu4Pb2Mg1	AlMg2	AlMg2.5	AlMg4.5Mn	AlMg5	AlSi1Mg	AlMgSi	AlMn1	AlZn4Mg1
Apparatus	X	X				X	X	X	X			X	X
Building structures						X	X			X	X	X	X
Panels	X	X										X	
Machined components				X	X				X		X		
Electro–mechanical		X	X							X	X		
Vehicles and accessories						X	X	X		X	X	X	X
Tools							X			X			X
Machine making				X						X			X
Metallic goods, hardware	X	X	X			X				X		X	
Furniture							X		X		X		
Foodstuffs industry	X	X	X			X	X			X		X	
Rivets	X			X			X		X	X		X	
Shipbuilding							X	X	X	X			
Packing		X	X			X	X					X	
Architecture										X	X		

The Pourbaix diagram for Al shown in Figure 3.11 is based upon passivation with $Al_2O_3 \cdot 3H_2O$. Other forms of Al_2O_3 that may exist in practice give a somewhat narrower passive region, but a common feature is that the passive region comprises neutral and slightly acid water, and that the passivity is very efficient for pure aluminium in reasonably pure water. The corrosion regions at higher and lower pH in the Pourbaix diagram extend to very low potentials, and the corrosion resistance depends therefore on a dense and continuous oxide film.

As mentioned in Chapter 3, the most significant deviations from the theoretical Pourbaix diagram in practice are that chloride or other aggressive species may destroy the passivity and that impurities in the metal may cause weak points in the oxide, with pitting as a possible consequence (Section 7.6). For pure aluminium, the corrosion resistance decreases considerably when the content of impurities in the metal increases from 0.01 to 1%. However, even 99% aluminium resists neutral atmospheres and chloride-free water very well. In seawater, pitting will usually occur, but the weight loss is low and the pits shallow (Section 7.6).

In severe (acidic) industrial atmospheres, a mixture of pitting and uniform corrosion may take place. The oxide film is partly dissolved under such conditions (compare the Pourbaix diagram).

For many environments, corrosion of aluminium cannot be explained by means of the Pourbaix diagram. Since Al, like Zn, shows very high hydrogen overvoltage, super-pure aluminium (99.99%) may corrode very slowly even in the active state (in an acid solution with little access of air). A 1% content of impurities will in this case accelerate the corrosion heavily.

The most corrosion-resistant commercial Al alloys are those containing Mg and/or Mn. These can be exposed to neutral and slightly acidic solutions, various atmospheres and seawater with good results. AlMgSi withstands marine atmosphere but is less suited for components immersed in seawater. Results of atmospheric exposure of different Al materials in a wide variety of testing sites around the world are listed by Roberge [10.3]. Characteristics of the resistance to SCC of various alloys in different conditions are also listed. When aluminium materials are immersed they are liable to intensive galvanic corrosion if coupled to more noble metallic materials. Corrosion of aluminium materials is further dealt with in Sections 7.3 and 7.6, and as coatings in Section 10.6.1. The corrosion resistance can be improved by anodizing.

In addition to the fields of application listed in Table 10.12, aluminium is used in helicopter decks and living quarters on offshore oil platforms, and in superstructures on ships.

Special aluminium alloys have been developed for sacrifical anodes (see Section 10.4 about cathodic protection).

10.1.9 Titanium and Its Alloys

Various aspects are dealt with in the proceedings from a conference concentrated upon application of titanium [10.18].

Titanium is an interesting material that, due to its high price, has not been particularly widespread in use, but the number and extent of applications are increasing. For demanding conditions, selection of titanium may be an economic solution because it has a series of good properties. Its primary field of application has been in the aerospace industry: compressor wheel discs, blades in gas turbines. etc. It has traditionally been used also in the chemical industry. Its use in the oil production industry has increased due to conditions dealt with in the following paragraphs.

The corrosion properties of titanium versus the actual environments are superior to the properties of competing materials. The experience with titanium in heat exchangers is excellent, both in nuclear power plants, offshore installations and process industry. The material resists seawater up to 130°C (some alloys even higher); generally it tolerates polluted seawater very well (contrary to Cu-based alloys). For oil production systems it is of interest that CO_2 is not aggressive versus titanium and that neither hydrocarbons nor sulphides or chlorides usually cause corrosion on this material. Some attacks have been detected on titanium exposed to solutions with high H_2S concentration, but on most oil fields this has not been a problem. The material is resistant to hypochlorites and wet chlorines, as well as nitric acids including hot acids. In the process industry, titanium has performed well in pumps, valves, gas purification plants and pipelines.

Titanium alloys constitute the group of metallic materials showing the highest strength/weight ratio. The density is 4.5 g/cm^3. A grade widely used under corrosive conditions (the one with the best corrosion properties) is commercially pure titanium, which has a yield strength in the range 300–400 MPa, but far stronger alloys exist. The fatigue properties are also beneficial, e.g. the fatigue strength of material without surface notches in air at 10^7 cycles may be about 60% of the ultimate tensile strength, a percentage that is high compared with steel.

Titanium is not particularly difficult to weld if important precautions are taken. Inert gas protection of the weld and the nearest material is required during the welding work. However, Ti cannot be welded to other materials. Practical precautions must also be taken under machining to avoid formation of microscopic cracks in the surface.

For use in heat exchangers, the thermal conductivity is of interest. In titanium it is at the same level as in austenitic stainless steels, but much lower than in copper-based alloys. Here, it must also be taken into account that the wall thickness of titanium can be less, partly because of higher strength and partly due to superior corrosion properties compared with Cu alloys.

It should also be noted that, at low and moderate flow velocities, the biological growth tendency is larger on titanium than on copper materials, which may, however, be counteracted by chlorination of the water or allowing higher velocities.

Titanium alloys are sensitive to absorption of hydrogen, which makes the materials brittle. Contact with any corroding material that can be the site for hydrogen evolution must therefore be avoided. Ti should not be used in pure hydrochloric, sulphuric and hydrofluoric acids or dry chlorine [10.1].

10.1.10 Other Metallic Materials

Magnesium is thermodynamically one of the less noble metals, and it can protect most other metals when used as sacrificial anodes (see Section 10.4). In the atmosphere the metal is covered by an oxide film. Therefore it resists rural atmospheres but is subject to pitting in marine atmospheres. Magnesium alloys are also liable to SCC and erosion corrosion, and are attacked by most acids. Mg alloys are used in automobile engines, aircraft, missiles and various movable and portable equipment, in all cases primarily because of their low density (1.76 g/cm^3).

Zinc is, with respect to corrosion, most interesting in the form of coatings and sacrificial anodes (Sections 10.6 and 10.4). Besides these applications it is used in die-cast components, e.g. in automobiles (door handles, grilles), but in the latter cases it is usually covered with a corrosion-resistant coating.

Tin and *cadmium* are also primarily interesting as coatings for protection of steel.

Lead as a corrosion-resistant material is particularly used versus dilute sulphuric acid, where it is protected by a passivating layer of lead sulphate. For this purpose it is alloyed with 0.06% copper.

Of very corrosion-resistant materials we have metals such as *tantalum* and *zirconium* (both resistant due to passivity), and *platinum* (immune, passive in strongly oxidizing environment) (see the Pourbaix diagrams for tantalum and platinum in Figure 3.11). Tantalum is attacked by some alkaline solutions, fluoric acid and hot, concentrated sulphuric acid. The metal is used, e.g. for repair of glass-lined equipment and for handling of pure chemicals. Zirconium is used in nuclear power plants. One of the applications of platinum is anodes for impressed current cathodic protection (Section 10.4).

Very high strength and stiffness are obtained in two relatively new groups of materials, namely *metallic glasses* and *metal composites*. The former are also very ductile, and their corrosion resistance is outstanding, but they may be liable to hydrogen embrittlement under certain conditions. Only very thin foils can be produced, which may be utilized in composites with a metal binding phase. The corrosion resistance of metal composites depends on the combination of materials. Often, this group shows improved resistance to crack growth compared with the actual metallic matrix [10.1].

10.1.11 Non-metallic Materials

The most important groups of non-metallic materials are plastics and rubbers, carbon, ceramics (including concrete), and wood.

Plastics (polymers) and *rubbers* (elastomers) may, in many cases with advantage, replace metallic materials in corrosive environments. Particularly, these organic materials resist chloride-containing media better than most metallic materials. For instance, rubber linings are used in metal pipes carrying hydrochloric acid. On the other hand, plastics are less resistant in concentrated sulphuric acid, oxidizing acids such as nitric acid, organic solvents, and at high temperatures.

Polymers possess the benefit of low density, and in the form of composites they can obtain a very high strength/weight ratio. They are therefore suitable in means of transport. The application of plastics on offshore platforms has been limited primarily because the demands on fire precautions are stricter than they are onshore, and for some purposes due to higher cost than the metallic alternatives. The use of polymers offshore has, however, increased in recent years. One field of application where these materials are interesting is pipe systems of flame-resistant polymers.

Soft rubbers are excellent against wear and erosion corrosion.

Plastics and rubbers are used as insulating and separating materials between different metallic materials to prevent galvanic corrosion. For this application specific requirements must be fulfilled, such as the necessary mechanical strength and thickness; they must be fireproof, and resistant to the environment (including heat-resistant to a specified extent). In addition the following are important:

1. The ohmic resistivity must be sufficiently high. By mistake, materials such as chloroprene (neoprene) containing carbon particles have sometimes been used for insulation purposes in spite of their relatively low resistivity. Such materials not only allow galvanic corrosion, they also act as efficient cathodes and will therefore increase the rate of galvanic corrosion. For insulating aluminium from steel a resistivity of minimum 10^{10} ohm cm has been recommended (see Section 7.3.2), which is satisfied by most polymers not containing conducting particles.
2. Insulating materials should not release species that are aggressive versus the present metallic materials. This may, for instance be the case for vulcanized rubber, which releases sulphur.
3. Water-absorbing materials must be avoided.

Carbon possesses high resistance to corrosion (except against halogens and oxidizing acids) and high electrical and thermal conductivity, but it is brittle. Graphite anodes are used in cathodic protection by impressed current.

Ceramics have excellent resistance to corrosion, wear and heat, but relatively low tensile strength, ductility and toughness. Fluoric acid and alkalis are environments that attack ceramic materials. Concrete is resistant to mildly corrosive surroundings, but has to be protected by coatings against strongly aggressive environments.

Somewhat higher ductility is obtained by combining ceramic particles with a metallic binder phase. Such *ceramic–metallic materials* (cermets) are excellent under erosive and abrasive conditions. A well-known material is WC Co. Better corrosion resistance is obtained by alloying Co with Cr or by using alloys of Ni, Cr and possibly Mo as the binding phase.

Wood is attacked by strong acids, oxidizing acids, dilute alkalis and biological environments.

10.2 Change of Environment

The environment may be changed in the following ways in order to reduce corrosion rates:

a) Decreasing (or increasing) the temperature
b) Decreasing (or increasing) the flow velocity
c) Decreasing (or increasing) the content of oxygen or aggressive species
d) Adding inhibitors

The first three items are not subjects for the present section. It should, however, be emphasized that corrosion rates most often are reduced by reducing temperature, flow rates, or content of oxygen or aggressive species, although examples of the opposite behaviour do exist for each of these items. For further study of the effects of these variables, reference is made to the previous chapters, particularly Chapter 6.

On the following pages, various groups of inhibitors, their effects, and examples of compositions and applications are presented. The inhibitors can be arranged in groups based on which reaction (anodic or cathodic) they affect and how they influence upon the polarization properties. Five different groups are characterized by the overvoltage curves shown in Figure 10.5a–e.

The *passivating inhibitors,* also called passivators (a and b in Figure 10.5), are usually inorganic. The *oxidizing* ones (a) act by depolarizing the cathodic reaction (making it more efficient), or more frequently by introducing an additional cathodic reaction. When the concentration of inhibitor becomes high enough (higher than a critical value c, see Figure 10.5a), the cathodic current density at the primary passivation potential becomes higher than the critical anodic current density, and consequently the metal is passivated.

Examples of oxidizing inhibitors are chromates and nitrites. Critical concentrations of these for steel in neutral water are 10^{-4}–10^{-3} M at room temperature [10.19]. At 70–90°C as much as 10^{-2} M is needed because the critical current density increases by increasing temperature. Chlorides in the water affect the critical concentration in the same direction as does the temperature. Chlorides and sulphates also lead to higher passive current density and reduce the efficiency of the inhibitor. Chromates and nitrite inhibitors may with advantage be combined with pH adjustment to values between 7.5 and 9.5, at which the critical concentration of inhibitor is lower than at lower pH values.

Both chromates and nitrites are often used in cooling systems. Chromates are poisonous, which is a serious drawback. Nitrites are also advantageous in cooling water with antifreeze liquid on an alcohol or ethylene glycol basis since they, contrary to chromates, do not react with the antifreeze liquid. Furthermore, nitrites are used in oil–water emulsions in machining, in oil storage tanks in houses and internally in gas and oil-carrying pipelines. A common nitrite inhibitor is $NaNO_2$.

When using oxidizing inhibitors it is very important that the inhibitor concentration is high enough everywhere in the system because a concentration below the critical value is worse than no inhibitor at all. At positions with stagnant liquid there may be a risk of lower supply than consumption of inhibitor and consequently too low a concentration, resulting in localized corrosion. In order to avoid this, the liquid should be kept moving, and crevices as well as deposits should

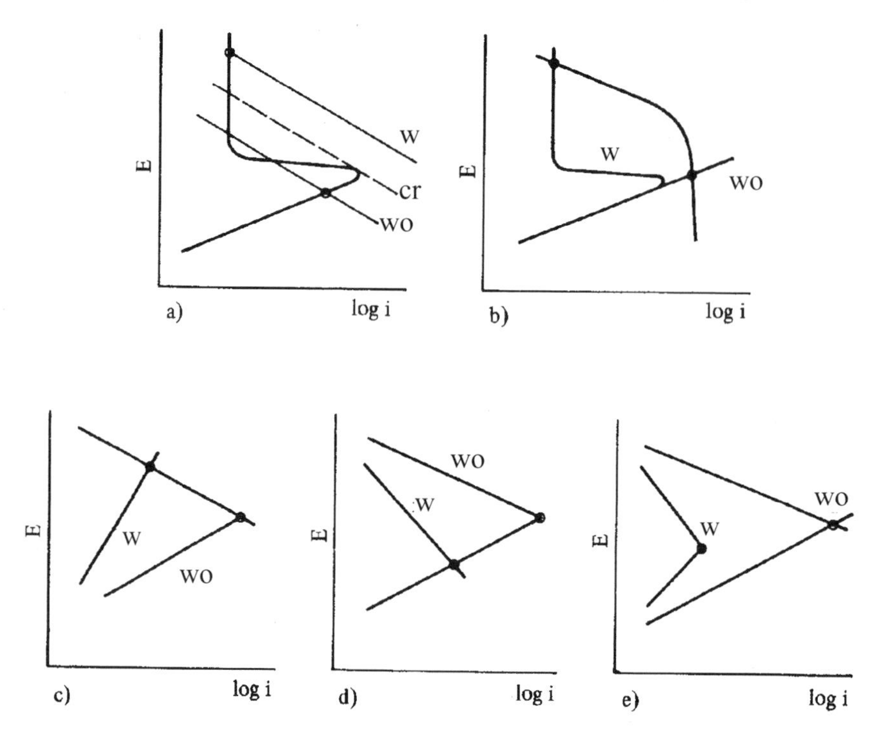

Figure 10.5 Effect of different types of inhibitor on cathodic and anodic overvoltage curves. w = with and wo = without inhibitor. cr = critical concentration of inhibitor. a) Oxidizing inhibitor (passivator), b) anodic, passivating inhibitor (passivator), c) anodic, not passivating, d) cathodic, and e) double acting inhibitor.

be avoided when using such inhibitors. It is also important to check the inhibitor concentration regularly.

The described effect of the oxidizing inhibitors implies that they are primarily suitable for active–passive materials, steel being the most common example. However, it has been demonstrated empirically and experimentally that, e.g. sodium chromate is efficient also on aluminium in water-containing chlorides, some acids and slightly alkaline solutions, and that chromates are useful for protection of copper alloys, soft solder joints and zinc in cooling water systems [10.20]. This indicates that chromates also affect the anodic reaction, i.e. partly as an *anodic passivating inhibitor.* Generally, this group can rather be characterized by the overvoltage curves in Figure 10.5b. It comprises alkaline compounds such as NaOH, Na_2PO_4, $Na_2O_nSiO_2$ and $Na_2B_4O_n$. Contrary to the oxidizing inhibitors, the anodic passivating inhibitors usually lead to reduced corrosion rate only in waters with dissolved oxygen. Within this group we have examples such as sodium benzoate and sodium polyphosphate. The latter may in some systems act by forming diffusion-limiting layers of deposits on cathodic areas (similar to some cathodic inhibitors, see below). The mentioned anodic passivating inhibitors are particularly suitable for iron and steel in near neutral waters.

Non-passivating inhibitors include both some anodic (Figure 10.5c) and the cathodic ones (Figure 10.5d). The latter are exemplified by those that remove free oxygen by a reaction, as is the case for hydrazine and sodium sulphite:

$$N_2H_4 + O_2 \rightarrow N_2 + 2H_2O, \tag{10.1}$$

$$2Na_2SO_3 + O_2 \rightarrow 2Na_2SO_4. \tag{10.2}$$

These inhibitors are effective in all environments where the oxygen reduction is the dominating cathodic reaction in the uninhibited state, such as in neutral natural waters. Conversely, other types must be used in strongly acid solutions. Typical inhibitors for strong acids are the ones used as pickling inhibitors (in pickling for removal of mill scale or for cleaning metal surfaces in other connections). Pickling inhibitors are most often organic substances that are adsorbed on the surface and are double acting, in the sense that both the cathodic and the anodic reaction are made less efficient (Figure 10.5e). They are often added in concentrations of 0.01–0.1%. Organic, adsorption species form the largest group of inhibitors. Typical examples in this group are organic amines. The group includes – among others – substances that are used in oil, grease or wax on machine parts during transport and storage.

Vapour phase inhibitors can also be considered as adsorption inhibitors. These are used for protection of wrapped components temporarily. The inhibitor is placed together with the component(s) and acts due to its suitable low saturation pressure, leading to a sufficiently durable inhibitor condensate on the metal surface. By this the effect of water and oxygen is prevented. It should be emphasized that these inhibitors may accelerate corrosion on some non-ferrous metals and alloys. Examples of vapour phase inhibitors are bicyclohexyl-ammonium nitrite and cyclohexyl-amine carbonate.

A large number of inhibitors for various metals and environments and their behaviour under different conditions are thoroughly dealt with in the literature, e.g. References [10.3, 10.5,10.20–10.22].

Inhibitors are very important for corrosion prevention in oil and gas production plants. The amounts, the cost, and the consequences of wrong treatment are so huge that selection and concentration of inhibitors must be decided by real specialists in the field. Companies specializing in inhibitors and other chemicals can be consulted. Oil companies have sponsored and carried out their own research on the subject, and may also hold a lot of knowledge about inhibitors themselves.

Use of inhibitors is generally most common in recirculation systems. For once-through flow systems the protection method is relatively expensive.

10.3 Proper Design

Design and materials selection are performed in connection with each other. In these processes both the individual components, the interactions between them and the relation to other structures and the surroundings have to be taken into account. The various phases of the life cycle of the construction, i.e. manufacturing, storing, transport, installation, operation and service, maintenance, and destruction should be considered. Materials selection has been dealt with in Section 10.1, and the following pages concentrate on design and arrangement.

The majority of the corrosion forms are affected more or less by the geometry: galvanic and thermogalvanic corrosion, crevice corrosion, erosion and cavitation corrosion, fretting corrosion, stress corrosion cracking and corrosion fatigue. Important design measures for prevention of these corrosion forms are dealt with in the respective sections of Chapter 7. In addition some general guidelines are useful:

1. Design with sufficient corrosion allowance. Pipes, tanks, containers and other equipment are often made with a wall thickness twice the corrosion depth expected during the desired lifetime. However, this must of course depend on load and maximum permissible stresses too.
2. Design such that the components that are most liable to corrosion are easy to replace. Special parts may be installed for "attracting" the corrosion. An example is shown in Figure 10.6: a "spool" of carbon steel is inserted between a heat exchanger and the adjacent copper pipeline for the purpose that copper shall be deposited *before* the water enters the heat exchanger (see Section 7.3.2).
3. For structures exposed to the atmosphere: the design should allow easy drainage with ample supply of air. Alternatively, the opposite: hinder air transport to cavities by complete sealing.

   For components immersed in aqueous solutions there are similar extremes: efficient aeration should be secured (when this will cause passivation), or aeration should be prevented as far as possible.
4. Design in a way that makes drainage, inspection and cleaning easy (Figure 7.26 and Figure 10.7). Use joints that do not cause corrosion problems, e.g. butt welds instead of overlap joints. However, in cases where the materials are

protected by a paint or zinc coating before joining, sometimes it is beneficial to avoid welding. Alternatively, the welds can be post-treated (blast cleaned and coated, e.g. with paint, tape or thermal spray metal).

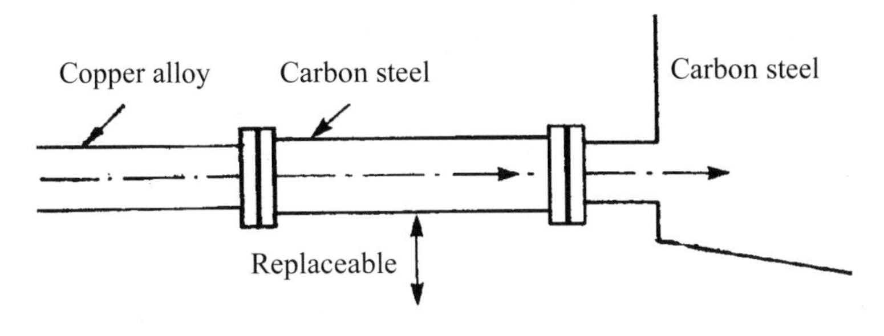

Figure 10.6 The part with highest corrosion rate should be easy to replace.

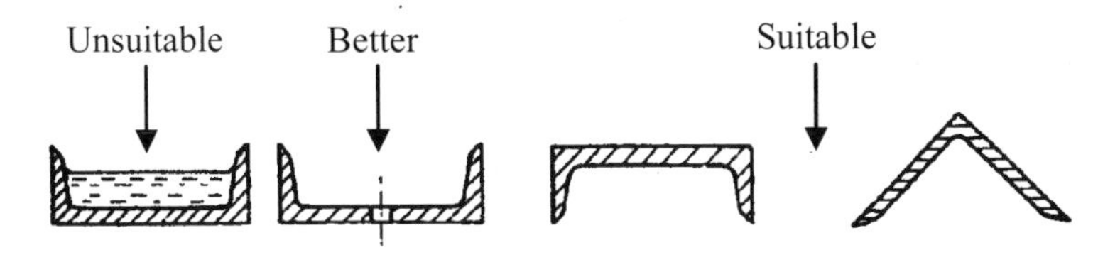

Figure 10.7 Sections should be arranged in such a way that they can be drained and kept clean.

5. Avoid hot as well as cold spots. Heat exchangers and other equipment where heat transport occurs should be so designed that the surface temperature varies as little as possible. On superheated spots, increased (possibly thermogalvanic) corrosion will occur. For systems containing vapour, cold spots leading to local condensation and corrosion should be avoided (Figure 10.8).

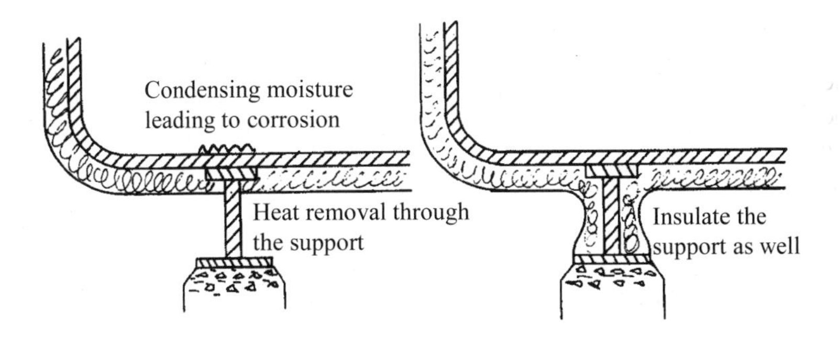

Figure 10.8 Heat insulation of a vessel containing a moist gas. (From Reference [10.23] with kind permission of Kluwer Academic Publishers.)

6. Take the surroundings into account: make arrangements for minimizing the consequences of corrosion, e.g. where it may cause leakage (Figure 10.9). Make sure that separate systems do not impair the environment for each other. An example of this is if moisture containing copper ions (originating from corrosion of a copper alloy pipe) can come in contact with aluminium components.

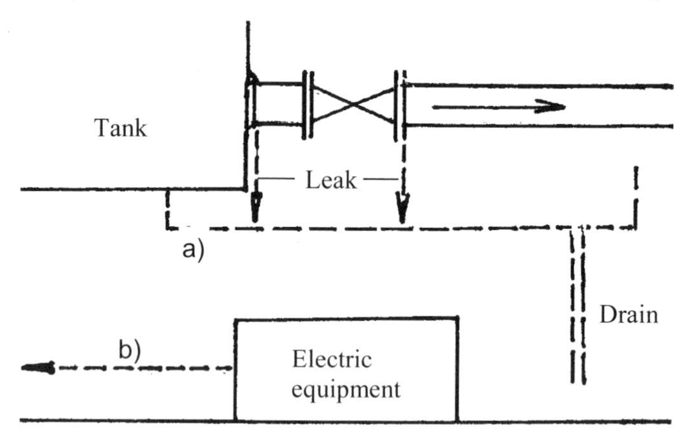

Figure 10.9 Avoiding serious damage due to leakage by a) using a tray for collecting drips or b) moving sensitive equipment to a safer possition.

7. Avoid high corrosion risk on load-bearing parts or on critical places by shifting the attack to less critical places (e.g. extending the inlet part of a tube or pipe as shown in Figure 7.46c).
8. Aim at simple geometry, and avoid heterogeneity and sharp changes in the system. Heterogeneity comprises different metals, uneven temperature and stress distribution, uneven dimensions etc. Abrupt dimension changes are unfortunate for stress distribution, temperature distribution, flow conditions, and possibly for the ease of cleaning. Sharp edges and irregularities on the surface impair the basis for painting, electrolytic plating and hot dipping (Figure 10.10). Welds may contribute to several such problems. Crevices may cause harm in more than one sense, for instance on parts that are going to be pickled, since pickling acid may be entrapped and cause corrosion. Complex geometry and narrow gaps impede surface treatment such as blast cleaning, painting and thermal spraying, increase the cost, and make it more difficult to keep the structure clean and dry. Certain rules for minimum gaps between profiles are given in ISO 12944-3:99 (Figure 10.11).
9. Design for flow velocities that are compatible with the selected materials.
10. Adjust pressure on gaskets (depending on gasket material) to avoid penetration of liquid into crevices.

Further reference is made to Pludek [10.24] who deals with a large number of precautions with respect to design.

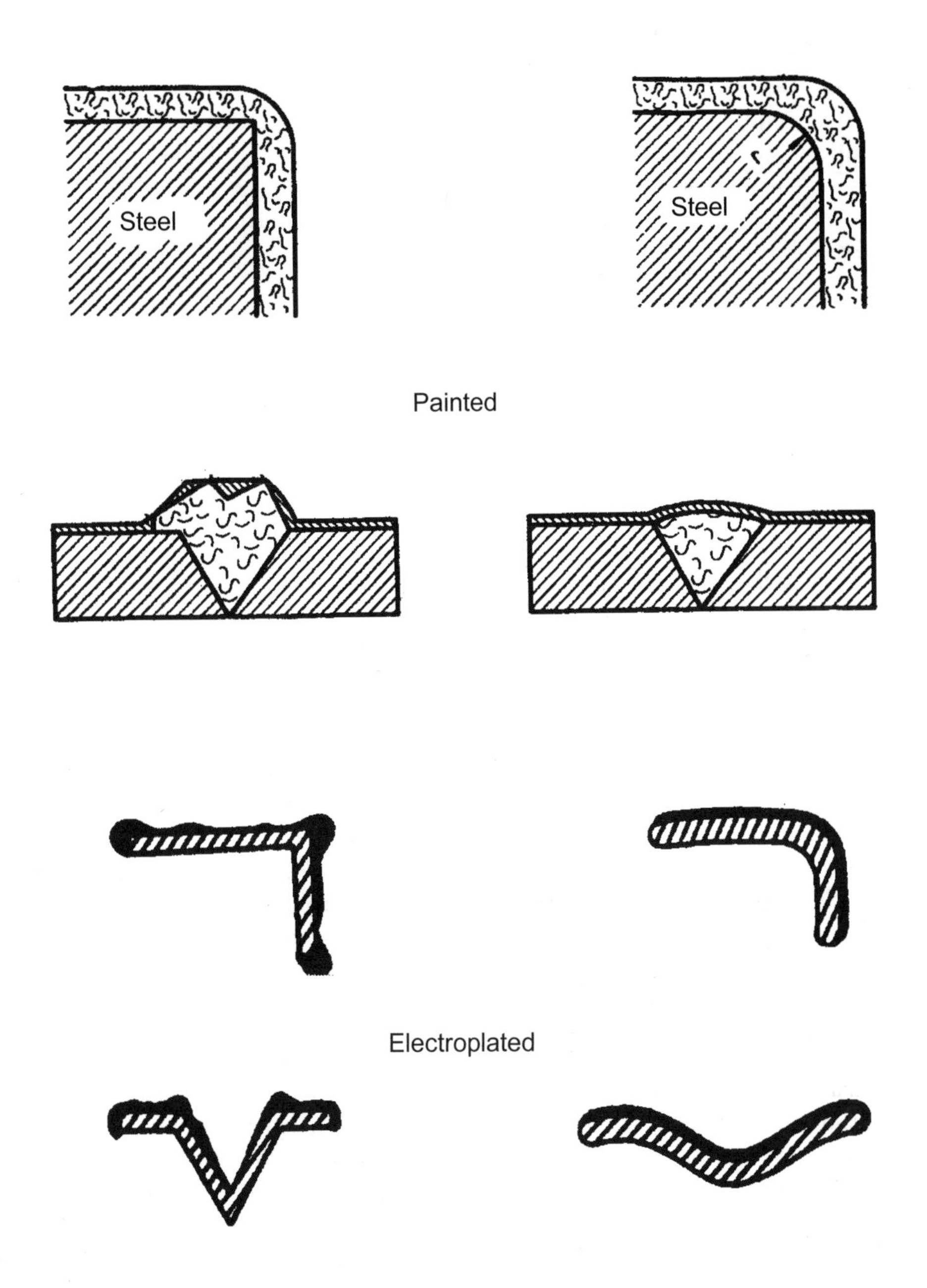

Figure 10.10 Sharp edges and irregularities on the surface are unsuitable as a basis for coating. (Adapted from ISO 12944-3, 1999, and Reference [10.24].)

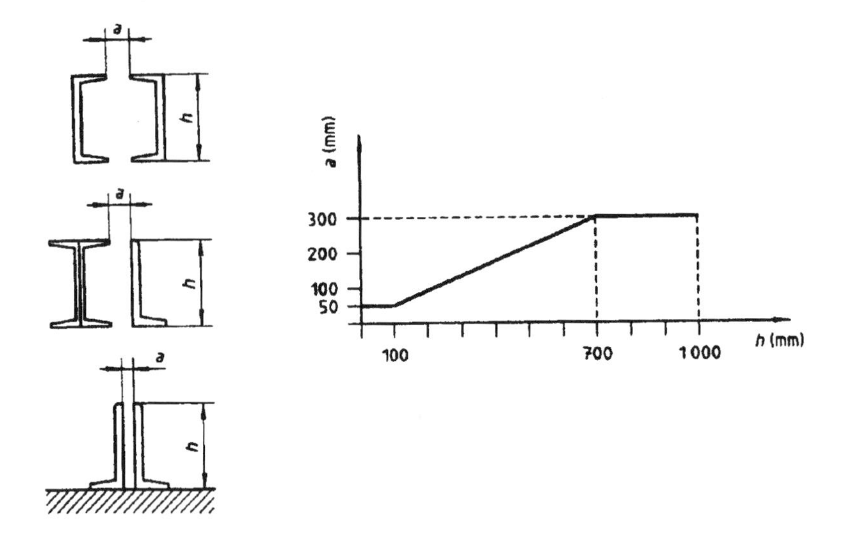

Figure 10.11 Minimum distances between sections according to ISO 12944-3: 1999.

10.4 Cathodic Protection

To get to the source, one must swim against the current.
STANISLAW JERZY LEC

Cathodic protection (CP) has been known for about 170 years. Primarily it has been used for protection of ordinary structural steel in soil and seawater, more seldom (and under special conditions) for steel exposed to fresh water. Other materials can also be protected by CP, for instance to prevent localized corrosion on stainless steel and aluminium. CP has in most cases been applied in combination with a coating, with the intention to protect the steel on "holidays" and damaged areas of the coating. In recent decades the application of this technology has increased considerably in connection with the expanding offshore oil and gas exploration and production. It is the dominating protection method for the large submerged parts of fixed oil installations, most often the only one for the parts freely exposed to seawater.

10.4.1 Principle

The main principle of CP is to impress an external current on the material, which forces the electrode potential down to the immune region, or, for protection against localized corrosion, below a protection potential. In other words, the material is

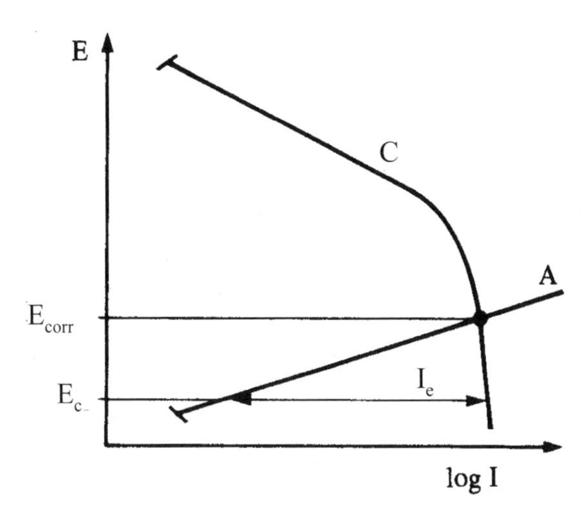

Figure 10.12 Shift of potential from the free corrosion potential E_{corr} to a lower potential E_c by means of an external current I_e.

made the cathode in an electrochemical cell; it is polarized cathodically as illustrated by potential–log current curves in Figure 10.12.

The external current I_e is the difference between the cathodic and the anodic current at the actual potential E_c. At this potential, the anodic current on the protected material is very low, i.e. the external current is nearly equal to the cathodic current.

The external current can be produced in two different ways (Figure 10.13):

a) By means of a less noble material in the form of sacrificial anodes, which are connected by metallic conductors to the structure to be protected.
b) By means of an external current source, usually a rectifier. A reference electrode may be used to control the rectifier potentiostatically.

A potential–log current diagram for a system with sacrificial anodes is shown in Figure 10.14. The flat anodic curve of the sacrificial anode is typical for a suitable anode material. There is a potential difference ΔE between the structure and the anode due to a potential drop in the water. When the structure is completely protected, the galvanic current $I_g = I_c = I_a$, where I_c is the cathodic current on the structure, and I_a is the anodic current on the sacrificial anode, implying that the relatively small cathodic current on the anode is disregarded.

When dealing with CP we operate with different definitions of the term potential. We have the *electrode potentials* of the structure and the sacrificial anode, i.e. the cathode and the anode potential, and we have a *varying electrical potential* in the system. To avoid confusion as regards the sign, plus or minus, of potential differences from one place to another it is useful to present the electrical potential variation in a potential–distance diagram. For a CP system with sacrificial anodes, such a diagram is shown in Figure 10.15. The system forms a galvanic element, and

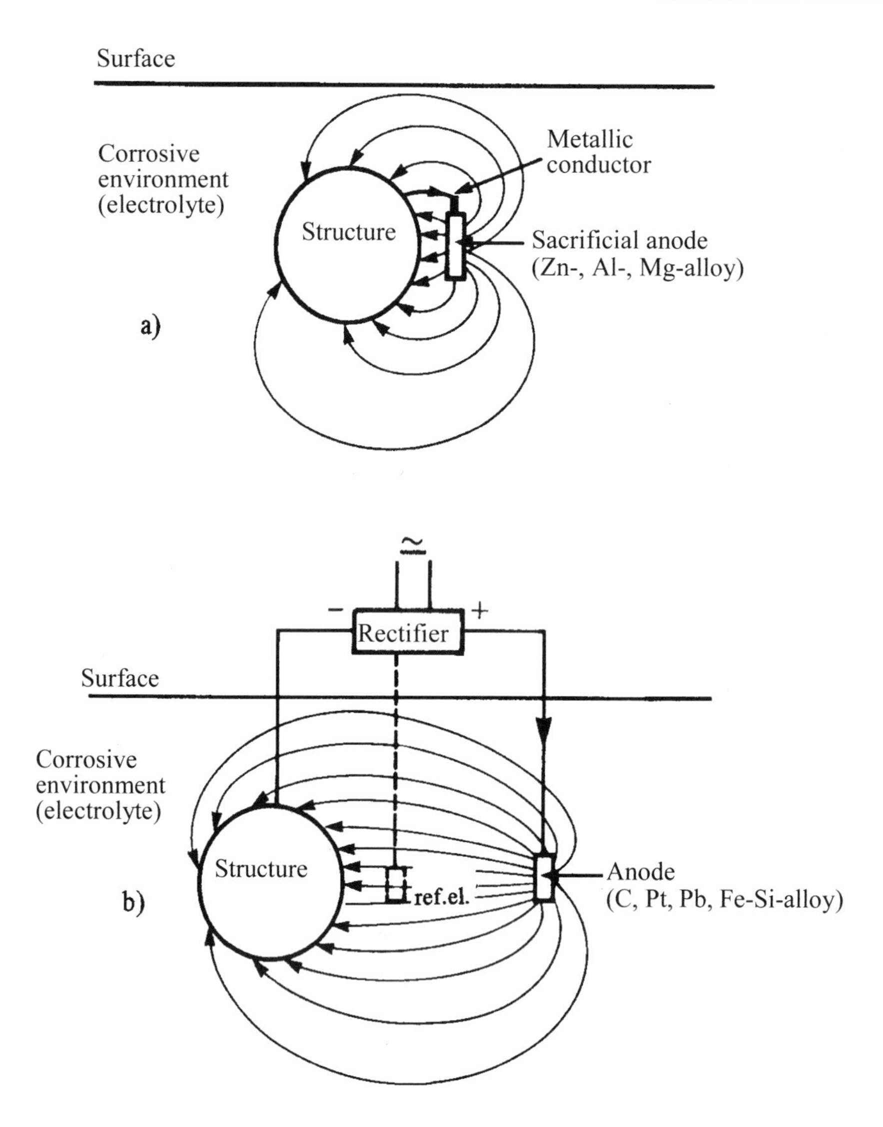

Figure 10.13 Cathodic protection by a) sacrificial anodes and b) impressed current.

similar diagrams are also useful in connection with other cases of galvanic corrosion. In Figure 10.15, any potential drop in metallic materials (in the structure and the connecting conductors) is disregarded since it is usually insignificant in comparison with the other potential differences. Therefore, the presentation involves a common electrialc potential P_m in the anode and the structure material. The electrode potential is the difference in electrical potential between the metal and the adjacent electrolyte (see Section 3.10). Versus common reference electrodes such as the standard hydrogen electrode, the saturated calomel electrode and the silver-silver chloride electrode, the electrode potentials on the steel structure and the sacrificial anodes (E_c and E_a) are negative, as illustrated in the diagram. The anode potential E_a is more negative than the electrode potential of the structure E_c, due to a potential drop ΔE in the water. It should also be emphasized that the electrical potential in the water decreases in the direction from the anode to the structure (positive ions move towards the structure, negative ions towards the anode). Most of the potential drop in the water is localized close to the anode because the surface of the anode in reality is much smaller than that of the structure, i.e. the electrolytic current density is much higher near the anode than elsewhere.

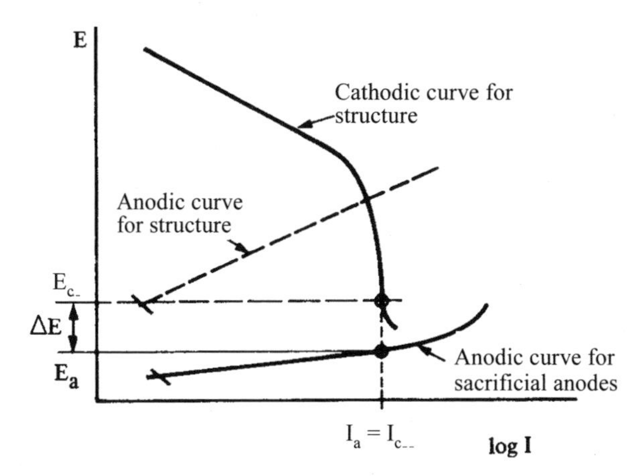

Figure 10.14 Potential-current diagram for cathodic protection with sacrificial anodes.

If an external current source is connected with its negative terminal to the structure an its positive terminal to the anode, we obtain a protection system with impressed current. The potential-distance diagram for this case is shown in Figure 10.16.

10.4.2 Protection Criteria and Specifications

In most cases it is appropriate to use a *potential criterion* for CP. The main criterion is $E_c < E_p$, where E_p is a protection potential defined mainly upon an empirical basis. For steel in seawater, the following protection potential has usually been recommended, referred to the various reference electrodes:

$Cu/CuSO_4$	Ag/AgCl/seawater	Zn	SCE	SHE
–850 mV	–800 mV	–250 mV	–780 mV	–530 mV

This potential is 90 mV above the equilibrium potential of the reaction $Fe^{2+} + 2e = Fe$ at an activity of $Fe^{2+} = 10^{-6}$. The stated value of the protection potential corresponds to the equilibrium potential at a ferrous ion activity of 10^{-3}. Theoretically, there is some corrosion at this potential, but experience shows that it is practically insignificant.

For steel buried in the seabed the corrosive environment may be affected by H_2S that is developed due to sulphate–reducing bacteria. Under these conditions the corrosion rate at the given value of E_p is higher, and therefore E_p = –900 mV Reference Ag/AgCl is recommended in this case [10.25].

With a properly designed CP system with sacrificial anodes of a Zn- or Al-alloy, usually a potential in the range –900 to –1000 mV is obtained after short time of exposure of a steel structure. This potential range also gives the lowest current consumption because a layer of calcareous deposits (dominated by $CaCO_3$) is soon formed. If an impressed current CP system is used, the potential may be considerably lower. Then, in addition to reduction of oxygen (Equation (4.4)) there will also be some hydrogen development (Equation (4.2b)). In certain cases this may cause damage [10.25]:

- Cathodic disbonding of paint coatings due to hydroxide ions (OH^-) and possibly hydrogen gas produced in the cathodic reactions. Common paint systems for marine applications usually tolerate potentials down to –1100 mV vs. Ag/AgCl. For lower potentials, a special documentation of the system's resistance to cathodic disbonding is required.
- Hydrogen embrittlement due to absorbed hydrogen atoms produced in the cathodic reaction (4.2b). The risk for this type of damage exists particularly at high tensile stresses and for materials with high tensile strength and hardness.

Special conditions for welding, cold working and maximum material hardness (HV_{10} = 350) are specified. For materials with a minimum yield strength above 550 MPa, qualification testing should be carried out when they are going to be used under critical conditions (e.g. at potentials < –1100 mV Ag/AgCl and when the structure is exposed to plastic deformation and/or fatigue load) or when sufficient experience from the actual conditions does not exist [10.25]. The tendency to increased fatigue crack growth rate due to cathodic polarization at low potentials has been dealt with in Section 7.13.

In order to obtain the specified potential limits, a current density (CD) that depends on the conditions is needed. In Tables 10.13 and 10.14, various CD values are stated for design of systems for protection of steel in seawater for different climatic zones and depths in the sea. Table 10.13 shows initial/final CD values chosen with the aim of securing sufficient polarization under unfavourable conditions in the first and the final period of the lifetime ($E_c \leq$ –800 mV Ag/AgCl). In Table 10.14, we can see the mean design values, which are based upon an assumption that a calcareous scale has been established on the surface; these values correspond to a more negative potential and they are lower than both the initial and final values given in Table 10.13. The application of these data is dealt with in connection with design of a sacrificial anode system in Section 10.4.3. The specified values are conservative, so that the real lifetime in most cases is longer than the design lifetime.

The limitation of the current demand to the order of 100 mA/cm^2 in most regions is, as mentioned in Chapter 6, caused by the formation of calcareous deposits, which is a result of the electrode reaction and the composition of the seawater. Under rough weather conditions and/or mechanical action of ice in arctic regions, the calcareous deposits may be worn off, which leads to a strong increase of the current demand.

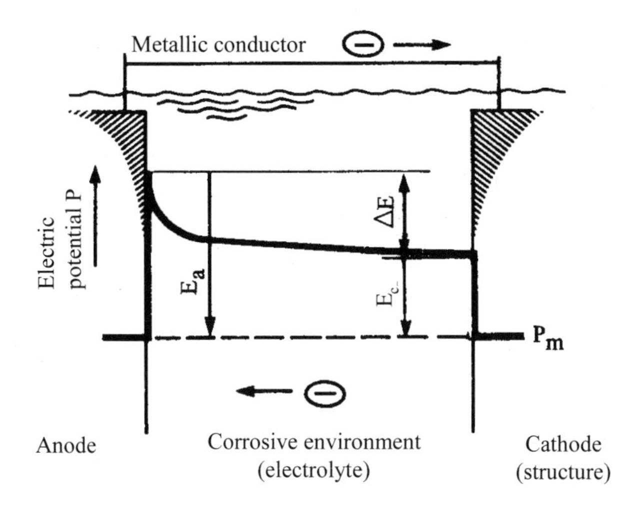

Figure 10.15 Schematic potential–distance diagram for cathodic protection with sacrificial anodes. P_m = electrical potential in the metallic materials (structure, anodes and connections), E_a = electrode potential of the anode, E_c = electrode potential of the structure (the cathode), $\Delta E = E_c - E_a$ = electrical potential drop in the corrosive medium (the water, the electrolyte) *from* the anode *to* the cathode.

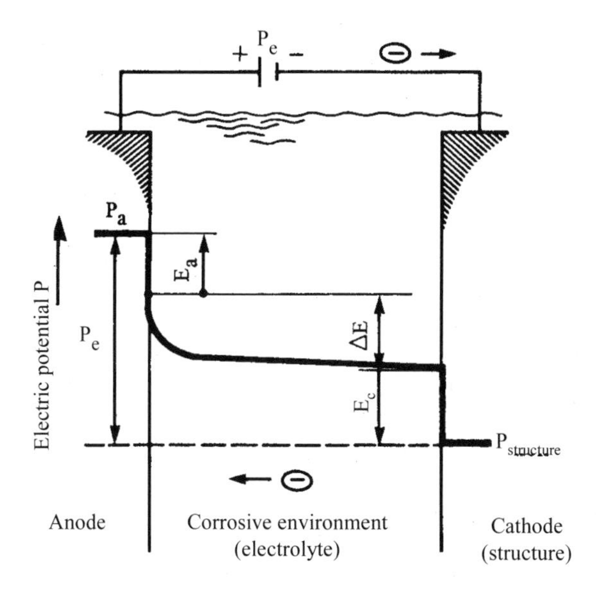

Figure 10.16 Schematic potential–distance diagram for cathodic protection with impressed current. Definition of E_a, E_c and ΔE as in Figure 10.15. P_a = electrical potential in the anode material, $P_{structure}$ = electrical potential in the structure (the cathode material), P_e = the terminal voltage of the external current source.

Table 10.13 Initial/final design values of current density for cathodic protection of steel in different climatic zones and at different ranges of seawater depth. The given temperatures are sea surface temperatures in average during the year [10.25].

Depth (m)	Design values of current density (initial/final) i_c (mA/m^2)			
	Tropical (>20 °C)	Subtropical (12–20 °C)	Temperated (7–12 °C)	Arctic (<7 °C)
0–30	150/90	170/110	200/130	250/170
>30	130/80	150/90	180/110	220/130

Table 10.14 Mean design values of current density for design of cathodic protection of steel in different climatic zones and at different water depths [10.25].

Depth (m)	Design values of current density (mean) i_m (mA/m^2)			
	Tropical (>20 °C)	Subtropical (12–20 °C)	Temperated (7–12 °C)	Arctic (<7 °C)
0–30	70	80	100	120*
>30	60	70	80	100

*Possible wear effects due to ice are not taken into account.

For instance, in Cook Inlet in Alaska they have experienced a current demand of 250–500 mA/cm^2 [10.26]. Under most conditions, however, a calcareous scale will be developed in the long run, so that the necessary current density decreases during the years, e.g. to values as low as 30 mA/m^2 [10.19].

Reference [10.25] specifies a current density of 10 mA/m^2 for thermally sprayed aluminium on steel in seawater (must possibly be adjusted for higher temperatures) and 20 mA/m^2 for bare steel buried in the seabed. For concrete structures they have specified 1–3 mA per m^2 surface of reinforcement steel depending on the climatic zone and the water depth. Contrary to the practice for steel structures, the highest design values of CD on concrete are specified for the warm climatic zones.

A traditional application of CP is the protection of ship hulls. Because these always are coated, a typical design value of the average CD is around 10 mA/m^2; in special cases, such as for icebreakers and oil platform supply vessels, 30 mA/m^2 is more appropriate [10.27]. CP is also used in ballast compartments of tankers, with CD in the order of 100 mA/m^2 on uncoated and, e.g. 5 mA/m^2 on coated surfaces.

For steel in other environments, recommended values are, e.g. 50 mA/m^2 for aerated hot water, 10–500 mA/m^2 in soils and 400,000 (!) mA/m^2 in hot sulphuric acid for pickling [10.19].

To check that the current density is high enough, potential measurements have to be carried out, with reference electrodes placed (permanently or at regular monitoring) on potentially critical sites.

A general requirement is that conductors are thoroughly fastened to the anodes and the structure.

For CP of steel in seawater, Zn and Al anodes give sufficient driving voltage. For CP in environments with low conductivity (soil, fresh water), however, either impressed current or Mg sacrificial anodes are usually necessary.

A series of recommended practices comprising various aspects and applications of CP has been published by NACE [10.28].

10.4.3 Cathodic Protection with Sacrificial Anodes

The majority of offshore installations, particularly in aggressive environments such as in the North Sea, are protected with sacrificial anodes. On submerged parts of fixed platforms, this is in most cases the only protection system. A few oil companies have used a combination of a paint coating system and sacrificial anodes, while a combination of thermally sprayed aluminium and sacrificial anodes have been used in special cases. Combinations of coatings and CP and/or a combination of sacrificial anodes and impressed current may be of particular interest for the reduction of weight and drag forces from the anodes on structures in deep water. Arguments for and against the different systems under various conditions are listed in Reference [10.27].

Floating platforms are generally protected by CP (sacrificial anodes or impressed current) combined with a high-quality paint system. Pipelines with an organic coating and an external concrete coating are also protected by sacrificial anodes.

For bare steel on offshore structures in seawater, the sacrificial anodes are commonly made of special aluminium alloys because these give the highest current output for a certain anode weight (Tables 10.15 and 10.16) as well as the lowest cost. Zn anodes are usually applied on coated and buried pipelines offshore, where the risk for passivation of Al anodes is higher due to a lower CD.

When sacrificial anodes (usually Zn or Al) are used on ship hulls they are combined with a paint system. The anodes are mounted close to each other on the area around the propeller, partly because the propeller "steals" some of the current and partly because of turbulence leading to higher demand of cathodic current density. Zn and Al anodes are also used in ballast tanks on tankers and bulk carriers.

For some other applications, such as on steel structures in soil, and in hot water tanks for fresh water, magnesium anodes are used to a higher extent. In these cases Mg anodes have a great benefit in higher driving voltage than the Al and Zn anodes. In high-resistivity environments, a combination of CP and coating is particularly beneficial with the aim to reduce the total current and thereby the potential drop near the anodes, and thus to obtain a better distribution of the current to the spots where it is needed.

For pipe systems of stainless steel carrying chlorinated seawater, internal localized corrosion can be very efficiently prevented by the application of *Resistance-controlled Cathodic Protection (RCP).* A resistance is simply inserted between the sacrificial anode and the pipeline, and this makes a system that is particularly suitable when there is a low diffusion-limiting cathodic current in the critical potential range [10.29]. Typical of the method is that the current output from the anode is kept low, which has the consequence that the voltage drops are low and the protected pipe length from each anode is long.

Crucial properties of sacrificial anodes are listed in Tables 10.15 and 10.16. In aluminium, Hg (or alternatively In to avoid pollution of the environment) contributes to reduced tendency to passivation. In Zn anodes, the content of Fe (among some other species) should be kept very low to avoid self-corrosion. As regards anode testing, see Reference [10.25].

Table 10.15 Typical properties of sacrificial anodes. (Extracts adapted from References [10.26, 10.30].) Design is based on somewhat higher potential and lower capacity, see Table 10.16

Material	ρ[1] ohm cm	i_a[2] A/m^2	Alloying elements	Potential E_a ref. Ag/AgCl, V	Utilization factor u	Capacity C Ah/kg
Al alloys	150	1–10	Si, Fe, Zn, Hg or In, Cu	–1.05 to –1.15	0.9–0.95	2700–2830
			Si, Zn, In, Mg, Cu	–1.15	0.95	2700
			(Fe), Zn, Sn, Cu	–1.10	0.5–0.8	Varying
Zn alloys	500	1–10	Cu, Al, Si, Fe, Pb, Cd	–1.05	0.95	780
			Cu, Fe, Hg	–1.05	0.95	780
Mg alloys	6000	4–11	Cu, Al, Si, Fe, Mn, Ni, Zn, (others)	–1.50	0.5	1230
			Cu, Al or Si, (Fe), Mn, Ni, (others)	–1.70	0.5	1230

[1] Maximum resistivity of the water. [2] Anode current density.

Table 10.16 Design data for Al- and Zn-based sacrificial anodes [10.25].

Anode material	Environment	Potential E_a, V ref. Ag/AgCl/seawater	Capacity C Ah/kg	Utilization factor u[3]
Al-based	Seawater	–1.05	2000[1]	0.75–0.90
	Sediment	–0.95		
Zn-based	Seawater	–1.00	700[2]	
	Sediment	–0.90		

[1] Anode temp. max. 25^{0}C. [2] Anode temp. max. 50^{0}C. [3] Depends on the type of anode.

Figure 10.17 shows various arrangements for fixing sacrificial anodes on tubular structures. Submarine pipelines are usually protected by bracelet anodes encompassing the pipe and flush with the outer concrete coating. Sacrificial anodes on ship hulls are commonly flush-mounted and welded directly to the hull.

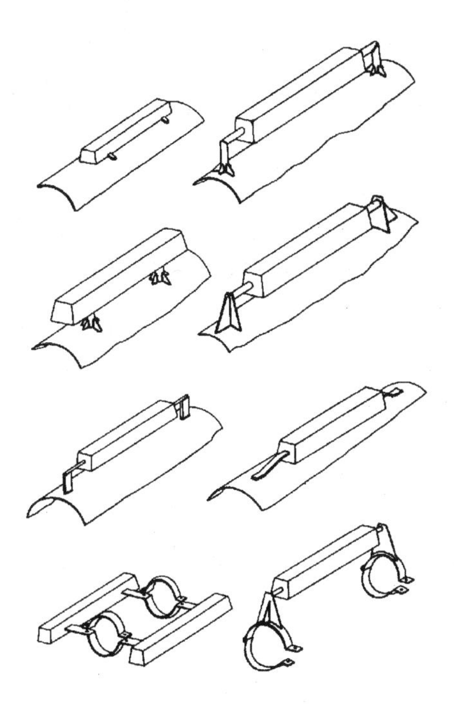

Figure 10.17 Typical arrangements for fixing sacrificial anodes on offshore tubular structures [10.30].

A sacrificial anode system can be designed in the following way. The exact description of the procedure is a little different in different sources, e.g. [10.25, 10.26], but the principles are the same:

1. The surface area A_c of the structure is calculated in a reasonably conservative way [10.25]. Areas with different current density demand must be calculated separately.

2. Design values of CD (I_c) on the steel surface are chosen (Table 10.13).

3. The total current demand, $I_t = I_c \cdot A_c$, is calculated. If the steel is coated, the current must be multiplied by a coating breakdown factor f_b [10.25].

4. Material, shape and size of the anodes are selected, and the anode resistance R_a is calculated by use of the appropriate formula (see Section 10.4.5).

5. The current output from each anode is given by

$$I_a = \frac{\Delta P_a}{R_a}, \tag{10.3}$$

where P_a is the potential difference that is available for overcoming R_a (Section 10.4.5).

6. The number of anodes is calculated:

$$n \geq \frac{I_t}{I_a}. \tag{10.4}$$

It should be noticed that both the current demand I_t and the output from each anode I_a is reduced during the service time. The number of anodes should therefore be calculated both for the initial and for the final stage. The largest of the calculated n-values is chosen.

7. The total weight of anodes necessary to protect the structure during the specified lifetime t^* is calculated:

$$G_t = \frac{I_g\, t}{C\, u}, \tag{10.5}$$

where I_m is the average (mean) current demand $= \Sigma\, i_m \times A_c$ (i_m from Table 10.14), C is the anode capacity and u the utilization factor (Tables 10.15 and 10.16). Then it is checked that

$$n \geq \frac{G_t}{G_a}, \tag{10.6}$$

where G_a is the weight of each anode.

E_a, I_a, C, u and other anode data are available from anode suppliers.

10.4.4 Cathodic Protection with Impressed Current

When such a system is used on fixed offshore platforms it may be preferable to combine it with sacrificial anodes (hybrid system). Then, the sacrificial anodes can protect the structure before the impressed current installation is ready for use, and also contribute to a better current distribution. The largest benefits of impressed current compared with sacrificial anodes are lower anode weight and lower drag forces from the sea. Also, relatively few anodes are necessary. Theoretically, it is a more economic solution than sacrificial anodes, but practical experience with cases of serious mechanical damage of the CP system has led to fewer applications of impressed current systems.

---

* Usually the lifetime of the anodes is set equal to the lifetime of the structure, but exceptions exist [10.25].

On large ship hulls and floating platforms, impressed current has often been used in combination with a paint coating. The same is the case for land-based buried structures.

Anodes of a corrosion-resistant material such as Pt, PbSbAg, graphite, magnetite or high-silicon iron are normally used in impressed current installations. Pt is often used as a thin layer on a substrate of another material, e.g. in the form of platinized titanium. A corroding material, such as scrap steel, can also be used, but additional anode material must be supplied regularly in this case.

The anodes can be arranged in different ways. On fixed offshore structures the anodes may be mounted on the structure itself (but of course insulated from it), they may be supported by special conduits more or less remote from the structure, or remote seabed anodes may be used.

On ship hulls protected by impressed current, flush-mounted anodes are used in order to avoid additional hydrodynamic resistance. Around the anode, the nearest steel surface is covered with a dielectric shield or coating with the aim of obtaining a better current distribution.

A land-based arrangement for CP in soil is shown in Figure 10.18. A backfill of a carbonaceous material such as coke breeze or graphite particles is used around the anode. The backfill is electron-conducting, which makes the effective size of the anode larger. This in turn reduces the resistance between the anode and the structure, and it reduces the consumption of anode material. If the resistivity of the soil is very high, buried pipelines may be protected by means of a continuous anode along the pipeline at a suitable distance from it.

For further study of arrangements and requirements the reader is referred to the literature [10.3, 10.5, 10.25, 10.27, 10.31, 10.32].

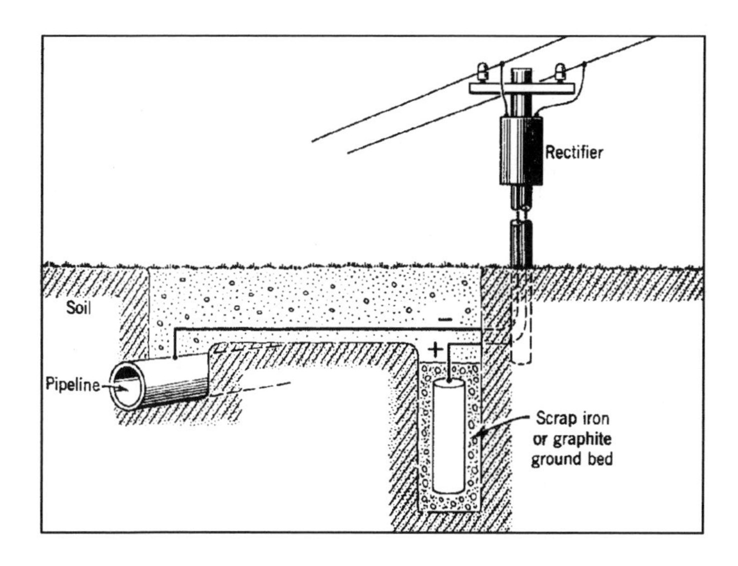

Figure 10.18 Cathodic protection of a buried pipeline by impressed current [10.19]. (Reproduced from Uhlig HH. Corrosion and Corrosion Control. John Wiley & Sons, 1971.)

10.4.5 Electrolyte Resistance, Potential Variation and Current Distribution in CP Systems and Galvanic Elements

The simplest possible geometry of a galvanic element is characterized by plane-parallel and equally large electrodes (anode and cathode) and between the electrodes an electrolyte volume with a cross-section identical with each electrode area, and consequently (provided a homogeneous electrolyte) parallel current paths. For this model, there is a potential drop in the electrolyte over the length l from the anode to the cathode given by

$$\Delta E = I\,R = i\,A\,\rho\,l/A = i\,\rho^*\,l \qquad (10.7)$$

and we can call it a linear model.

Other simple models are the cylindrical and the spherical model, where the two electrodes are concentric shells with radii r_i and r_o, respectively, and the current paths are radial. For the cylindrical model, the resistance of an arbitrary shell (between the electrodes) with radius r, area A and thickness dr, is given by the following equation (when we consider a long cylinder and disregard end effects):

$$dR = \rho\,dr/A = \rho\,dr/2\pi rL$$

and the total resistance is

$$R = \frac{\rho}{2\pi L}\int_{r_i}^{r_o}\frac{dr}{r} = \frac{\rho}{2\pi L}\times \ln\frac{r_o}{r_i}. \qquad (10.8)$$

The potential drop from r_i to r_o:

$$\Delta E = I\,R = \frac{I\,\rho}{2\pi L}\ln\frac{r_o}{r_i} = \frac{I_u\,\rho}{2\pi}\ln\frac{r_o}{r_i} = \rho\,i_i\,r_i\ln\frac{r_o}{r_i}, \qquad (10.9)$$

where I_u is the current per unit length of the cylinder and i_i is the current density at the inner electrode.

For the spherical model we have correspondingly

$$dR = \rho\,dr/4\pi r^2 \quad \text{and}$$

$$R = \frac{\rho}{4\pi}\left(\frac{1}{r_i} - \frac{1}{r_o}\right). \qquad (10.10)$$

---

* As an example, the resistivity ρ in seawater is about 30 ohm cm.

With intelligent combinations of these simple models it is possible to make approximate estimations for other geometries. Typical of the three models dealt with above is that both the anode and the cathode potential are constant over the respective electrode surfaces. Analytical solutions exist also for some other geometries. On the basis of such solutions (and partly on an empirical basis) several formulae for the resistance of the electrolyte volume near the anode – the anode resistance R_a – have been developed. Such formulae for various anode geometries are used in design of cathodic protection systems [10.25–10.34].

Recommended formulae for different types of anodes are listed in Table 10.17 (see also Figure 10.19). The definitions of the symbols in the table are:

ρ (ohm m) = resistivity,
L (m) = anode length,
R (m) = $C/2\pi$, where C is the circumference of the anode cross-section,
S (m) = arithmetic mean value of the length and width of the anode,
A (m^2) = exposed surface area of the anode.

Table 10.17 Formulae for anode resistance [10.25].

Type of anode	Resistance	
Slender, stand-off $L \geq 4r$ *	$R_a = \frac{\rho}{2\pi L}\left(\ln\frac{4L}{r} - 1\right)$	(10.11)
Slender, stand-off $L < 4r$ *	$R_a = \frac{\rho}{2\pi L}\left[\ln\left\{\frac{2L}{r}\left(1 + \sqrt{1 + \left(\frac{r}{2L}\right)^2}\right)\right\} + \frac{r}{2L} - \sqrt{1 + \left(\frac{r}{2L}\right)^2}\right]$	(10.12)
Long, close to surface	$R_a = \frac{\rho}{2S}$	(10.13)
Short, close to surface. Bracelet	$R_a = \frac{0.315\rho}{\sqrt{A}}$	(10.14)

* The formula is valid for distances between anode and steel surface ≥ 0.3 m. For distances from 0.15 to 0.3 m the same formula can be used with a correction factor of 1.3.

The driving voltage used in CP design is

$$\Delta P = E_c - E_a, \qquad (10.15)$$

where usually $E_c = -0.8$ V Reference Ag/AgCl and E_a is taken from Table 10.16.

In connection with modern offshore oil and gas activities, numerical methods have been applied which make it possible to carry out more accurate and extensive calculations of potential variation and current distribution, in which the total system of structure and anodes is covered. This can be done for more complex geometries and realistic electrochemical boundary conditions, partly based on polarization properties in the form of a relationship between the electrode potential of the steel and the current density.

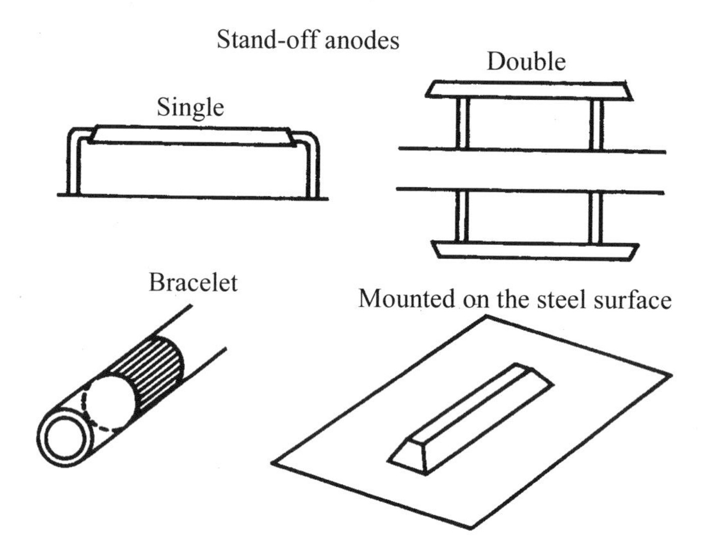

Figure 10.19 Types of sacrificial anodes for offshore oil installations.

For the linear model, which can be considered as a model with one-dimensional, stationary charge transport, the potential variation in the electrolyte can be expressed by a simple differential equation (see Equation (10.7)):

$$\frac{dP}{dx} = \frac{\Delta E}{l} = i\,\rho\,. \tag{10.16}$$

This relationship between the potential gradient and the transport of charge (i.e. current density i) is analogous to the relationship between concentration gradient and mass transport (the first Fick's law) and between temperature gradient and heat transport under stationary conditions. Since the current density i (in the x-direction) in the electrolyte volume is constant in the one-dimensional, stationary case, Equation (10.16) can be expressed by

$$\frac{\partial^2 P}{\partial x^2} = 0\,. \tag{10.17}$$

For a three-dimensional system (x, y, z), the potential variation in the electrolyte volume can correspondingly be expressed by the Laplace equation

$$\frac{\partial^2 P}{\partial x^2} + \frac{\partial^2 P}{\partial y^2} + \frac{\partial^2 P}{\partial z^2} = 0\,, \tag{10.18}$$

or generally

$$\nabla^2 P = 0. \tag{10.19}$$

Both the analytic and numeric calculation of potential variation and current distribution is directed to the solution of Equation (10.19). The numerical method that seems most suitable for calculation of potential and current distribution on offshore steel structures is the boundary element method [10.35], but the finite difference method has also been used.

10.5 Anodic Protection

Anodic protection can be applied on materials with a well-defined and reliable passive region and low passive current density, i.e. with an anodic overvoltage curve like that shown in Figure 10.20. The material is polarized in the anodic direction so that the potential is lifted to the passive region. As indicated in the figure, the current density necessary to hold the potential E_a at a suitable value is low, and this makes the method economic in use. However, the initial current density i_{emax} may be relatively high. The low current density in the established protection state leads to a high throwing power of the current, i.e. areas "in shadow", internal corners etc. are adequately covered, keeping the whole surface of the component in the passive state. Figure 10.21 shows in principle the arrangement for internal protection of a steel tank by means of a potentiostat.

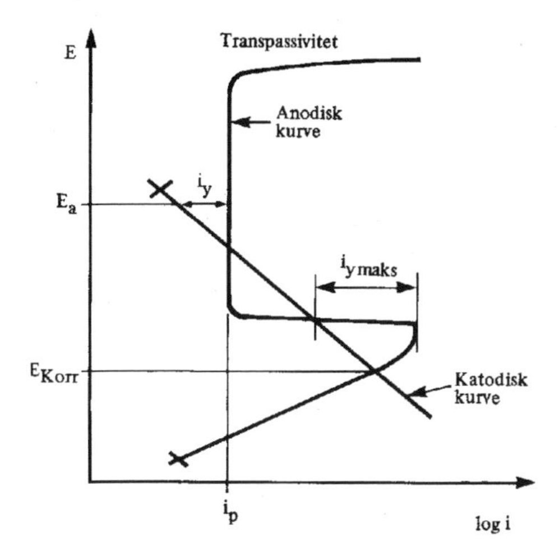

Figure 10.20 Potential shift from the corrosion potential E_{corr} to a higher potential E_a by means of an external current with a final value $I_e = i_e \cdot A$

Anodic protection is used on objects such as steel tanks for storing and transport of sulphuric acid, apparatus made of stainless steels and titanium for treatment of various acids and salt solutions, and for aluminium exposed to water at high temperature. The method cannot be used in aggressive liquids that may cause localized corrosion or high passive current density. It may be used on 18/8 CrNi stainless steel exposed to 30% H_2SO_4 + 1% NaCl [10.24] and on titanium exposed to hydrochloric acid solutions [10.20]. Pitting is avoided, in the former case because the sulphate ions counteract the chloride, in the latter case because the pitting potentials of titanium in the actual solutions are very high.

Anodic protection might be utilized much more than it has been so far, but the method must not be used under unfavourable conditions, because the anodic polarization may cause a strong increase of the corrosion rate. If there is any doubt, laboratory experiments may clarify if anodic protection can be recommended.

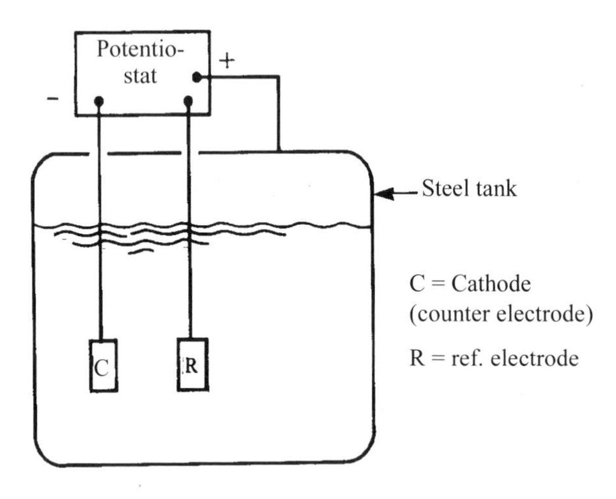

Figure 10.21 Internal anodic protection of a steel tank.

10.6 Corrosion Protection by Coatings

Through the application of coatings, corrosion is prevented by one of the following three main mechanisms or by combination of two of them:

i) Barrier effect, where any contact between the corrosive medium and the metallic material is prevented.

ii) Cathodic protection, where the coating material acts as a sacrificial anode.

iii) Inhibition/passivation, including cases of anodic protection.

In the following sections, a survey of different types of coating is given, which includes properties, mode of action, fields of application, and coating methods. Without regard to the type of coating and the application method, a proper pre-treatment of the material to be protected is necessary for obtaining a good result. The most important pre-treatment methods are common to some types of coatings, and pre-treatment is therefore dealt with in a separate section at the end of the chapter.

10.6.1 Metallic Coatings

In most cases of corrosion protection by metallic coatings, the purpose is to protect unalloyed or low-alloy steel, but there also exist many cases of other metals to be protected this way. Metallic coatings can be divided in two groups, the *cathodic* coatings, which are more noble than the substrate, and the *anodic* ones, which are less noble than the substrate, i.e. the coatings that have, respectively, a higher and a lower corrosion potential than the substrate in the environment in question. The cathodic coatings will most often act by the barrier effect only, but for some combinations of substrate and environment the substrate can also be anodically protected (on uncovered spots). The anodic coatings will in addition to the barrier effect provide cathodic protection of possible "holidays", i.e. spots or parts of the surface where the coating is imperfect and the substrate is exposed to the corrosive environment. The normal major difference between a cathodic and an anodic coating is just the behaviour at such a defect. This is illustrated in Figure 10.22. In the case of a cathodic coating (a) the substrate is subject to galvanic corrosion in the coating defect. The corrosion may be rather intensive because the area ratio between the cathodic coating and the anodic spot of bare substrate usually is very high. In the other case (b), only a cathodic reaction occurs on the bare substrate (it is protected cathodically), while the coating is subject to a corresponding galvanic corrosion distributed over a larger surface area. In order to protect the substrate, low porosity, high mechanical strength and continuous adhesion are even more necessary for a cathodic than for an anodic coating. Various coating metals are briefly described on the following pages, with reference particularly to the sources [10.36–10.40]. In relation to steel, Ag, Ni, Cr and Pb are cathodic, while Zn and Cd are anodic in most environments. The polarity of Sn and Al referred to steel varies from one environment to another.

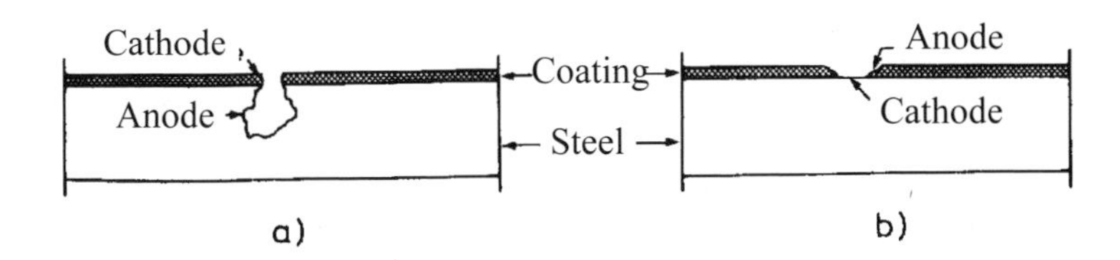

Figure 10.22 Localization of corrosion at a defect in a metal coating on steel. a) Cathodic coating. b) Anodic coating.

Silver is used as a coating material for decoration purposes; furthermore its chemical and electrical properties are utilized, respectively, in various equipment and environments in the process industry and in electric contacts and other small components in electrical and electronic equipment. Silver coatings are commonly applied by electrolytic plating (see p. 288). A coating of copper or nickel is often used under the silver top coat. Silver may be attacked by sulphides in industrial atmospheres. To prevent attack, sometimes a thin layer of rhodium is used on top of the silver coating; alternatively, a passivating chemical treatment is carried out.

Nickel has often been used for preventing corrosion and for providing a nice surface on steel parts in contact with ordinary atmosphere, e.g. bicycle parts, car parts and fittings. In order to make the surface brighter and more corrosion resistant the nickel plate is covered by a thin coating of chromium. Depending on the service conditions and the type of nickel and chromium coating, the recommended thickness of nickel can vary from 5 to 40 μm and of chromium from 0.3 to 0.8 μm [10.39]. The chromium coating has a large number of micropores or microcracks, which is advantageous because they make the corrosion of the underlying nickel well distributed and shallow. The chromium layer is usually deposited on top of a thin layer of bright nickel, which is anodic in relation to a thicker basic layer of (semi) matt nickel. Because of its less noble nature, the bright nickel layer gives a certain cathodic protection effect on the basic layer in existing pores.

A basic layer of copper may be used on iron and steel substrates, which allows a thinner nickel coating. Nickel–chromium coatings can also be used on zinc (with a basic layer of copper), on copper alloys and aluminium alloys [10.39].

For industrial use, nickel coatings are applied without chromium on top, and the coatings are most often thicker, usually 25–250 μm, and up to 500 μm for some applications. Such coatings are most suitable for environments that tend to cause general corrosion with reaction products of low solubility (environments without chlorides). However, it is also stated that nickel performs well in seawater provided that the flow rate is high. Furthermore, the experience with nickel in various alkalis, neutral and alkaline salts and organic chemicals is good.

Electrochemical deposition (electroplating) has traditionally been, and is still, the dominating technique for application of nickel coatings. In recent years, the number of applications of chemical (electroless) deposits of nickel has been growing (parts in pumps, valves, compressors, storage vessels for chemicals etc).

Nickel coatings can increase the resistance to corrosion fatigue, but stresses set up by the coating may reduce the pure mechanical fatigue strength of the base material. The effect can be counteracted by compression stresses in the coating, which can be obtained with certain electrolytic baths, or by applying a soft basic coating (tin, lead), which can prevent cracks in the coating propagating into the base material.

Thick or so-called hard *chromium* coatings are used directly on steel, cast iron and light alloys to improve resistance to wear and corrosion in hydraulic cylinders, pump shafts, bearings, dies etc. Thermodynamically, chromium is less noble than iron, but under atmospheric and many other conditions it is practically more noble due to the strong tendency to passivation.

Lead is particularly suitable in coatings exposed to dilute sulphuric acid and industrial atmospheres.

Tin shows rather complicated corrosion properties in contact with steel. In the atmosphere and in solutions of natural inorganic salts, Sn is cathodic in relation to Fe. In contact with fruit juice, meat, milk and alkaline solutions, tin is anodic. Tin plating is most often carried out by electrochemical deposition, but hot dipping is also common.

Zinc. The Pourbaix diagram for zinc is shown in Figure 10.23. The boundaries of the passivity region shown by solid lines are valid for water without CO_2. For pH < 8–9, the following reactions occur:

Anodic: $Zn \rightarrow Zn^{2+} + 2e^-$, (4.1)

Cathodic: $2H^+ + 2e^- \rightarrow H_2$, (4.2)

$O_2 + 2H_2O + 4e^- \rightarrow 4OH^-$. (4.4.b)

The equilibrium potentials of the three reactions as functions of pH are drawn in the diagram. For pH values between 8.5 and 11, zinc is passivated by formation of $Zn(OH)_2$. When CO_2 is dissolved in the corrosive medium the passive region is enlarged so that it reaches from pH = 6 to pH = 11 (dashed lines). The reason for this is the formation of a protective film of alkaline zinc carbonate. The effect of this film is strong in rural atmospheres, in mildly aggressive city atmospheres and in neutral and slightly acidic rain water. Also the calcium content in hard waters leads to deposits that prevent corrosion. Conversely, in various salt solutions the formation of soluble corrosion products may counteract the passivation process. Chlorides act in this way, but in marine atmospheres and seawater, the detrimental effect of chloride is at least partly balanced by protective calcareous deposits. Corrosion of zinc in seawater is therefore mainly under cathodic control, which is also indicated by the level of the corrosion potential (marked C in the diagram in Figure10.23). In industrial atmospheres with high SO_2 content, zinc sulphate and zinc sulphite are formed, which are easily washed off by rain. Under such conditions, the corrosion rate will be relatively high, in agreement with the indication given by the Pourbaix diagram for acid environments. The effects of the mentioned substances are reflected by the data in Table 8.2.

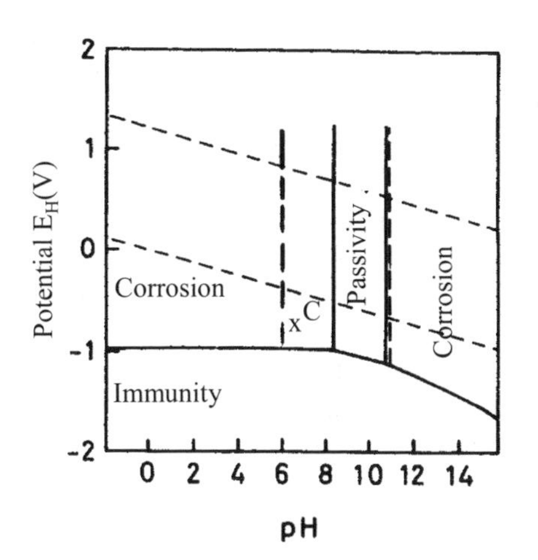

Figure 10.23 Pourbaix diagram for Zn in water at 25°C. Metal ion activity = 10^{-6}. Corrosion potential in seawater marked "C".

Another omportant factor is the temperature. The corrosion rate at 70°C may be 50 times that at 20°C. Above 60–70°C, zinc is cathodic in relation to iron and carbon steel (see below).

Zinc is easy to apply electrochemically, by hot dipping, and by thermal spraying. A fourth method is sherardizing [10.37]. Electrochemically applied coatings are relatively thin, usually up to 25 μm, and they are in most cases post-treated with a converting process, such as chromating, which gives a yellow, blue, drab or colourless appearance. Hot-dipped zinc coatings are often 80–100 μm thick, and a typical thickness range for thermal spray Zn is 100–200 μm, and these two methods can therefore provide more long-term protection under suitable conditions.

Contact between zinc and steel. In spite of its electronegative character, zinc corrodes relatively slowly when separated from other metallic materials. The reasons are that the reaction $2H^+ + 2e^- \rightarrow H_2$ does not go fast on Zn, and that (in the atmosphere) protective carbonate is formed. As a coating on a more electropositive material such as carbon steel (at $t < 60°C$), the corrosion rate of zinc is increased near the bare steel regions. The potential at these places will increase to some extent compared with the normal potential on zinc, but it will still remain below a practical limit for corrosion of steel. For larger uncovered regions, the following should be noted: with increasing distance from the zinc, the potential of the steel surface increases, and at a certain distance significant corrosion of steel will occur. This critical distance increases with increasing conductivity of the liquid and its corrosivity versus zinc. However, at the same time as an increased area of bare steel can be protected, the corrosion rate of zinc may be so high that protection of steel with zinc is not an economic method. In such cases, coating the steel with aluminium may be a better alternative, but possible bare steel regions must then be kept small.

On the basis of the electrochemical nature of corrosion, it is clear that metallic zinc must be in direct contact with the electrolyte to protect bare steel regions. Paint or other coatings on top of the zinc coating will therefore eliminate or at best reduce the cathodic protection effect.

The change of polarity between the steel substrate and zinc coating when the temperature is increased to 60–70°C may cause corrosion of steel on bare regions. Therefore a continuous coating is required at higher temperatures.

Aluminium. Corrosion and other properties of aluminium are dealt with in Section 10.1.8. As a coating material it is applied in thermal spraying (metal spraying) both on land-based and on offshore steel structures. Hot dipping is a more sensitive process with aluminium than with zinc as coating material, and it has not obtained the same widespread application for Al as it has for Zn. Hot-dipped coatings produced of mixtures of aluminium and zinc have shown good all-round properties for atmospheric exposure. Aluminium can also be plated by electrochemical as well as chemical deposition. Cladding of surfaces can be carried out by rolling, extrusion, drawing, and explosive bonding [10.37].

Aluminium in contact with structural steel. Although steel corrodes faster than aluminium in environments such as seawater, marine and industrial atmospheres, water containing SO_2, and soft water, aluminium is the material that primarily corrodes when these two materials are galvanically coupled in the mentioned environments. The reason for this apparent paradox is that the corrosion potential of steel is higher than on aluminium when the materials are separated. By the coupling a mixed potential is established which is lower than the separate steel potential and above the corrosion potential of aluminium. Particularly in seawater, the corrosion of aluminium is accelerated considerably by the galvanic contact (see further explanation in Section 7.3). But the relationship may be more complicated in some other environments; aluminium may in one case be anodic and in another cathodic to steel. In pores in thermally sprayed aluminium on steel in moist atmospheres, the following phenomena may occur successively:

1. The coating acts as a cathode due to the oxide, which prevents conductance of ions but allows electronic conductance to some extent.
2. Due to increased pH at the cathode, the solubility of the aluminium oxide increases (see the Pourbaix diagram in Figure 3.11). Therefore, metallic aluminium is exposed to the solution, the coating becomes less noble than steel and begins to act as an anode.
3. The corrosion products of Al gradually block the pores in the coating, the galvanic element is made ineffective and the coating becomes stable.

Comparison between zinc and aluminium as coatings on steel. Due to the oxide, aluminium has not the same remote effect as zinc for protection of a bare steel area. In acid solutions (pH = 4–5) and severe industrial atmospheres the consumption of zinc may, however, be so large that an Al coating is a more economic solution. It has been claimed that the lifetime of aluminium in industrial atmosphere may be three times that of a Zn coating. Also exposed to soft water, aluminium is considered superior to zinc. Zinc is preferable in hard water and alkaline solutions, but for two different reasons. In hard chloride-free water, Al corrodes too slowly to be able to protect the steel in coating defects. Conversely, in alkaline solutions aluminium corrodes too fast, which results in too short a lifetime of the coating. Aluminium normally corrodes more slowly than zinc in marine atmospheres and seawater. The remote effect of aluminium is satisfactory in seawater under most conditions. Unfavourable localized corrosion effects may occur on a sprayed aluminium coating if it is galvanically coupled to a large bare steel surface.

Coatings of Zn and Al can be combined with an external paint coating, with good results when the work is done properly. An alternative is to use a thin external organic sealer, which blocks the pores in the metal coating. Because Al and Zn coatings are relatively inefficient as cathodes, they can also be combined with cathodic protection, particularly on structures that are going to have a long life in the sea. Aluminium is favourable with regard to current density demand and potential. In some cases, thermal spraying of aluminium has therefore been applied on submerged parts of offshore steel structures. Compared with bare steel, this protection system gives a strong reduction of the cathodic current demand (see Section 10.4).

In the selection of a coating method, there are usually more real possibilities with zinc than with aluminium. For steel structures, aluminium has mainly been competitive on price only in the form of thermal spray coatings.

Cadmium has been a very useful coating material, but because of its toxic nature the numbers of applications have been reduced during recent decades.

For further study of metallic coating materials, see Shreir et al. [10.37].

Coating Methods

A few characteristic features of three of the methods for applying metal coatings, namely electrolytic plating, hot dipping and thermal spraying, will be summarized here. Other methods are only mentioned at the end.

Electrolytic plating [10.36–10.38]. Compared with other coating methods, electrolytic deposition is favoured by low process temperatures. Hence, any heat effect, which might cause changes in structure, shape and mechanical properties of the substrate, is avoided. As for other methods, both local and overall geometry of the surface to be coated must be carefully considered as part of the design work. Cavities, where air or liquid may be enclosed, cannot be accepted, and sharp corners should be avoided (Figure 10.10). Since the coating usually is thin (5–40 μm), factors such as roughness and porosity in the substrate may cause problems that do not exist for the methods where thicker coatings are common.

For electrolytic deposition of zinc it is easier to control the coating thickness than it is for hot dipping and thermal spraying. Any brittle alloy, like the Zn–Fe layer that is typical in hot-dip galvanized coatings, does not exist in electroplated Zn coatings, i.e. the latter is more ductile. But the lifetime of zinc coatings is nearly proportional to the thickness, independent of coating method, and this limits the numbers of applications of electroplated Zn.

Hot dipping. Zinc, tin and tin/lead coatings are commonly applied by hot dipping, i.e. dipping of the object to be coated in molten coating metal. Aluminizing by this method is less widespread, because Al makes the process more difficult to operate.

Hot-dip galvanizing is the most economical method in cases where a relatively thick Zn coating (up to 100–200 μm) is needed or desirable on components that can be placed in a bath without much difficulty. A common size limit is 12–13 m in length and 1–1.5 m in width, but baths that can take structures of dimensions up to 18 m × 2 m × 5 m exist.

The normal temperature of the galvanizing bath is 445–465°C. For thin uniform coatings, particularly on threaded components, a higher temperature is used (up to 560°C). At these temperature levels an alloy layer of Fe and Zn is formed, which constitutes a proportion of the coating thickness that depends on the steel type. As shown in Figure 10.24, addition of Si to the steel affects the whole structure of the coating to a great extent. It is also used to regulate the coating thickness. The Fe–Zn alloy is hard and brittle. In galvanizing thin sheets a method is used that gives an alloy layer thickness of only a few micrometres, and the sheets can therefore undergo metal-forming processes after galvanizing.

With respect to requirements on design, effects on mechanical and mechanical–technological properties (e.g. toughness, fatigue strength, weldability) and painting of zinc coatings, the zinc coating industry can give adequate information, like that in Reference [10.40].

For other metals applied by hot dipping, see References [10.36, 10.37, 10.41].

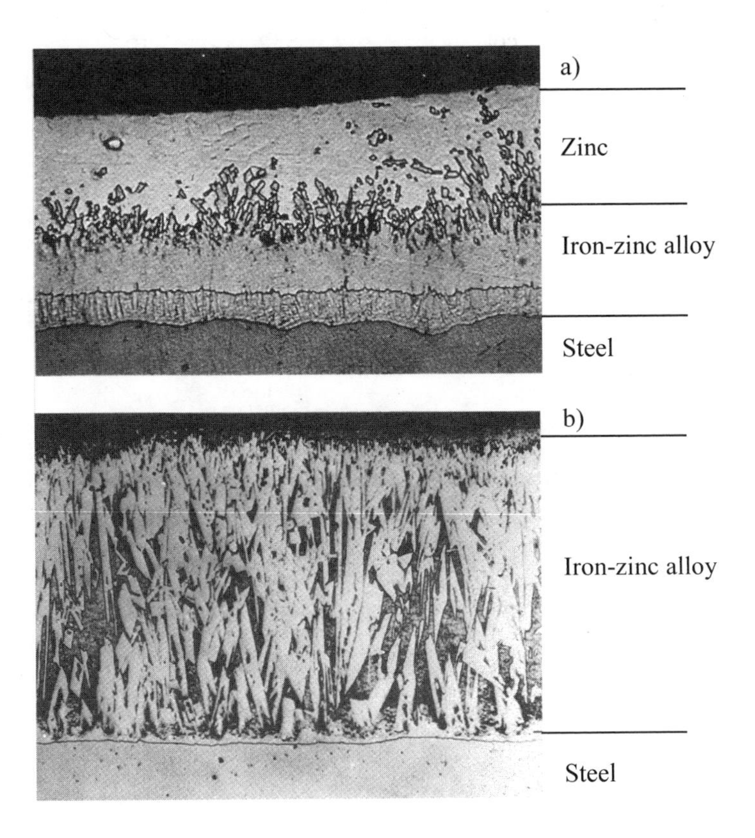

Figure 10.24 Cross-sections of hot-dip zinc coatings a) on rimming steel (a similar coating is obtained on Al-killed steel) and b) on Si-killed steel [10.40].

Thermal spraying. The principles of thermal spray methods are shown in Figure 10.25. Flame and arc spraying are most common for prevention of low-temperature corrosion. These classical methods have been thoroughly described by Ballard [10.42]. Coated test samples have been studied after long-term exposure [10.43].

Most metals can be sprayed. For corrosion protection of steel in waters and the atmosphere, aluminium and zinc are of particular importance. These coatings are usually from 100 to 300 μm thick. Other materials such as stainless steel are also sprayed with good results and have important application, e.g. on machine parts and on cylinders in the pulp and paper industry. For these purposes the coatings are considerably thicker, e.g. up to a few millimetres. The special advantages of thermal spraying compared with other coating methods are just that thicker coatings can be applied, and particularly that the spraying equipment is portable, so that the coating work can be carried out almost anywhere. For large and fixed structures that cannot be hot-dip galvanized, thermal spraying is often the only alternative to painting.

The drawbacks of thermal spraying are considered to be relatively high cost, relatively low adhesion between substrate and coating, high porosity in the coating and rough surface. The porosity and roughness of a sprayed zinc coating are shown

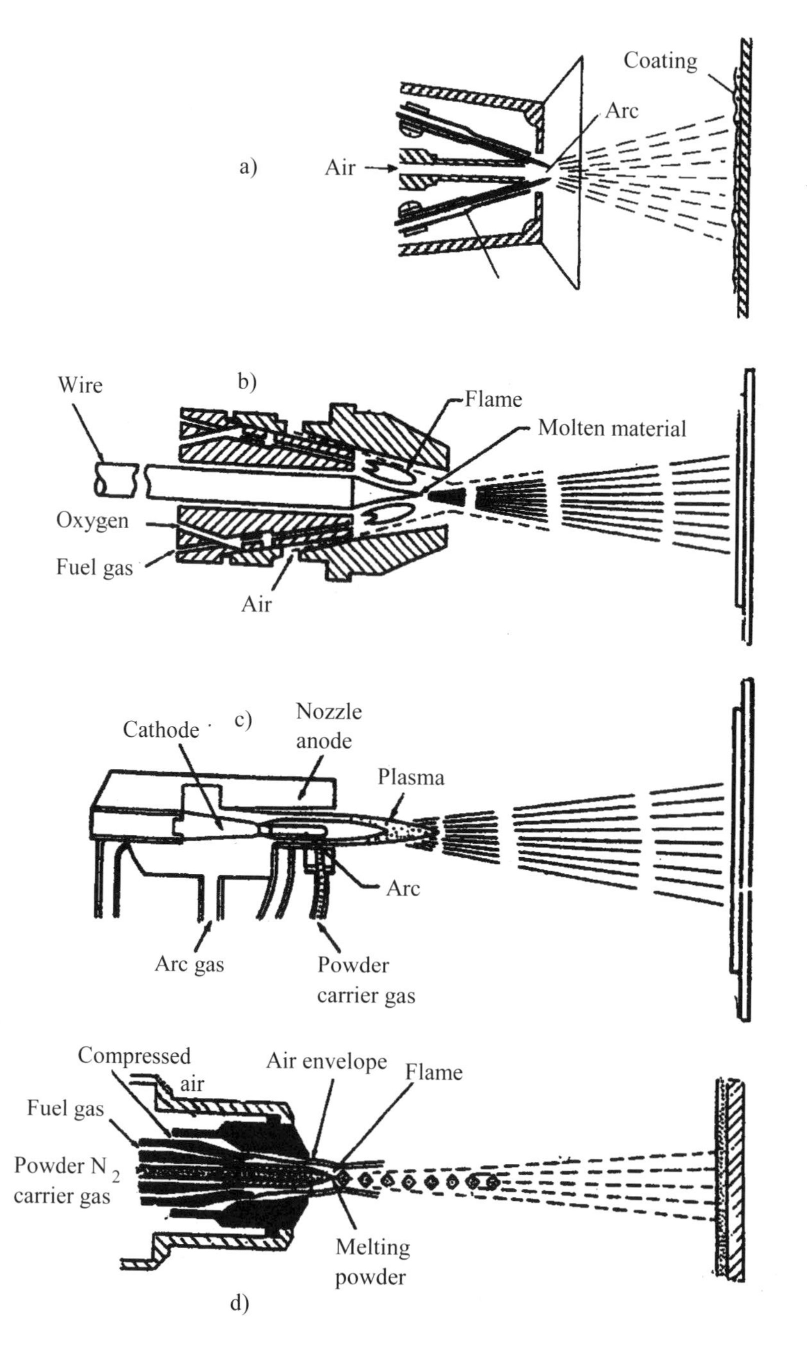

Figure 10.25 Thermal spray guns for a) arc spraying, b) flame spraying with wire, c) plasma spraying, and d) high-velocity oxygen fuel (HVOF) spraying (the most common method of high-velocity flame spraying).

in Figure 10.26. In addition, it can be commented that thermal spraying is more expensive than hot dipping in cases where the latter method is possible. The initial cost of thermal spray coatings is higher than for good paint coating systems, but for structures with long lifetime, for instance more than 25 years, lower maintenance cost will probably more than compensate for the higher initial cost. When it is favourable to have a paint coating on the outside, the basic zinc or aluminium coating will increase the lifetime of the paint film because the metal coating prevents rusting underneath the paint. With arc spraying of aluminium, considerably higher adhesion is obtained than what is possible with flame spraying of aluminium and zinc. However, with thorough pre-treatment and professional spraying, the adhesion of flame-sprayed coatings as well is satisfactory. The porosity of aluminium and zinc coatings should not cause any problem, and the roughness forms a good basis for paint, commonly better than provided by hot-dip galvanizing. In spite of the large difference in structure and surface between a hot-dip and a spray zinc coating (compare Figs. 10.25 and 10.26), the protection ability and lifetime of coatings with the same thickness may be about the same for the two coating methods.

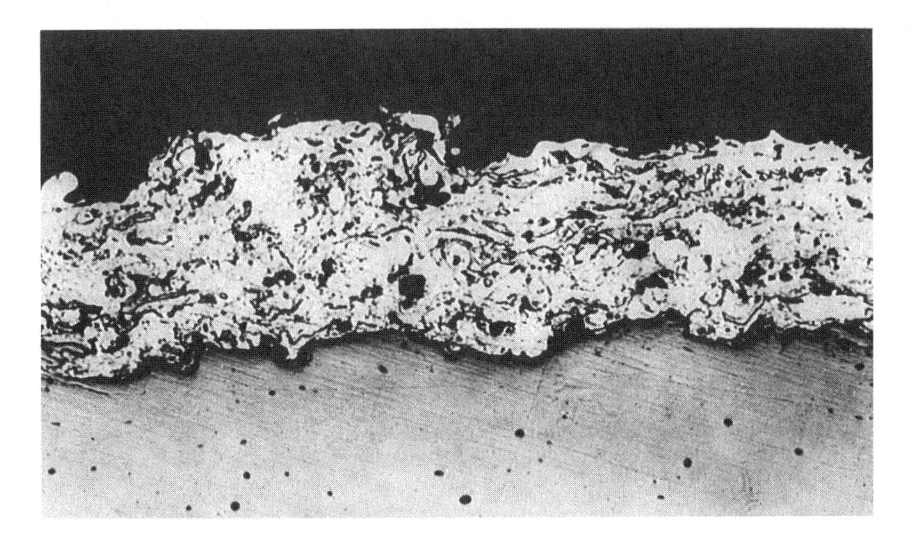

Figure 10.26 Cross-section of a flame-sprayed zinc coating.

In order to obtain a good result with thermal spraying it is quite necessary that the base material is clean, dry and rough all over prior to spraying. Therefore blast cleaning to a grade of Sa3 after the Swedish standard SS 055900 (ISO 8501–1) is specified (see also Section 10.6.4).

Thermal spray aluminium and zinc coatings with a paint system on top provide first class corrosion protection with very long lifetime under most natural conditions, and have been extensively used on steel structures, e.g. road bridges [10.44].

In addition to the methods just described, several other principles for applying metal coatings exist, such as *cladding* (e.g. application by rolling nickel or stainless steel on carbon steel, or pure aluminium on a strong but less corrosion-resistant aluminium alloy), *spot welding, overlay welding, chemical deposition, diffusion coating* (performed at high temperature), and advanced methods such as *physical* and *chemical vapour deposition (PVD and CVD), ion plating*, and *laser treatment.*

10.6.2 Other Inorganic Coatings

Coatings in this group prevent corrosion by their barrier effect. The group comprises, among others, various types of *enamel and glass linings.* The coating material, which must have a suitable thermal expansion coefficient, is applied in the form of a powder (after pickling or other cleaning of the base material) and heated in a furnace until the lining material becomes soft and is bound to the the metal surface [10.19]. High resistance to numerous chemicals and other environments can be obtained by variation of the composition of the coating material. The linings are brittle and can tolerate very little deformation, and some of them are sensitive to thermal shock.

Coatings of ordinary *mortar* (from Portland cement) have nearly the same thermal expansion coefficient as steel, and they are cheap and easy to repair. They can be applied with a thickness in the range 0.5–2.5 cm and more to protect pipelines and tanks, internally or externally. Thicker coatings are often reinforced with a net. On internal cylindrical surfaces the coatings can be applied by centrifugal casting, or by spraying and manual finishing. Imperfections and damaged parts in the coating may cause concentrated attacks, particularly if there is a galvanic contact with more noble materials than steel, which can be the case in pipe systems including valves, pumps etc. The problem is avoided if the coating is combined with cathodic protection.

Ceramic coatings of oxides, nitrides, borides etc. can be applied by thermal spraying. Some of the ceramic materials can be sprayed by the classical flame spraying method, while others must be plasma sprayed because the melting points are too high for the former method. Ceramic coatings are particularly used to prevent high-temperature corrosion and wear, and to provide thermal barriers (thermal insulation). Important alternatives in many cases are ceramic–metallic coatings that most often are composed of hard carbide particles bound together by a metallic material. These materials are also called hardmetals. A common composition is WC with a Co binding phase. However, the corrosion properties of cobalt are rather poor, and it is preferably replaced by an alloy of Co and Cr or an alloy of Ni, Cr and possibly some other elements for application under corrosive conditions [10.45]. These coatings are among the best at resisting abrasion and erosion possibly combined with corrosion. With modern high-velocity flame spraying (HVOF) denser and better coatings are obtained.

Conversion coatings. This group comprises phosphate, chromate and oxide coatings. *Phosphate treatment* of steel is usually carried out by exposing the steel surface to phosphoric acid containing phosphates of Zn and Mn, either by spraying or by dipping after prior acid pickling or blast cleaning. Phosphate coatings that are only 1–10 μm thick do not provide a complete protection by themselves, and the process is mainly used as pre-treatment before painting or treatment with wax or oil. Metals such as Zn, Cd, Al and Mg can also be phosphated. *Chromate treatment* has been widely used on non-ferrous metals, particularly Al and Zn, sometimes as the only treatment and sometimes as pre-treatment before painting. Since chromates are highly toxic the environmental regulations connected with chromate processes have become more stringent, and several efforts have been made to find more environment-friendly conversion alternatives. *Oxidation* of steel by dipping in hot concentrated alkaline solutions containing persulphates, nitrates and chlorates is used as a base for oil and wax. An important process is oxidation of aluminium and aluminium alloys. It is achieved by anodic polarization, and is usually called anodizing. The oxide film can be dyed in the anodizing process by various additions to the bath, or it can be dyed afterwards. It is sealed by treatment in boiling water, steam or possibly in a hot, dilute chromate solution. Anodizing provides in itself some improvement of the corrosion resistance of aluminium, and furthermore it gives a good base for painting. More information on conversion coatings and the actual processes can be found in Reference[10.37].

10.6.3 Paint Coatings

The use of paint coatings is the most common method for corrosion prevention.

Compositions and types of paint. An anticorrosive paint is composed of a binder, pigments, a solvent/diluent, extenders and a variable number of other additives such as antioxidants, surface-active agents, driers, thickeners and antisettling agents. A paint is primarily characterized by its pigment or by its binder. We distinguish between primers, which usually contain pigments causing some inhibition or cathodic protection of the substrate, and paints for finishing coats, which contain colour pigments and extenders, which may improve the barrier effect of the coating system. The coats in thicker paint systems may be divided into primer, intermediate or body coats and topcoat. Recent development of thick film paints implies less need for distinction between intermediate coats and topcoat.

An actual inhibiting type of pigment in primers is zinc phosphate, while red lead and zinc chromate, which were earlier in widespread use, are seldom applied nowadays because of the risk of health injuries. Metallic zinc powders provide cathodic protection of the substrate if the zinc concentration is high enough. The pigments in the finishing coats provide colour and protect the binder from being damaged by ultraviolet sunlight. The most efficient barrier-acting pigment consists of aluminium flakes. The aluminium flakes increase the length of the diffusion path and are therefore assumed to increase the resistance to diffusion of water, oxygen and ions, but more recent investigations have shown that the effect and the mechanism depend strongly upon the type of binder. Other commonly used pigments in intermediate coats and topcoats are various forms of iron oxide.

The binder may, for instance be bitumen (coal tar or asphalt), or linseed oil (natural materials), alkyd, chlorinated rubber, epoxy, vinyl, polyurethane etc. (synthetic organic materials), or silicate (inorganic). Some paints are hardened by reaction with oxygen in the air (oil and alkyd paints), others by evaporation of the solvent (e.g. chlorinated rubber) and a third group by a chemical reaction between two components (epoxy and polyurethane paints). Recently, a binder of polysiloxine has been introduced. There is also a growing interest for application of water-based paints, e.g. in marine environments [10.46].

Properties and selection of paint systems for protection of steel in various environments [10.37, 10.46–10.48]. The primary questions that have to be asked before selecting a paint system are:

- Which pre-treatment is possible, and what will the condition of the substrate be before painting?
- How will the environment around the painted structure change during various periods of the lifetime? Which mechanical and chemical attacks will the coating be exposed to?
- What are the conditions for application and drying/hardening of the paint, particularly the temperature and the humidity?
- What are the initial and maintenance costs of paint and painting work?

Several so-called rust-stabilizing or rust-converting products are commercially available, but the practical efficiency of these products depends on the conditions. They are made for coating on rusty substrates, but it is still important to remove loose rust, which can be done by thorough steel brushing. Some products are based on phosphoric acid. These must be applied in appropriate thickness, and excessive acid must be washed away with fresh water before the coating is overpainted.

Regarding selection of paint on steel in different atmospheric environments, some proposals are given in Table 10.18. More extensive guides for selection of paint systems are given in References [10.37, 10.48]. Particularly for ships and marine structures, Chandler [10.27] has presented a clear survey comprising structure parts above water as well as submerged and in the splash zone.

As indicated in Table 10.18, a zinc-rich primer is often recommended. It can be an organic zinc-epoxy or an inorganic zinc-ethyl-silicate primer. Zinc-rich primers are also used as so-called *shop primers,* or prefabrication primers, for temporary protection of semi-manufactured steel goods. After fabrication, e.g. of welded steel structures, the shop primer surface must be cleaned (degreased), and possible shop primer defects and weld joints have to be blast cleaned and coated with a primer before the whole structure is painted. Iron oxide is also used as a pigment in some shop primers. These must not be overpainted with a zinc-rich paint.

The main rule for painting previously painted steel is to use the same type of paint. If the old paint coating has turned out to be unsuitable in the actual environment, it should be removed before the structure is coated with a different type of paint.

The cost of the paint itself forms often 15-20% of the total cost of the painting operations including pre-treatment and application. It is important that the pre-treatment and the type of paint are compatible. The advanced paints depend upon good pre-treatment (as indicated in Table 10.19) to obtain the necessary adhesion to the substrate. In Table 10.19, the pre-treatment quality is denoted with cleanliness grades which are defined in a Swedish standard, and in recent years included in the

Table 10.18 Proposed paint systems (Reference [10.47]) for various categories of corrosivity defined in ISO standard 12944-2 [10.48].

Corrosivity category[1]	Environment	Pre-treatment[2]	Paint system	No. of coats	Thickness μm
C2	Unheated interiors where condensation may occur. Exteriors with little pollution. Mainly rural atmosphere.	Sa 2½	Alkyd	2–3	120
C3	Interiors with high relative humidity, some air pollution. Exteriors in urban and industrial atmosphere, moderately SO_2 polluted. Coast climate with little salt content.	Sa 2½	Zinc–rich primer. Finishing coats: vinyl, chlorinated rubber or epoxy plus polyurethane	3–4	175–240
C4	Interiors in chemical industry, swimming pools, shipyards. Exteriors in industrial and marine atmosphere with moderate content of salt.	Sa 2½	Zinc rich primer. Finishing coats: vinyl, chlorinated rubber or epoxy plus polyurethane	4–5	250–275
C5-M (marine)	Interiors with almost continuous condensation, heavily polluted. Exteriors in coast and offshore climate with high salt content.	Sa 2½	Zinc-rich primer. Finishing coats: vinyl, chlorinated rubber, epoxy, polyurethane, coal tar vinyl or coal tar epoxy.	4–5	325–400
C5-I (industry)	Interiors with almost continuous condensation, heavily polluted. Exteriors in aggressive industrial atmosphere with high relative humidity.	Sa 2½	Glass flake polyester or vinyl ester	1	1000–1500

[1] Categories: C2: Low corrosivity. C5: Very high corrosivity.
[2] See also Table 10.19.

international standard (ISO 8501-1) [10.49]. The table shows also the lower limits of application temperature, recommended time before application of the next coat, and for which corrosion class of environment (defined in Table 10.18) the paint types are suitable.

The rate of chemical reactions normally decreases with decreasing temperature. Therefore, two-component paints and oxidative hardening paints require an application temperature above a certain minimum. The steel temperature is more important than the air temperature. Temperature limits often used are: epoxy paints:

10°C, polyurethane: 0°C, oxidative hardening paints: 0°C. If it is strictly necessary to paint in cold weather, physically drying paints (i.e. those drying by evaporation of the solvent) should preferably be used, since these dry relatively fast also at low temperatures.

Moisture is detrimental to the application of paint, and condensation may sometimes be a problem. All paints give the best result when they are applied on a clean and dry substrate. However, special paints based on an alcohol solvent are more tolerant than others versus moisture. Furthermore, zinc-ethyl-silicate has to absorb water from the air in order to dry, and in this case the relative humidity of the air should not be too low either. Conversely, pure vinyl paints are particularly sensitive to high humidity. Painting a moist surface should be done by brush rather than by spraying. Special paints that can be applied and that harden submerged in water have also been produced.

A suitable film thickness and appropriate periods between application of the successive coats are important, but depend upon the type of paint. Data sheets from the paint producers give information about this. The thickness should be checked during the painting work. Other important properties are adhesion between old and new paint as well as resistance to detergents and mechanical wear. It can be mentioned that hardened two-component paints may be less suitable for over painting unless they are rubbed mechanically, but on the other hand they are resistant to detergents and mechanical wear.

Table 10.19 Type of paint, required pre-treatment, application temperature, time limits for over-painting, and actual environmental categories as defined in the ISO standard [10.48].

Type of paint	Pre-treatment[1]	Application temperature	Can be over-painted after, min/max	Corrosivity category[2]
Alkyd	St 2–3, Sa 2½		8 h/∞	C1–C4
Vinyl	Sa 2½		2 h/∞	C4–C5
Chlorinated rubber	Sa 2–Sa 2½	min 10–15°C	4 h/∞	C4–C5
Epoxy	Sa 2½–Sa 3	min 10°C	18 h/3 d	C4–C5
Coal tar epoxy	St 2, Sa 2½		16 h/3 d	C4–C5
Epoxy mastic	Sa 2½–Sa 3			C4–C5
Polyurethane	Sa 2½–Sa 3			C4–C5
Polyester	Sa 2½			C5
Zinc silicate	Sa 2½–Sa 3		24 h	C4

1 ISO 8501-1: St = wire brushing, Sa = blast cleaning.
2 Categories: C0: very low corrosivity, C5: very high corrosivity.

Protection and deterioration mechanisms of paints are described by Munger [10.50], Stratmann et al. [10.51], and others. For barrier coatings, the resistance to transport of water, ions and oxygen is of crucial significance for protection of the substrate, and transport and absorption of these substances are also important factors in the deterioration of the coating. The film resistance and the potential determine the cathodic reaction rate underneath the paint coating [10.52], which interacts closely with two of the main deterioration mechanisms, namely blistering and cathodic disbonding [10.53]. Examples of such defects are shown in Figs. 10.27 and 10.28. A good barrier coating is generally characterized by low uptake of water, low conductivity, and low permeability of water and oxygen, but these properties do not give a direct expression of the durability of the coating. A correlation between water permeability/water uptake and tendency to blistering of paints in fresh water has been shown in Reference [10.54].

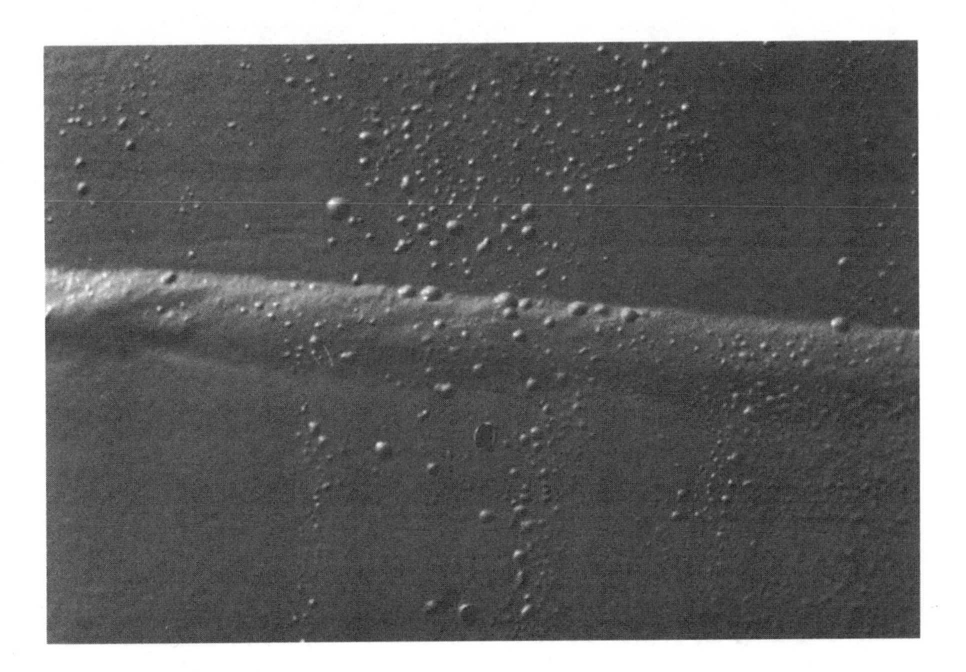

Figure 10.27 Blisters in a chlorinated rubber paint coating. (Photo: T.G. Eggen, SINTEF Corrosion Centre.)

Various thermal, mechanical, chemical and biological conditions can cause degradation of coatings. UV radiation is an important factor in the atmosphere. Corrosion underneath the coatings is also a common cause of paint degradation, which may result in phenomena such as blistering, anodic undermining and cathodic disbonding. Anodic undermining means that the coating is undermined by the anodic dissolution of the substrate, often at the edge of a defect in the coating. Cathodic disbonding implies that the coating loses its adhesion to the substrate because of the cathodic reaction. The formation of a local alkaline environment is a crucial step in the process. The cathodic reaction may be part of a corrosion process, or a result of cathodic protection. The attack starts at a weak point in the coating and develops radially, most often forming a circular defect. An interesting example of this is shown in Figure 10.28, where we can see a few such circular defects that have been

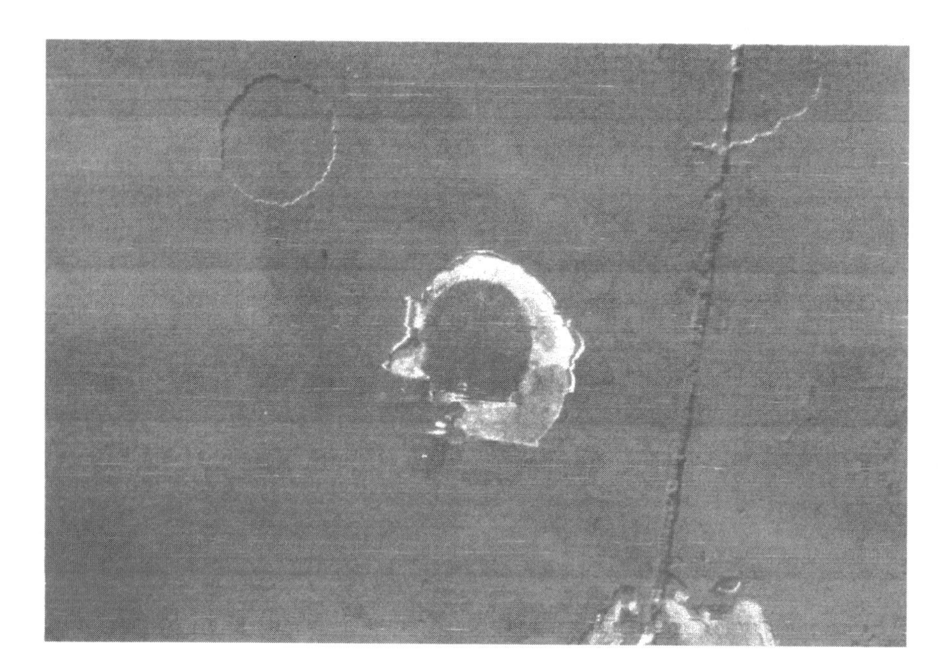

Figure 10.28 Cathodic disbonding of a paint coating on a cathodically protected ship hull. The defects developed during the first period have been overpainted. (Photo: T.G. Eggen, SINTEF.)

overpainted. At one of them (in the centre) a new defect has developed from the boundary of the overpainted defect.

As regards cathodic protection, the condition or state of the coating is important for the current requirement. In order to express the expected increase in current demand during the paint degradation, a degradation factor is used varying from 0 (for a completely insulating coating) to 1 (when the coating provides no protection effect) [10.25].

Painting zinc and aluminium makes special demands on the pre-treatment of the substrate. Chromate treatment provides a good basis for paint, and it is still in widespread use in spite of the toxicity of chromate. Phosphate treatment or wash-priming can also be used. Alternatives for aluminium are light blast cleaning or anodizing.

Application of paint. In order to get a good result the application of the paint must be done professionally with suitable methods and equipment. Common application methods are:

a) *Spraying*, which can be done by the conventional method with compressed air, or by high-pressure airless spraying. With the latter method, less scattering of paint droplets to the surrounding occurs during spraying, and it is possible to apply thicker coats by this method (100 μm with thick film paint, while a common film thickness for conventional spraying is 30–40 μm).

Electrostatic spraying, in common widespread use in industrial lacquering plants, gives a uniform coating on external surfaces. In another industrial method, suitable for plane panels, the object passes through a curtain of paint (curtain coating).

Powder of polymers such as epoxy and polyester is sprayed and subsequently heated to obtain a continuous film.

b) *By brush*, more tolerant than other methods for a substrate that is not completely dry or clean, although an optimum result is not possible in any case under such conditions.

c) *Dipping*, which is suitable for small articles and for priming in industrial plants. A special dipping method implies electrolytic deposition of paint.

Application equipment and processes (in addition to pre-treatment, types of paints etc.) are thoroughly described by Kjernsmo et al. [10.47]. This textbook, which soon (2003) will be available in English, is used in the widely recognized Frosio courses [10.55] for education and certification of inspectors in the surface treatment industry.

By properly performed painting work we can avoid defects such as "holidays" (areas left uncoated), flaking (e.g. due to inadequate pre-treatment of the substrate), too thin or too thick coats (insufficient checking of thickness) and blistering (which may, for instance occur because of short hardening periods between application of successive coats combined with low temperature). In Figure 10.28 an example of blistering in a chlorinated rubber coating is shown. The blistering has in this case been accelerated heavily due to heating by sunshine after completed painting. Regarding coating thickness, particular attention should be paid to edges, welds and other irregularities on the base material. Thick-film paints give better coverage on such places.

10.6.4 Other Forms of Organic Coatings

Rubber coatings provide excellent corrosion protection in seawater and several chemicals, and are used internally in tanks and pipelines. Externally on buried pipelines, asphalt or coal tar has commonly been used, often reinforced with a textile or mineral net, and during the last generation, also tape made from materials such as polyvinyl chloride or polyethylene with adhesive and primer on one side. Various factory methods for direct application of polyethylene or powder epoxy coatings on pipelines have been developed and used. During the last few decades, glass-flake-filled unsaturated polyester or vinyl-ester coatings have been applied to an increasing extent, particularly on offshore installations, on ships and in industry. These are fast-hardening coatings with high mechanical wear resistance and with good ability to protect against corrosion in most environments. The coatings are applied by high-pressure spraying in 1–2 coats giving a thickness of 0.75–1.5 mm.

Depending on the environment, all the mentioned types of coating can be combined with cathodic protection to prevent corrosion in/at coating defects.

10.6.5 Pre-treatment Before Coating

This is the most important step in the surface treatment. The purpose is primarily to get a surface with high cleanliness and adequate roughness. The pre-treatment can mainly be divided in two groups: a) degreasing and b) removal of mill scale and rust, and simultaneously roughening the surface.

a) *Degreasing* is carried out with i) organic solvents, such as white spirit and paraffin, ii) alkaline cleaning agents, e.g. solutions containing sodiumphosphate, silicate, carbonate or hydroxide, together with soap or some other wetting agent, iii) a combination of organic and alkaline cleaning agents (an emulsion), or iv) high-pressure steam containing small amounts of a cleaning agent.

   Alkaline cleaning give the cleanest surface and are often used in the last step of the pre-treatment, e.g. just before electrolytic metal plating. Steam can beneficially be used after degreasing with emulsions and after removal of old paint by alkaline cleaning before application of new paint [10.24]. All coatings and application methods require a surface free from grease before coating.

b) *Removal of rust and mill scale* is preferably done chemically by pickling or mechanically by blast cleaning.

Pickling of steel is carried out by immersing the object in a bath of pickling acid containing a pickling inhibitor (see Section 10.2). A commonly used acid is 3–10% by weight of H_2SO_4 in water, and with this solution the pickling is normally carried out at 65–90°C for 5–20 min. For various purposes, other pickling acids are also used, namely hydrochloric acid (cold), phosphoric acid, nitric acid or a mixture of different acids [10.37].

Pickling causes removal of scale partly by dissolution of rust and mill scale, and partly by dislodging oxide scale due to hydrogen gas that is produced by corrosion beneath the scale. Inhibitors are added to prevent intensive corrosion of the metal. Pickling is used before application of metallic coatings (by electroplating or hot dipping), inorganic coating material such as enamel, and organic coatings.

Blast cleaning is the best pre-treatment for painting (in addition to necessary degreasing). The process can be carried out by centrifugal blasting (sling-cleaning), or by compressed air blasting with a dry blasting agent, or possibly a mixture of sand and water (preferably with an inhibitor) that is blown against the object under high pressure. As blasting agents the following are used: steel shot or wire cuts (in centrifugal blasting plants), grit of aluminium oxide, steel or white iron (primarily when the blasting agent can be recirculated), olivine sand, iron slag or copper slag, iron silicate or aluminium silicate. (Silica sand is forbidden in some countries because of the risk of silicosis.) For large structures such as offshore platforms, blasting of water at a very high pressure (without solid blasting media) has become usual in recent years.

As mentioned previously, demands have to be made with respect to the quality of a blast-cleaned surface. The Swedish Standard SS 055900 (ISO 8501-1) is in widespread use for characterizing blasting quality. The quality grades are Sa 1, Sa 2, Sa 2½ and Sa 3, the last being the best one. As a tool for judging the blasting quality in real life (as well as rust grade before blasting and quality of wire-brushed and scraped surfaces) colour pictures of the various grades are collected in a book [10.49]. For modern, advanced paints, a blasting quality of at least Sa 2½, in some cases 2½–3 is required (see Table 10.19).

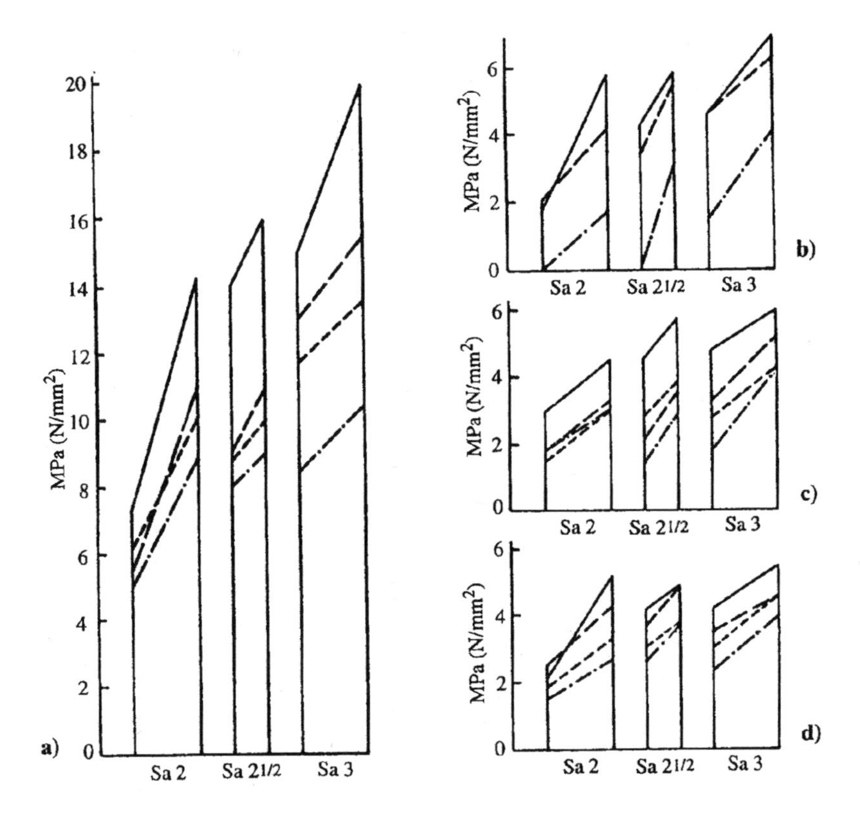

Figure 10.29 Bond strength of thermal spray coatings as a function of blasting quality for different types of grit and sand.
a) Arc-sprayed aluminium, b) flame-sprayed aluminium,
c) flame-sprayed zinc, and d) arc-sprayed zinc.
——— iron grit (0.1–0.6, 0.1–0.9, 0.5–0.9 mm)
– – – – copper slag (0.3–2.5 mm)
— · — silica sand (0.6–1.5 mm)
--------- silica sand (0.1–1.0 mm)

Complete cleaning of heavily rusty steel surfaces is difficult to obtain by blast cleaning, particularly if the rust has absorbed salts from the sea or from marine or industrial atmospheres. Pickling is better for removing salt remnants. The surface must be carefully washed with water afterwards. When blast cleaning is recommended for removal of rust, it should be taken into account that a relatively fine-grained blasting agent is most efficient.

In most cases, blast cleaning is also the best pre-treatment for thermal spraying of metals as well as ceramic or ceramic–metallic coating materials. For certain purposes rough grinding is used. As an example of the significance of blasting quality in general, and for thermally sprayed coatings specially, Figure 10.29 shows how the adhesion of arc-sprayed and flame-sprayed aluminium and zinc on steel depends on Sa-grade and blasting medium, the latter characterized by material and grain size [10.56].

In addition to pickling and blast cleaning, the following methods are also used for removing rust and mill scale from steel surfaces: hammering, scraping, wire-brushing, grinding and tumbling (mechanical methods), as well as flame cleaning and induction heating with subsequent quenching (thermal methods).

References

10.1 Fontana MG Greene ND. Corrosion Engineering. New York: McGraw-Hill, 1978, 1986.

10.2 Rabald E. Corrosion Guide. Amsterdam-Oxford-New York: Elsevier, 1968.

10.3 Roberge PR. Handbook of Corrosion Engineering. New York: McGraw-Hill, 1999.

10.4 Corrosion data survey, Metals section, 5th ed. and Non-metals sections, 5th ed. Houston, Texas: NACE, 1974 and 1975.

10.5 Shreir LL, Jarman RA, Burstein GT. Corrosion Vol 1, 3rd Ed. Oxford: Butterworth-Heinemann, 1994.

10.6 Metals Handbook, 9th Ed. Vol. 13 Corrosion. Metals Park, Ohio: ASM International, 1987.

10.7 Corrosion Tables of Stainless Steels. Stockholm: Jernkontoret, 1979 (in Swedish).

10.8 Corrosion handbook: Stainless Steels. Sandviken, Sweden: Sandvik Steel & Avesta Sheffield, 1994.

10.9 Alfonsson E. Corrosion of stainless steels. General introduction, Avesta Sheffield Corrosion Handbook, 1994; 9–17.

10.10 NORSOK Standard M-001. Materials selection. Oslo: Norwegian Technology Standard Institution. 1997.

10.11 Publications in Norwegian at NIF/NTH conferences on materials technology for the petroleum industry. Trondheim, Jan. 1989, Jan. 1993, Norwegian Association of Chartered Engineers, (NIF), 1989, 1993.

10.12 Moniz BJ, Pollock WJ, editors. Process Industries Corrosion, Houston, Texas: NACE, 1986.

10.13 Sedriks A. Corrosion of Stainless Steel. John Wiley & Sons, 1980.

10.14 Friend WZ. Corrosion of Nickel and Nickel-base Alloys. John Wiley & Sons, 1980.
10.15 Scarberry RC, editor. Corrosion of Nickel-base Alloys. Proceedings of Conference Oct. 1984, Cincinnati, Ohio: ASM, 1985.
10.16 NVS Metalliske materialer. Håndbok P 170, Del 3–Metaller. Oslo: Norwegian Technology Standard Institution, 1978 (in Norwegian).
10.17 Hatch JE. Aluminium – Properties and Physical Metallurgy. Metals Park, Ohio: American Society of Metals (ASM), 1984.
10.18 Titanium in Practical Applications. International conferance, Trondheim, 19-20 June 1990. The Norwegian Association of Corrosion Engineers, 1990.
10.19 Uhlig HH. Corrosion and Corrosion Control. John Wiley & Sons, 1971.
10.20 Nathan CC, editor. Corrosion Iinhibitors. Houston, Texas: NACE, 1973.
10.21 Rozenfeld IL. Corrosion inhibitors. McGraw-Hill, 1981.
10.22 Corrosion Inhibitors, Book B559, EFC 11. London: Institute of Materials, 1994.
10.23 Wranglen G. An Introduction to Corrosion and Protection of Metals. Chapman and Hall, 1985.
10.24 Pludek VR. Design and Corrosion Control. The MacMillian Press, 1977.
10.25 Det norske Veritas Industri Norge AS. Recommended practice, RP B401, Cathodic protecting design, 1993.
10.26 Strømmen R. Cathodic Protection of offshore structures. A review of present knowledge. SINTEF-report STF 16 F80009. Trondheim, 1980.
10.27 Chandler KA. Marine and Offshore Corrosion. London: Butterworths, 1985.
10.28 Recommended practice. RP-series on Cathodic Protection, NACE, Houston.
10.29 Gartland PO, Drugli JM. Methods for evaluation and prevention of local and galvanic corrosion in chlorinated seawater pipelines. Corrosion/92. Nashville. Paper no. 408, 1992.
10.30 Mackay WB. Sacrificial anodes, Section 11.2. In: Shreir LL. Corrosion. London: George Newnes, 1976.
10.31 Morgan JH. Cathodic Protection, 2nd Ed. Houston: NACE, 1987.
10.32 Baeckmann W, Schwenk W. Handbook of Cathodic Protection. Redhill, UK.: Portocullis Press, 1975.
10.33 Willis AD. Cathodic protection of novel offshore structures. In: Ashworth V, Googan C, editors. Cathodic Protection Theory and Practice. Ellis-Horwood,1993.
10.34 Strømmen R. Evaluation of anode resistance formulas by computer analysis. Corrosion/84. Paper no. 253, Houston, Texas: NACE, 1984.
10.35 Gartland PO, Johnsen R. Comcaps – mathematical modelling of cathodic protection systems. Corrosion/84. Paper no. 319. Houston, Texas: NACE 1984.
10.36 Tödt F. Korrosion und Korrosionsschutz. Berlin: W. De Gruyter & Co., 1961.
10.37 Shreir LL, Jarman RA, Burnstein GT, editors. Corrosion, Vol. 2, 3rd Ed. Oxford: Butterworth Heinemann, 1994.
10.38 Handbook of Electroplating. Birmingham: W. Canning & Co., 1966.
10.39 ISO-1456: 1988 (E). Metallic coatings – electro-deposited coatings of nickel plus chromium and of copper plus nickel plus chromium.

10.40 Thomas R, Wallin T. Corrosion Protection by Hot Dip Galvanizing. Stockholm: Nordisk försinkningsförening, 1989. English ed., 1992.
10.41 Burns RM, Bradley WW. Protective Coatings for Metals. New York: Reynolds Publ. Corp., 1967.
10.42 Ballard WE. Metal Spraying. London: Griffin, 4th Ed., 1963.
10.43 Corrosion tests in Flame-sprayed Coated Steel. 19 Years Report. Miami: American Welding Society, 1974.
10.44 Klinge R. Sprayed zinc and aluminium coatings for the protection of structural steel in Scandinavia. Proceedings 8th International Thermal Spray Conferance, American Welding Society, 1976; 203–213.
10.45 Rogne T, Solem T, Berget J. Effect of metallic matrix composition on the erosion corrosion behavior of WC-coatings. Proceedings of the United Thermal Spray conferance ASM Thermalspray Society, sept. 1997.
10.46 Knudsen OØ. Private communication. SINTEF, Trondheim, 2003.
10.47 Kjernsmo D, Kleven K, Scheie J. Overflatebehandling mot korrosjon. Oslo: Universitetsforlaget, 1997. (An English edition is being prepared, 2003.)
10.48 ISO-Standard 12944 – 2, 1988. Paint and varnishes. Corrosion protection of steel structures by protective paint systems. Part 2, 1998.
10.49 ISO-Standard 8501-1. Preparation of steel substrates before application of paint and related products. Part 1, 1988/suppl. 1994.
10.50 Munger CG. Corrosion Prevention by Protective Coatings. Houston, Texas: NACE,1984.
10.51 Stratmann M, Feser R, Leng A. Corrosion protection by organic films. Electro-chimica Acta, 39, 8/9, 1994: 1207.
10.52 Steinsmo U, Bardal E. Factors limiting the cathodic current on painted steel. Journal of the Eletrochemical Society 136, 12, 1989; 3588-3594.
10.53 Knudsen OØ, Steinsmo U. Effect of cathodic disbonding and blistering on current demand for cathodic protection of coated steel. Corrosion,. 56,. 3, 2000:
10.54 Steinsmo U. An evaluation of corrosion protective organic coatings for steel structures in fresh water. Final report. SINTEF report STF 34 A 92229, Trondheim, 1992.
10.55 Korsøen L. Presentation of NS 476 and Frosio. Proceedings Eurocorr’97, Vol. 2, Trondheim 1997; 423–427.
10.56 Bardal E, Molde P, Eggen TG. Arc and flame sprayed aluminium and zinc coatings on mild steel. Bond strength, surface roughness, structure and hardness. British Corrosion Journal, 8, January 1973: 15–19.

EXERCISES

1. Explain why the following materials are not suitable for use in the given environments, and mention at least one acceptable material for each environment:

   - Copper in acid, oxidizing environment
   - 18Cr8Ni stainless steel in 6% $FeCl_3$ solution

- 17Cr12Ni2.5Mo stainless steel in natural seawater at 10°C
- High-strength carbon steel in unprocessed oil/gas from a well containing H_2S
- 18Cr8Ni steel in aqueous solutions containing chloride at 90°C
- Copper–nickel in polluted or stagnant seawater
- Aluminium alloys in contact with steel in environments containing chloride
- 22Cr5Ni3Mo ferritic–austenitic stainless steel in seawater with some sand content at 20°C a) in stagnant condition, and b) flowing at a velocity of 20 m/s
- Titanium in strongly reducing acids
- Zinc and aluminium in alkaline environments (e.g. at pH = 12)

2. For design of structures, equipment and single components it is, from a corrosion technology point of view, a good rule to aim at simple geometry, and to avoid heterogeneity and abrupt changes. Heterogeneity comprises different materials, varying temperature, uneven stress distribution and varying dimensions. Give examples of problems which can occur when this rule is neglected. As many corrosion forms (Chapter 7) and as many corrosion prevention methods (Chapter 10) as possible should be represented in the solution.

3. Give examples of design that help to localize corrosion at preferable places.

4. A cylindrical container of steel, 1.5 m in diameter, is filled with seawater to a height of 2 m. It is planned to protect the container for a period of 10 years with a cylindrical sacrificial anode of an Al alloy positioned in the centre, as shown in Figure 1. The bottom is covered with a thick organic coating, so that no current

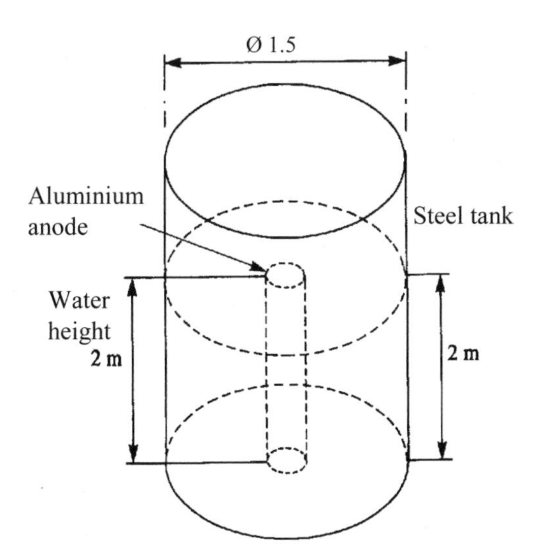

Figure 1. Steel container with seawater, cathodically protected by an aluminium anode.

goes to the bottom area. The cylindrical mantel surface is not coated. The resistivity of the seawater is ρ_r = 30 ohm cm. The cathodic reaction on the steel surface is reduction of oxygen. The reaction is assumed to be controlled by concentration polarization in the whole potential range in question, and the limiting diffusion current density is found to be 15 μA/cm^2.

Unprotected steel is dissolved by the reaction $Fe \rightarrow Fe^{2+} + 2e^-$, the equilibrium potential of which under the actual conditions is estimated to be E_o =–-850 mV vs. the saturated calomel electrode (SCE)

Figure 2 shows the anodic polarization curve of the sacrificial anode. It is assumed that the anode is dissolved by the reaction $Al \rightarrow Al^{3+} + 3e^-$.

Atomic weight of Al M = 27
Density of Al ρ_d = 2.7 g/cm^3
Faraday's number F = 96,500 C/mol e$^-$

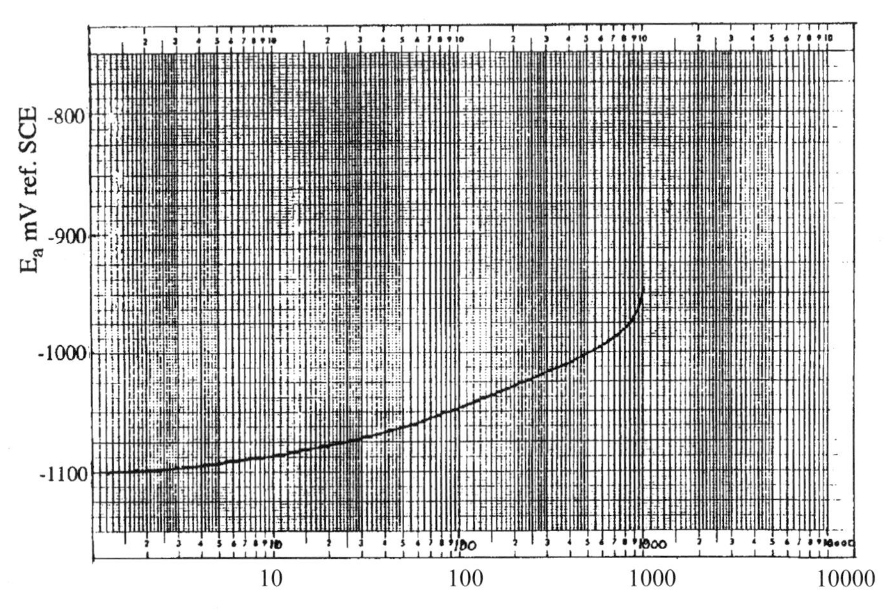

Figure 2 Polarization curve for the anode i, μA/cm^2.

a) Determine the total current demand for full protection of the cylinder wall.

For solution of the following tasks b)– e) we assume complete protection of the container continuously for all the 10 years, we disregard self-corrosion of the anode, and the calculation is to be based on the data given in Figure 2 and in the text.

b) Assume for simplicity that the anode dissolves uniformly. Suppose a minimum diameter (anode diameter after t = 10 years) of 3 cm, and determine the current density and potential of the anode when this stage is reached.
c) Determine the necessary original diameter of the anode.
d) Derive formulae for i) total (ohmic) resistance R in the seawater between the anode and the container wall as a function of the radius r_a of the anode, ii) the radius r_a as a function of time t.
e) Draw diagrams that show i) current density on the anode as a function of time; ii) electrode potential of the anode as a function of time; iii) electrode potential of the container wall as a function of time. (It is sufficient to determine the respective values for t = 0, t = 10 years and two moments in between.)
f) Is the assumption of complete protection satisfied all the time?
g) Why are potentials up to –780 mV vs SCE accepted for steel structures?

5. A steel structure for seawater exposure is to be protected by means of sacrificial anodes, such that the potential of the steel surface becomes ≤ – 0.8 V referred to Ag/AgCl/seawater. The sea depth is 60 m. The total surface area of steel exposed to seawater is 10,000 m^2, half of which is at depth 0–30 m. Average temperature during the year is 10°C.

   a) Select a suitable anode material (alloys of Al, Zn or Mg). State the reasons for the selection
   b) Determine the total current demand for full protection at the beginning and the end, respectively, of the lifetime. (Use data from the tables in Section 10.4.2.)
   c) For the most suitable anodes, the supplier has stated an "anode resistance" R_a = 0.05 ohm for new anodes. Because of the change in dimensions, this resistance is assumed to increase 50% during the lifetime. The potential of the anode can be taken from Table 10.15. Determine the current delivered from each anode at the beginning as well as at the end, assuming that the potential of the structure is at its highest allowable value.
   d) Calculate the necessary number of anodes.
   e) Table 10.15 gives the value of anode capacity C. The supplier of the anodes has stated a utilization factor u = 0.85 for the chosen type of anode. Use relevant current density data given in Section 10.4.2, assume a lifetime of t = 20 years and determine the necessary total weight G_t of the anodes.
   f) With a weight G_a of each of the anodes (as stated by the supplier), explain how we can ensure that the calculated number of anodes really is large enough.

Subject Index

A

Abrasion corrosion 138, 139, 150
Activation
  potential drop 61
Activation polarization 37–38
Active–passive metals, definition 54
Activity 17–18
Aluminium and aluminium alloys
  applications 255
  as coatings 286–287, 291
  atmospheric corrosion 196
  exfoliation corrosion 135
  galvanic corrosion 97–99
  in general 254–256
  intergranular corrosion 135
  mechanical properties 254
  pitting corrosion 123, 124, 125, 127–129, 131
  product forms 254
  sacrificial anodes 273–274
  stress corrosion cracking 157
Anodic control 7
Anodic protection 281–282
Anodic reaction 5, 6
Anodising 293
API RP14E 80–81
Atmospheric corrosion
  environmental factors 193–196
  function of water film thickness 74
  on various materials 196–198
  prevention 197

B

Bacteria
  effects 67, 77–78, 96, 98, 202–203
  iron bacteria 202–203
  sulphate-reducing 77–78, 202, 207
Blast cleaning 296, 295, 300–302

C

Cadmium 99, 287
Carbon 259
Cast iron
  high-alloy 242–243
  unalloyed/low alloy 241–242
Cathodic control 6
Cathodic protection 266
  anode material 274
  anode resistance 278–279
  arrangement of sacrificial anodes 275, 280
  current demand 270, 272–273
  design 275–276
  effect on fatigue 172, 176, 178–180
  in soils 277
  monitoring 227–229
  potential–log current diagram 267, 269
  potential–distance diagram 271
  potential variation and current distribution 278–281
  principles 266–269
  protection criteria 269–273
  with impressed current 271, 276–277
  with sacrificial anodes 271, 273–276, 279–280
Cathodic reactions
  dependence on oxidizer concentration 65–66
  in general 65–66
Cavitation corrosion 152–154
Cell voltage 15
  standard 19
Ceramic materials 259
  as coatings 292
Ceramic–metallic materials 259
  as coatings 292
Chlorine 83–84
Chromate treatment 292–293
Chromium
  as a coating 283, 284
  as an alloying element 241–251
Cladding 292
CO_2 corrosion 78, 212–215
  cathodic polarization curves 79
  corrosion rates 80–81
  mechanism 78–80

Coatings
inorganic 283, 292–293
metallic 283–292, 301
organic 293, 299
paints 293–299
pretreatment 300–302
protection mechanisms 282, 283, 296–297
Combined polarization 43–44
Concentration polarization 38, 40–43
Concrete structures 210–212
Copper and copper alloys
applications 251–253
atmospheric corrosion 197
corrosion resistance 250–253
erosion corrosion 138, 144, 145, 146, 149
in general 250–253
in seawater 145, 149, 205–206
mechanical properties 252–253
selective corrosion 135–137
stress corrosion cracking 157
tables 252–253
Corrosion
importance 1–3, 91–92
cost 2–3, 211
Corrosion allowance 262
Corrosion current density 8–9
Corrosion failures
frequency of various forms 91, 92
Corrosion fatigue 170
beach marks 170–171
characteristical features 170
crack growth 173
crack growth rate 176–179
definition 170
influencing factors 171–172, 175–179
initiation 173, 175
lifetime calculation 180
mechanisms 173–175
prevention 180–181
S–N curves 171–172
stages 173–174
Wöhler curves 171–172
Corrosion forms 89
classification, main groups 90–91
frequency of occurrence 91, 92
Corrosion monitoring 226
cathodic protection 227–229
cost 229
current density measurement 227–228
diver-carried measuring equipment 227–228
in process plants 229–232
in soil 232
of reinforcement in concrete 232–233
potential measurement 227–228
Corrosion potential
determination 44–47
Corrosion prevention by change of environment 259
Corrosion rates
conversion factors 9
determination 44–50
expressions and units 8–9
influencing factors 50
of steel in seawater 9
Corrosion science 3
Corrosion technology 3–4
Corrosion testing 219
electrochemical methods 223–226
in the atmosphere 232, 233
methods 220
objectives 219
procedures 221–222
standardized methods 118, 130, 134, 135, 169, 222
Corrosion types
defined by cathodic reaction 65
Corrosive (corrosive–abrasive) wear 139, 149–150
Crevice corrosion 108
anodic polarization curves 112, 116
calculation model 113–117
cases 119–120
conditions 108, 113
critical crevice temperature 118–119, 248
initiation 109, 111–112
initiation potential 118, 119

incubation period 110, 118
mechanism 109–113
potental–log current diagram 112
prevention 121–122
protection potential 112
testing 115–119

D

Dealuminization 137
Degreasing 300
Deposit corrosion 108, 119, 120
at welds 119
Deposits
properties 56, 70
Design 262–266
DeWaard and Milliams equation 80
nomogram 81
Dezincification 136–137
Diffusion
for oxygen reduction 40–41
Diffusion boundary layer 41, 68, 71

E

Electrochemical cell 6
Electrode kinetics 35
Electrode potential 13
definition 19–21
standard 21, 23, 24
Electrode reactions
definition 19–20
Electrolytic plating 284, 285, 286, 288
Enamel 292
Equilibrium potential 20
hydrogen reaction 75
iron 25–26
oxygen reaction 67
standard 24
ER (electric resistance) probes 230–231
Erosion 138–144
Effect of particles 140–144
Erosion corrosion 71–72
cases 145–149
characteristic features 138–139
corrosive–erosive wear 149
critical velocities 146–149
effect of particles 140–144
influencing factors 144–146
mechanisms 139–140
overvoltage curves 146–149
prevention 150–152
sensitive materials 138, 149
synergy effects 143
Exchange current density 37
table 39
Extrapolation of Tafel region 49–50, 223–224

F

Faraday's constant 9, 15
Faraday's laws 9, 15
Fatigue fracture 170–171
Fatigue limit 172
Fatigue strength 171–172
Filiform corrosion 108–109
Flowline 146, 147
Flow velocity
effects 70–74, 145 (see also erosion corrosion, erosioin)
Free enthalpy 13, 14, 16, 17, 18, 19
standard 16
Fretting corrosion 154–156

G

Galvanic corrosion
area ratio 7, 100–101
atmospheric 105
cases 105
cathodic efficiency 97–100
conditions 94
corrosion rates of aluminium alloys 97–99
element of carbon steel and stainless steel 103–104
insulation materials 105–106
joint between iron and copper 106
ohmic potential drop 102–103
potential–log current diagram 102
prevention 105–107
titanium in contact with steel 100
total corrosion rate 97

utilization in engineering 107
Galvanic crevice corrosion 120–121
Gavanizing, see Zinc, hot dipping
General corrosion, see Uniform corrosion
Geometric effects, design 262–266
Glass linings 292
Grain boundary corrosion, see Intergranular corrosion

H

H_2S corrosion 82–83, 215
cathodic polarization curves 82
Half-cell potential, definition 20
Hastelloy 121, 250, 251
Hot-dip zinc coatings 288, 289
Hot dipping 288
Hydrogen electrode
standard 21
Hydrogen embrittlement 161–162
Hydrogen-induced (hydrogen-assisted) cracking 161–162
Hydrogen reduction 65
corrosion under h. r. 75–77
reversible (equilibrium) potential 75
polarization curves for iron 76

I

Immunity 27
Impingement 139
Inhibitors 259–262
Inspection, methods 229–230
Insulators
to prevent galvanic corrosion 105, 106, 258
Intergranular corrosion
causes, general features, 131–132
in Ni-based alloys 135
in stainless steel 132–134
in welded stainless steel 133–134
potential–log current diagram 133
prevention 133–135
testing 134, 135

L

Laboratory testing 220
electrochemical 223–226
procedure 221–222
Lead 257
Limiting current density 40–42, 68–75
with deposits present 70
without deposits 68
Linear polarization 224–225

M

Magnesium 251
sacrificial anodes 273, 274
Materials selection
in general 237–240
in the offshore industry 240–241
Metal deposition 284, 285, 286, 288
Metal spraying, see Thermal spraying
Mixed potential 45

N

Nernst's equation 23
Nickel
as a coating metal 283, 284
as an alloying element 241, 243–245, 247, 253
electrolytic deposition 284
Nickel alloys
crevice corrosion 113
in general 250
in seawater 121, 205, 206
intergranular corrosion 135
table 251
Ni–Resist 243
Nitric acid 84

O

Oil and gas 212
corrosion attacks 214
corrosion rates 80–81, 213
unfavourable factors 212, 214–215
Organic coatings 299–300
Overvoltage
definition 36
Overvoltage curves 43, 44, 48, 49
Oxide films
properties 56

Oxygen reduction 5, 6, 65, 66–67
 corrosion under o. r. 66
 effect of flow velocity 70–74
 effect of surface deposits 70
 effect of temperature 68–69
 equilibrium potential 67
 exchange current density 39, 67
 in acid solutions 66
 in neutral and alkaline solutions 66–67
 Tafel constants 39, 67

P

Paint coatings 293–299
 application 298–299
 compositions and types 293, 295, 296
 ISO standard 295, 296
 pretreatment 300–302
 properties 294–298
 protection and deterioration mechanisms 296–298
 selection of paint system 294–296
Passivation 7, 53–63
 critical current density 54
 Faraday's experiment 58–59
 steel in sulphuric acid 53–54
Passivation potential
 definition 53–54
 steel in 1 N sulphuric acid 53–54
Passivators (passivating inhibitors) 259–261
Passive current density
 definition 54
 steel in sulphuric acid 54
Passivity 27, 53–63
 breakdown 57–58
 causes 54–56
 conditional (practically unstable) 58–60
 stable 59–61
Phosphate treatment 292
Pickling 247, 300
Pitting corrosion 122
 characteristic features 122–124
 conditions and occurrence 122–124
 critical pitting temperature 130, 248
 influencing factors 126
 mechanisms 124–125
 pitting resistance equivalent 130
 prevention 131
 testing 130
 time dependence 127–130
Plastics 258
 as coating materials 299–300
 as insulation materials 105, 106, 258
Platinum, Pt 21, 22, 24, 28, 257
Polarization 36
Polarization curves
 recording 47, 47–50
Polarization resistance
 measurement 224–225, 231
Potential measurement 20–22, 223, 227–228
Potential series
 corrosion potentials in seawater 96
 standard euilibrium potentials 23–24
Potentiostat 47–49
Pourbaix diagram 26–29
 prevention 150–152
Propellers
 cavitation corrosion 143

R

Reduction potential 22
Reference electrodes 269
 saturated calomel (SCE) 48
 standard hydrogen (SHE) 21
Reinforcement corrosion 210
Resistance (ohmic) polarization 37, 44
Resistivity
 in soils 207, 208
 in seawater 278
Reversible cell 15
Reversible potential, see equilibrium p.
Rubber 258
 as a coating material 299
Rust 6

S

Sacrificial anodes 273–276, 280
Sand blasting, see blast cleaning

Sea cable corrosion 206, 207
Seawater
composition 204
corrosion rates 205
galvanic series 96
Selective corrosion 135–138
Silicon iron 242–243
Silver 28, 284
Soils 206
biological factors 207
corrosion rates 208–210
density 206–207
resistivity 207, 208
stray currents 207–208
Spontaneous reaction 14, 15
Stainless steels 243
alloying elements 243–245
anodic protection 281–282
austenitic245, 248
brushing 246
crevice corrosion 113–120, 247–249
ferritic 245
ferritic–austenitic 245, 248
for oil production 249
gas protection under welding 119, 245, 246
in NaCl+Na_2SO_4 solutions 59
in 1 N H_2SO_4 solution 57, 58
in 0.1 N NaCl solution 57, 59
intergranular corrosion 132–134
in water of various compositions 246–249
martensitic 245
mechanical properties 244, 245
pickling 245–246
pitting corrosion 59, 127, 130, 131, 247, 248
standard numbers 244
stress corrosion cracking 157, 156, 164, 166–169, 247
survey 243–245
tables 244, 245, 247
Standard electrode potentials 24
Standard state 16, 18
Steel and iron
atmospheric corrosion 194, 195, 196
CO_2 corrosion 146–147, 212–214
corrosion fatigue 170–172, 175–171
in fresh water and other waters 198–203
in general 241–242
in seawater 73, 145, 205, 206
in sulphuric acid 53–54, 242
in water at various temperatures 69
stress corrosion cracking 157, 160, 161, 162, 164–165
weathering steels 196–197
Stern–Geary's equation 224–225
Stress corrosion cracking 156
cases 164–169
characteristic features 156
data/diagrams 165–168
fracture mechanics quantities 163
hydrogen induced cracking 161–162
intergranular 162, 164
material/environment coimbinations 157
mechanisms 157–162
prevention 170
transgranular 156–162, 164

T

Tafel constant 38
table 39
Tafel equation 38
Tantalum 28, 257
Thermal spraying 289–291, 301
Thermodynamics 13
Thermogalvanic corrosion 107
Thickness reduction 9
Tin 257, 285
Titanium and its alloys
anodic protection 282
in general 239, 256–257
stress corrosion cracking 157
Transpassivity 54, 57, 58
Turbulence corrosion 139

U

Uniform corrosion 89, 91–94
conditions 89

overvoltage curves 94, 95
prevention 92

W

Water 198
bacteria 202–203
chlorination 202
deposition of salts 199–201
dissolved copper 202
dissolved salts 200–202
hardness 199–201
organic matter 202–203
Welds 105, 119, 121, 133–134, 170, 180, 246, 247, 248, 264, 265
Wet corrosion 5–6
Wood 259

Z

Zero-resistance amperemeter 226
Zinc 99, 257, 285–286
as a coating material 283, 285–286, 287, 288–289, 291
atmospheric corrosion 197, 198
corrosion rates in water 198, 205
electrode 19–20
electrolytic plating 286, 288
hot dipping 286, 288–289
sacrificial anodes 273, 274
thermal spraying 286, 288–289, 291, 301, 302
Zirconium 28, 257–258